CAMBRIDGE STUDIES IN ADVANCED MATHEMATICS 117

GALOIS GROUPS AND FUNDAMENTAL GROUPS

Ever since the concepts of Galois groups in algebra and fundamental groups in topology emerged during the nineteenth century, mathematicians have known of the strong analogies between the two concepts. This book presents the connection starting at an elementary level, showing how the judicious use of algebraic geometry gives access to the powerful interplay between algebra and topology that underpins much modern research in geometry and number theory.

Assuming as little technical background as possible, the book starts with basic algebraic and topological concepts, but already presented from the modern viewpoint advocated by Grothendieck. This enables a systematic yet accessible development of the theories of fundamental groups of algebraic curves, fundamental groups of schemes, and Tannakian fundamental groups. The connection between fundamental groups and linear differential equations is also developed at increasing levels of generality. Key applications and recent results, for example on the inverse Galois problem, are given throughout.

CAMBRIDGE STUDIES IN ADVANCED MATHEMATICS

All the titles listed below can be obtained from good booksellers or from Cambridge University Press. For a complete series listing visit: http://www.cambridge.org/series/sSeries.asp?code=CSAM

Already published

61 J. D. Dixon *et al. Analytic pro-p groups*
62 R. P. Stanley *Enumerative combinatorics, II*
63 R. M. Dudley *Uniform central limit theorems*
64 J. Jost & X. Li-Jost *Calculus of variations*
65 A. J. Berrick & M. E. Keating *An introduction to rings and modules*
66 S. Morosawa *et al. Holomorphic dynamics*
67 A. J. Berrick & M. E. Keating *Categories and modules with K-theory in view*
68 K. Sato *Lévy processes and infinitely divisible distributions*
69 H. Hida *Modular forms and Galois cohomology*
70 R. Iorio & V. Iorio *Fourier analysis and partial differential equations*
71 R. Blei *Analysis in integer and fractional dimensions*
72 F. Borceux & G. Janelidze *Galois theories*
73 B. Bollobás *Random graphs*
74 R. M. Dudley *Real analysis and probability*
75 T. Sheil-Small *Complex polynomials*
76 C. Voisin *Hodge theory and complex algebraic geometry, I*
77 C. Voisin *Hodge theory and complex algebraic geometry, II*
78 V. Paulsen *Completely bounded maps and operator algebras*
79 F. Gesztesy & H. Holden *Soliton equations and their algebro-geometric solutions, I*
81 S. Mukai *An introduction to invariants and moduli*
82 G. Tourlakis *Lectures in logic and set theory, I*
83 G. Tourlakis *Lectures in logic and set theory, II*
84 R. A. Bailey *Association schemes*
85 J. Carlson, S. Müller-Stach & C. Peters *Period mappings and period domains*
86 J. J. Duistermaat & J. A. C. Kolk *Multidimensional real analysis, I*
87 J. J. Duistermaat & J. A. C. Kolk *Multidimensional real analysis, II*
89 M. C. Golumbic & A. N. Trenk *Tolerance graphs*
90 L. H. Harper *Global methods for combinatorial isoperimetric problems*
91 I. Moerdijk & J. Mrcun *Introduction to foliations and Lie groupoids*
92 J. Kollár, K. E. Smith & A. Corti *Rational and nearly rational varieties*
93 D. Applebaum *Lévy processes and stochastic calculus*
94 B. Conrad *Modular forms and the Ramanujan conjecture*
95 M. Schechter *An introduction to nonlinear analysis*
96 R. Carter *Lie algebras of finite and affine type*
97 H. L. Montgomery & R. C. Vaughan *Multiplicative number theory, I*
98 I. Chavel *Riemannian geometry (2nd Edition)*
99 D. Goldfeld *Automorphic forms and L-functions for the group GL(n,R)*
100 M. B. Marcus & J. Rosen *Markov processes, Gaussian processes, and local times*
101 P. Gille & T. Szamuely *Central simple algebras and Galois cohomology*
102 J. Bertoin *Random fragmentation and coagulation processes*
103 E. Frenkel *Langlands correspondence for loop groups*
104 A. Ambrosetti & A. Malchiodi *Nonlinear analysis and semilinear elliptic problems*
105 T. Tao & V. H. Vu *Additive combinatorics*
106 E. B. Davies *Linear operators and their spectra*
107 K. Kodaira *Complex analysis*
108 T. Ceccherini-Silberstein, F. Scarabotti & F. Tolli *Harmonic analysis on finite groups*
109 H. Geiges *An introduction to contact topology*
110 J. Faraut *Analysis on Lie groups*
111 E. Park *Complex topological K-theory*
112 D. W. Stroock *Partial differential equations for probabilists*
113 A. Kirillov *An introduction to Lie groups and Lie algebras*
114 F. Gesztesy *et al. Soliton equations and their algebro-geometric solutions, II*
115 E. de Faria & W. de Melo *Mathematical tools for one-dimensional dynamics*
116 D. Applebaum *Lévy processes and stochastic calculus (2nd Edition)*

Galois Groups and Fundamental Groups

TAMÁS SZAMUELY

Alfréd Rényi Institute of Mathematics,
Hungarian Academy of Sciences, Budapest

CAMBRIDGE
UNIVERSITY PRESS

University Printing House, Cambridge CB2 8BS, United Kingdom

Cambridge University Press is part of the University of Cambridge.

It furthers the University's mission by disseminating knowledge in the pursuit of education, learning and research at the highest international levels of excellence.

www.cambridge.org
Information on this title: www.cambridge.org/9780521888509

First published 2009
Reprinted 2010

A catalogue record for this publication is available from the British Library

Library of Congress Cataloguing in Publication data
Szamuely, Tamás.
Galois groups and fundamental groups / Tamás Szamuely.
p. cm. – (Cambridge studies in advanced mathematics ; 117)
Includes bibliographical references and index.
ISBN 978-0-521-88850-9 (hardback)
1. Galois theory. I. Title. II. Series.
QA214.S95 2009
512′.32–dc22 2009003670

ISBN 978-0-521-88850-9 Hardback

Contents

Preface

Ever since the concepts of the Galois group and the fundamental group emerged in the course of the nineteenth century, mathematicians have been aware of the strong analogies between the two notions. In its early formulation Galois theory studied the effect of substitutions on roots of a polynomial equation; in the language of group theory this is a permutation action. On the other hand, the fundamental group made a first, if somewhat disguised, appearance in the study of solutions of differential equations in a complex domain. Given a local solution of the equation in the neighbourhood of a base point, one obtains another solution by analytic continuation along a closed loop: this is the monodromy action.

Leaving the naïve idea of substituting solutions, the next important observation is that the actions in question come from automorphisms of objects that do not depend on the equations any more but only on the base. In the context of Galois theory the automorphisms are those of a separable closure of the base field from which the coefficients of the equation are taken. For differential equations the analogous role is played by a universal cover of the base domain. The local solutions, which may be regarded as multi-valued functions in the neighbourhood of a base point, pull back to single-valued functions on the universal cover, and the monodromy action is the effect of composing with its topological automorphisms.

In fact, the two situations are not only parallel but closely interrelated. If our complex domain D is just the complex plane minus finitely many points, a local solution of a linear holomorphic differential equation that becomes single-valued on a cover with *finite* fibres lies in a finite algebraic extension of the field $\mathbf{C}(t)$ of meromorphic functions on D. We actually obtain a one-to-one correspondence between finite extensions of $\mathbf{C}(t)$ and finite covers of D, provided that we allow finitely many exceptional points called branch points. This opens the way for developing a unified theory within the category of algebraic curves, first over the complex numbers and then over a general base field. Algebraic geometers working in the 1950s realized that the theory generalizes without much effort to higher dimensional algebraic varieties satisfying the *normality* condition.

A further step towards generality was taken by Grothendieck. He gave a definition of the fundamental group not as the automorphism group of some space or field but as the automorphism group of a *functor*, namely the one that assigns to a cover its fibre over the fixed base point. This point of view permits a great clarification of earlier concepts on the one hand, and the most general definition of the fundamental group in algebraic geometry on the other. One should by no means regard it as mere abstraction: without working in the general setting many important theorems about curves could not have been obtained.

Grothendieck's concept of the algebraic fundamental group gives a satisfactory theory as far as finite covers of algebraic varieties or schemes are concerned, but important aspects of previous theories are lost because one restricts attention to finite covers. According to the fruitful motivic philosophy of Grothendieck and Deligne that underlies much of current research, this can be remedied by considering the algebraic fundamental group as only one incarnation, the 'étale realization' of a more general object. Other incarnations are the 'topological realization' where not necessarily finite covers of topological spaces are brought into play, and the 'de Rham realization' which is an algebraic formalization of the theory of differential equations. For the definition of the latter Grothendieck envisioned the algebraic formalism of Tannakian categories worked out in detail by Saavedra and Deligne. The various realizations of the fundamental group are related by comparison theorems. One instance of these is the correspondence between covers and field extensions mentioned above. Another one is the Riemann–Hilbert correspondence relating linear differential equations to representations of the fundamental group.

We have to stop here, as we have reached the viewpoint of present-day algebraic geometers and algebraic analysts on the subject. The future may well bring further unifications highlighting hitherto neglected aspects. Still, we feel that the time is ripe for a systematic discussion of the topic starting from the basics, and this is the aim of the present book. A glance at the table of contents shows that we shall be following the line of thought sketched above. Along the way we shall also mention a number of applications and recent results.

The first three chapters may be read by anyone acquainted with basic field theory, point set topology and the rudiments of complex analysis. Chapter 4 treats algebraic geometry, but is meant to be accessible to readers with no previous knowledge of the subject; the experts will skip a few introductory sections. The last two chapters are of a slightly more advanced nature. Nevertheless, we give a detailed summary of the basics on schemes at the beginning of Chapter 5, while most of Chapter 6 assumes only basic algebra and is largely independent of previous chapters.

Acknowledgments

In the course of the preparation of this book I have received valuable advice and comments from a number of colleagues and friends. It is a pleasure to thank Marco Antei, Gábor Braun, Antoine Chambert-Loir, Jean-Louis Colliot-Thélène, Mátyás Domokos, Hélène Esnault, Philippe Gille, Phùng Hô Hai, Luc Illusie, János Kollár, Andrew Kresch, Klaus Künnemann, Florian Pop, Wayne Raskind, Jean-Pierre Serre, Olivier Wittenberg, and especially Jakob Stix and Burt Totaro. Diana Gillooly was again a marvellous editor, and the remarks of two anonymous referees on a draft version were also appreciated.

1

Galois theory of fields

This first chapter is both a concise introduction to Galois theory and a warmup for the more advanced theories to follow. We begin with a brisk but reasonably complete account of the basics, and then move on to discuss Krull's Galois theory for infinite extensions. The highlight of the chapter is Grothendieck's form of Galois theory that expresses the main theorem as a categorical anti-equivalence between finite étale algebras and finite sets equipped with a continuous action of the absolute Galois group. This theorem is a prototype for many statements of similar shape that we shall encounter later.

1.1 Algebraic field extensions

In this section and the next we review some basic facts from the theory of field extensions. As most of the material is well covered in standard textbooks on algebra, we shall omit the proof of a couple of more difficult theorems, referring to the literature instead.

Definition 1.1.1 Let k be a field. An extension $L|k$ is called *algebraic* if every element α of k is a root of some polynomial with coefficients in k. If this polynomial is monic and irreducible over k, it is called the *minimal polynomial* of α.

When L is generated as a k-algebra by the elements $\alpha_1, \dots, \alpha_m \in L$, we write $L = k(\alpha_1, \dots, \alpha_m)$. Of course, one may find many different sets of such α_i.

Definition 1.1.2 A field is *algebraically closed* if it has no algebraic extensions other than itself. An *algebraic closure* of k is an algebraic extension $\bar{k}$ that is algebraically closed.

The existence of an algebraic closure can only be proven by means of Zorn's lemma or some other equivalent form of the axiom of choice. We record it in the following proposition, along with some important properties of the algebraic closure.

Proposition 1.1.3 *Let k be a field.*

1. *There exists an algebraic closure $\bar{k}$ of k. It is unique up to (non-unique) isomorphism.*
2. *For an algebraic extension L of k there exists an embedding $L \to \bar{k}$ leaving k elementwise fixed.*
3. *In the previous situation take an algebraic closure $\overline{L}$ of L. Then the embedding $L \to \bar{k}$ can be extended to an isomorphism of $\overline{L}$ onto $\bar{k}$.*

For the proof, see Lang [48], Chapter V, Corollary 2.6 and Theorem 2.8, or van der Waerden [106], §72.

Thus henceforth when speaking of algebraic extensions of k we may (and often shall) assume that they are embedded in a fixed algebraic closure $\bar{k}$.

Facts 1.1.4 A finite extension L of k is algebraic. Its *degree* over k, denoted by $[L : k]$, is its dimension as a k-vector space. If L is generated over k by a single element with minimal polynomial f, then $[L : k]$ is equal to the degree of f. For a tower of finite extensions $M|L|k$ one has the formula $[M : k] = [M : L][L : k]$. All this is proven by easy computation.

Definition 1.1.5 A polynomial $f \in k[x]$ is *separable* if it has no multiple roots (in some algebraic closure of k). An element of an algebraic extension $L|k$ is *separable* over k if its minimal polynomial is separable; the extension $L|k$ itself is called *separable* if all of its elements are separable over k.

Separability is automatic in characteristic 0, because a well-known criterion implies that an *irreducible* polynomial has no multiple roots if and only if its derivative f' is nonzero (see [106], §44). However, the derivative can be zero in characteristic $p > 0$, e.g. for a polynomial $x^p - a$, which is irreducible for $a \in k^\times \setminus k^{\times p}$.

In the case of finite extensions there is the following important characterization of separability.

Lemma 1.1.6 *Let $L|k$ be a finite extension of degree n. Then L has at most n distinct k-algebra homomorphisms to $\bar{k}$, with equality if and only if $L|k$ is separable.*

Proof Choose finitely many elements $\alpha_1, \dots, \alpha_m$ that generate L over k. Assume first $m = 1$, and write f for the minimal polynomial of α_1 over k. A k-homomorphism $L \to \bar{k}$ is determined by the image of α_1, which must be one of the roots of f contained in $\bar{k}$. The number of distinct roots is at most n, with equality if and only if α is separable. From this we obtain by induction on m using the multiplicativity of the degree in a tower of finite field extensions that L has at most n distinct k-algebra homomorphisms to $\bar{k}$, with equality if the α_i are separable. To prove the 'only if' part of the lemma, assume $\alpha \in L$ is not separable over k. Then by the above the number of k-homomorphisms $k(\alpha) \to \bar{k}$ is strictly less than $[k(\alpha) : k]$, and that of $k(\alpha)$-homomorphisms from L to $\bar{k}$ is at most $[L : k(\alpha)]$. Thus there are strictly less than n k-homomorphisms from L to $\bar{k}$. □

The criterion of the lemma immediately implies:

Corollary 1.1.7 *Given a tower $L|M|k$ of finite field extensions, the extension $L|k$ is separable if and only if $L|M$ and $M|k$ are.*

In the course of the proof we have also obtained:

Corollary 1.1.8 *A finite extension $L|k$ is separable if and only if $L = k(\alpha_1, \ldots, \alpha_m)$ for some separable elements $\alpha_i \in L$.*

We now show that there is a largest separable subextension inside a fixed algebraic closure $\bar{k}$ of k. For this recall that given two algebraic extensions L, M of k embedded as subfields in $\bar{k}$, their *compositum* LM is the smallest subfield of $\bar{k}$ containing both L and M.

Corollary 1.1.9 *If L, M are finite separable extensions of k, their compositum is separable as well.*

Proof By definition of LM there exist finitely many separable elements $\alpha_1, \ldots, \alpha_m$ of L such that $LM = M(\alpha_1, \ldots, \alpha_m)$. As the α_i are separable over k, they are separable over M, and so the extension $LM|M$ is separable by the previous corollary. But so is $M|k$ by assumption, and we conclude by Corollary 1.1.7. □

In view of the above two corollaries the compositum of all finite separable subextensions of $\bar{k}$ is a separable extension $k_s|k$ containing each finite separable subextension of $\bar{k}|k$.

Definition 1.1.10 The extension k_s is called the *separable closure* of k in $\bar{k}$.

From now on by 'a separable closure of k' we shall mean its separable closure in some chosen algebraic closure.

The following important property of finite separable extensions is usually referred to as the *theorem of the primitive element.*

Proposition 1.1.11 *A finite separable extension can be generated by a single element.*

For the proof, see Lang [48], Chapter V, Theorem 4.6 or van der Waerden [106], §46.

A field is called *perfect* if all of its finite extensions are separable. By definition, for perfect fields the algebraic and separable closures coincide.

Examples 1.1.12

1. Fields of characteristic 0 and algebraically closed fields are perfect.
2. A typical example of a non-perfect field is a rational function field $\mathbf{F}(t)$ in one variable over a field $\mathbf{F}$ of characteristic p: here adjoining a p-th root ξ of the indeterminate t defines an inseparable extension in view of the decomposition $X^p - t = (X - \xi)^p$.

   This is a special case of a general fact: a field k of characteristic $p > 0$ is perfect if and only if $k^p = k$ ([48], Chapter V, Corollary 6.12 or [106], §45). The criterion is satisfied by a finite field $\mathbf{F}_{p^r}$ as its multiplicative group is cyclic of order $p^r - 1$; hence finite fields are perfect.

1.2 Galois extensions

Now we come to the fundamental definition in Galois theory. Given an extension L of k, denote by $\mathrm{Aut}(L|k)$ the group of field automorphisms of L fixing k elementwise. The elements of L that are fixed by the action of $\mathrm{Aut}(L|k)$ form a field extension of k. In general it may be larger than k.

Definition 1.2.1 An algebraic extension L of k is called a *Galois extension* of k if the elements of L that remain fixed under the action of $\mathrm{Aut}(L|k)$ are *exactly* those of k. In this case $\mathrm{Aut}(L|k)$ is denoted by $\mathrm{Gal}\,(L|k)$, and called the *Galois group* of L over k.

Though the above definition is classical (it goes back to Emil Artin), it may not sound familiar to some readers. We shall now make the link with other definitions. The first step is:

Lemma 1.2.2 *A Galois extension $L|k$ is separable, and the minimal polynomial over k of each $\alpha \in L$ splits into linear factors in L.*

Proof Each element $\alpha \in L$ is a root of the polynomial $f = \prod(x - \sigma(\alpha))$, where σ runs over a system of (left) coset representatives of the stabilizer of α in $G = \mathrm{Gal}\,(L|k)$. The product is indeed finite, because the $\sigma(\alpha)$ must be roots of the the minimal polynomial g of α. In fact, we must have $f = g$. Indeed, both polynomials lie in $k[x]$ and have α as a root, hence each $\sigma(\alpha)$ must be a root of both. Thus f divides g but g is irreducible. Finally, by construction f has no multiple roots, thus α is separable over k. □

The converse also holds. Before proving it, we consider the 'most important' example of a Galois extension.

Example 1.2.3 A separable closure k_s of a field k is always a Galois extension. Indeed, to check that it is Galois we have to show that each element α of k_s not contained in k is moved by an appropriate automorphism in $\mathrm{Aut}(k_s|k)$. For this let $\alpha' \in k_s$ be another root of the minimal polynomial of α, and consider the isomorphism of field extensions $k(\alpha) \xrightarrow{\sim} k(\alpha')$ obtained by sending α to α'. An application of the third part of Proposition 1.1.3 shows that this isomorphism can be extended to an automorphism of the algebraic closure $\bar{k}$. To conclude one only has to remark that each automorphism of $\mathrm{Aut}(\bar{k}|k)$ maps k_s onto itself, since such an automorphism sends an element β of $\bar{k}$ to another root β' of its minimal polynomial; thus if β is separable, then so is β'.

The group $\mathrm{Gal}\,(k_s|k)$ is called the *absolute Galois group* of k.

We can now state and prove the following important characterization of Galois extensions.

Proposition 1.2.4 *Let k be a field, k_s a separable closure and $L \subset k_s$ a subfield containing k. The following properties are equivalent.*

1. *The extension $L|k$ is Galois.*
2. *The minimal polynomial over k of each $\alpha \in L$ splits into linear factors in L.*
3. *Each automorphism $\sigma \in \mathrm{Gal}\,(k_s|k)$ satisfies $\sigma(L) \subset L$.*

Proof The proof of (1) $\Rightarrow$ (2) was given in Lemma 1.2.2 above. The implication (2) $\Rightarrow$ (3) follows from the fact that each $\sigma \in \mathrm{Gal}\,(k_s|k)$ must map $\alpha \in L$ to a root of its minimal polynomial. Finally, for (3) $\Rightarrow$ (1) pick $\alpha \in L \setminus k$. As k_s is Galois over k (Example 1.2.3), we find $\sigma \in \mathrm{Gal}\,(k_s|k)$ with $\sigma(\alpha) \neq \alpha$. By (3), this σ preserves L, so its restriction to L yields an element of $\mathrm{Aut}(L|k)$ which does not fix α. □

Using the proposition it is easy to prove the main results of Galois theory for finite Galois extensions.

Theorem 1.2.5 (Main Theorem of Galois theory for finite extensions) *Let $L|k$ be a finite Galois extension with Galois group G. The maps*

$$M \mapsto H := \mathrm{Aut}(L|M) \quad \text{and} \quad H \mapsto M := L^H$$

yield an inclusion-reversing bijection between subfields $L \supset M \supset k$ and subgroups $H \subset G$. The extension $L|M$ is always Galois. The extension $M|k$ is Galois if and only if H is a normal subgroup of G; in this case we have $\mathrm{Gal}\,(M|k) \cong G/H$.

In the above statement the notation L^H means, as usual, the subfield of L fixed by H elementwise.

Proof Let M be a subfield of L containing k. Fixing a separable closure $k_s|k$ containing L, we see from Proposition 1.2.4 (3) that $L|k$ being Galois automatically implies that $L|M$ is Galois as well. Writing $H = \mathrm{Gal}(L|M)$, we therefore have $L^H = M$. Conversely, if $H \subset G$, then L is Galois over L^H by definition, and the Galois group is H. Now only the last statement remains to be proven. If $H \subset G$ is normal, we have a natural action of G/H on $M = L^H$, since the action of $g \in G$ on an element of L^H only depends on its class modulo H. As $L|k$ is Galois, we have $M^{G/H} = L^G = k$, so $M|k$ is Galois with group G/H. Conversely, if $M|k$ is Galois, then each automorphism $\sigma \in G$ preserves M (extend σ to an automorphism of k_s using Proposition 1.1.3 (3), and then apply Proposition 1.2.4 (3)). Restriction to M thus induces a natural homomorphism $G \to \mathrm{Gal}(M|k)$ whose kernel is exactly $H = \mathrm{Gal}(M|k)$. It follows that H is normal in G. □

Classically Galois extensions arise as *splitting fields* of separable polynomials. Given an irreducible separable polynomial $f \in k[x]$, its splitting field is defined as the finite subextension $L|k$ of $k_s|k$ generated by all roots of f in k_s. This notion depends on the choice of the separable closure k_s.

Lemma 1.2.6 *A finite extension $L|k$ is Galois if and only if it is the splitting field of an irreducible separable polynomial $f \in k[x]$.*

Proof The splitting field of an irreducible separable polynomial is indeed Galois, as it satisfies criterion (3) of Proposition 1.2.4. Conversely, part (2) of the proposition implies that a finite Galois extension $L|k$ is the splitting field of a primitive element generating L over k. □

Corollary 1.2.7 *A finite extension $L|k$ is Galois with group $G = \mathrm{Aut}(L|k)$ if and only if G has order $[L : k]$.*

Proof If $L|k$ is Galois, it is the splitting field of a polynomial by the proposition, so G has order $[L : k]$ by construction. Conversely, for $G = \mathrm{Aut}(L|k)$ the extension $L|L^G$ is Galois by definition, so G has order $[L : L^G]$ by what we have just proven. This forces $L^G = k$. □

Remark 1.2.8 An important observation concerning the splitting field L of a polynomial $f \in k[x]$ is that by definition $\mathrm{Gal}(L|k)$ acts on L by permuting the roots of f. Thus if f has degree n, we obtain an injective homomorphism from $\mathrm{Gal}(L|k)$ to S_n, the symmetric group on n letters. This implies in particular that $L|k$ has degree at most $n!$. The bound is sharp; see for instance Example 1.2.9 (3) below.

In the remainder of this section we give examples of Galois and non-Galois extensions.

Examples 1.2.9

1. Let $m > 2$ be an integer and ω a primitive m-th root of unity. The extension $\mathbf{Q}(\omega)|\mathbf{Q}$ is Galois, being the splitting field of the minimal polynomial of ω, the m-th *cyclotomic polynomial* Φ_m. Indeed, all other roots of Φ_m are powers of ω, and hence are contained in $\mathbf{Q}(\omega)$. The degree of Φ_m is $\phi(m)$, where ϕ denotes the Euler function. The Galois group is isomorphic to $(\mathbf{Z}/m\mathbf{Z})^\times$, the group of units in the ring $\mathbf{Z}/m\mathbf{Z}$. When m is a prime power, it is known to be cyclic.
2. For an example of infinite degree, let $\mathbf{Q}(\mu)|\mathbf{Q}$ be the extension obtained by adjoining all roots of unity to $\mathbf{Q}$ (in the standard algebraic closure $\overline{\mathbf{Q}}$ contained in $\mathbf{C}$). Every automorphism in Gal $(\overline{\mathbf{Q}}|\mathbf{Q})$ must send $\mathbf{Q}(\mu)$ onto itself, because it must send an m-th root of unity to another m-th root of unity. Thus by criterion (3) of Proposition 1.2.4 we indeed get a Galois extension. We shall determine its Galois group in the next section.

   By the same argument we obtain that for a prime number p the field $\mathbf{Q}(\mu_{p^\infty})$ generated by the p-power roots of unity is Galois over $\mathbf{Q}$.
3. Let k be a field containing a primitive m-th root of unity ω for an integer $m > 1$ invertible in k (this means that the polynomial $x^m - 1$ splits into linear factors over k). Pick an element $a \in k^\times \setminus k^{\times m}$, and let $\sqrt[m]{a}$ be a root of it in an algebraic closure $\bar{k}$. The extension $k(\sqrt[m]{a})|k$ is Galois with group $\mathbf{Z}/m\mathbf{Z}$, generated by the automorphism $\sigma : \sqrt[m]{a} \to \omega\sqrt[m]{a}$. This is because all roots of $x^m - a$ are of the form $\omega^i\sqrt[m]{a}$ for some $0 \le i \le m-1$.
4. When k does not contain a primitive m-th root of unity, we may not get a Galois extension. For instance, take $k = \mathbf{Q}, m = 3$ and $a \in \mathbf{Q}^\times \setminus \mathbf{Q}^{\times 3}$. We define $\sqrt[3]{a}$ to be the unique real cube root of a. The extension $\mathbf{Q}(\sqrt[3]{a})|\mathbf{Q}$ is nontrivial because $\sqrt[3]{a} \notin \mathbf{Q}$, but $\mathrm{Aut}(\mathbf{Q}(\sqrt[3]{a})|\mathbf{Q})$ is trivial. Indeed, an automorphism in $\mathrm{Aut}(\mathbf{Q}(\sqrt[3]{a})|\mathbf{Q})$ must send $\sqrt[3]{a}$ to a root of $x^3 - a$ in $\mathbf{Q}(\sqrt[3]{a})$, but $\sqrt[3]{a}$ is the only one, since $\mathbf{Q}(\sqrt[3]{a}) \subset \mathbf{R}$ and the other two roots are complex. Thus the extension $\mathrm{Aut}(\mathbf{Q}(\sqrt[3]{a})|\mathbf{Q})$ is not Galois. The splitting field L of $x^3 - a$ is generated over $\mathbf{Q}$ by $\sqrt[3]{a}$ and a primitive third root of unity ω that has degree 2 over $\mathbf{Q}$, so L has degree 6 over $\mathbf{Q}$.
5. Finally, here is an example of a finite Galois extension in positive characteristic. Let k be of characteristic $p > 0$, and let $a \in k$ be an element so that the polynomial $f = x^p - x - a$ has no roots in k. (As a concrete example, one may take k to be the field $\mathbf{F}_p(t)$ of rational functions with mod p coefficients and $a = t$.) Observe that if α is a root in some extension $L|k$, then the other roots are $\alpha + 1, \alpha + 2, \ldots, \alpha + (p-1)$, and therefore f splits in distinct linear factors in L. It follows that f is irreducible over k, and that the extension $k(\alpha)|k$ is Galois with group $\mathbf{Z}/p\mathbf{Z}$, a generator sending α to $\alpha + 1$.

Remark 1.2.10 There exist converse statements to Examples 3 and 5 above. The main theorem of *Kummer theory* says that for a field k containing a primitive m-th root of unity every cyclic Galois extension with group $\mathbf{Z}/m\mathbf{Z}$ is generated by an m-th root $\sqrt[m]{a}$ for some $a \in k^\times \setminus k^{\times m}$. This further generalizes to Galois extensions with a finite abelian Galois group of exponent m: they can be generated by several m-th roots.

According to *Artin-Schreier theory*, in characteristic $p > 0$ every cyclic Galois extension with group $\mathbf{Z}/p\mathbf{Z}$ is generated by a root of an 'Artin–Schreier polynomial' $x^p - x - a$ as above. There are generalizations to extensions with a finite abelian Galois group of exponent p, but also to extensions with group $\mathbf{Z}/p^r\mathbf{Z}$; the latter uses the theory of Witt vectors. For details and proofs of the above statements, see e.g. [48], Chapter VI, §8.

Our final example gives an application of the above ideas outside the scope of Galois theory in the narrow sense.

Example 1.2.11 Let k be a field, and $K = k(x_1, \dots, x_n)$ a purely transcendental extension in n indeterminates. Make the symmetric group S_n act on K via permuting the x_i. By definition the extension $K|K^{S_n}$ is Galois with group S_n. It is the splitting field of the polynomial $f = (x - x_1)\dots(x - x_n)$. As f is invariant by the action of S_n, its coefficients lie in K^{S_n}. These coefficients are (up to a sign) the *elementary symmetric polynomials*

$$\begin{aligned} \sigma_1 &= x_1 + x_2 \cdots + x_n, \\ \sigma_2 &= x_1x_2 + x_1x_3 + \cdots + x_{n-1}x_n, \\ &\vdots \\ \sigma_n &= x_1x_2\cdots x_n. \end{aligned}$$

But by definition K is also the splitting field of f over the field $k(\sigma_1, \dots, \sigma_n)$. As $k(\sigma_1, \dots, \sigma_n) \subset K^{S_n}$ and $[K : K^{S_n}] = n!$, Remark 1.2.8 shows that $K^{S_n} = k(\sigma_1, \dots, \sigma_n)$.

With a little commutative algebra one can say more. The x_i, being roots of f, are in fact *integral* over the subring $k[\sigma_1, \dots, \sigma_n] \subset k(\sigma_1, \dots, \sigma_n)$ (see Section 4.1 for basic facts and terminology). Therefore the subring $k[x_1, \dots, x_n]^{S_n} = k[x_1, \dots, x_n] \cap K^{S_n}$ of $k[x_1, \dots, x_n]$ is an integral ring extension of $k[\sigma_1, \dots, \sigma_n]$. But as $K \supset k(\sigma_1, \dots, \sigma_n)$ is a finite extension containing n algebraically independent elements, the σ_i must be algebraically independent over k. Thus $k[\sigma_1, \dots, \sigma_n]$ is isomorphic to a polynomial ring; in particular, it is integrally closed in its fraction field K^{S_n}. It follows that $k[x_1, \dots, x_n]^{S_n} = k[\sigma_1, \dots, \sigma_n]$. This is the *main theorem of symmetric polynomials*: every symmetric polynomial in n variables over k is a polynomial in the σ_i. For more traditional proofs, see [48], Chapter IV, Theorem 6.1 or [106], §33.

Remark 1.2.12 The above example also shows that each finite group G occurs as the Galois group of some Galois extension. Indeed, we may embed G in a symmetric group S_n for suitable n and then consider its action on the transcendental extension $K|k$ of the above example. The extension $K|K^G$ will then do. However, we shall see in the next section that the analogous statement is false for most infinite G.

1.3 Infinite Galois extensions

We now address the problem of extending the main theorem of Galois theory to infinite Galois extensions. The main difficulty is that for an infinite extension it will no longer be true that all subgroups of the Galois group arise as the subgroup fixing some subextension $M|k$. The first example of a subgroup that does not correspond to some subextension was found by Dedekind, who, according to Wolfgang Krull, already had the feeling that *'die Galoissche Gruppe gewissermaßen eine stetige Mannigfaltigkeit bilde'*. It was Krull who then cleared up the question in his classic paper [47]; we now describe a modern version of his theory.

Let $K|k$ be a possibly infinite Galois extension. The first step is the observation that K is a union of finite *Galois* extensions of k. More precisely:

Lemma 1.3.1 *Each finite subextension of $K|k$ can be embedded in a Galois subextension.*

Proof By the theorem of the primitive element (Proposition 1.1.11), each finite subextension is of the form $k(\alpha)$ with an appropriate element α. We may embed $k(\alpha)$ into the splitting field of the minimal polynomial of α which is Galois over k. □

This fact has a crucial consequence for the Galois group $\mathrm{Gal}(K|k)$, namely that it is determined by its finite quotients. We shall prove this in Proposition 1.3.5 below, in a more precise form. To motivate its formulation, consider a tower of finite Galois subextensions $M|L|k$ contained in an infinite Galois extension $K|k$. The main theorem of Galois theory provides us with a canonical surjection $\phi_{ML} : \mathrm{Gal}(M|k) \twoheadrightarrow \mathrm{Gal}(L|k)$. Moreover, if $N|k$ is yet another finite Galois extension containing M, we have $\phi_{NL} = \phi_{ML} \circ \phi_{NM}$. Thus one expects that if we somehow 'pass to the limit in M', then $\mathrm{Gal}(L|k)$ will actually become a quotient of the infinite Galois group $\mathrm{Gal}(K|k)$ itself. This is achieved by the following construction.

Construction 1.3.2 A *(filtered) inverse system* of groups $(G_\alpha, \phi_{\alpha\beta})$ consists of:

- a partially ordered set $(\Lambda, \leq)$ which is directed in the sense that for all $(\alpha, \beta) \in \Lambda$ there is some $\gamma \in \Lambda$ with $\alpha \leq \gamma, \beta \leq \gamma$;
- for each $\alpha \in \Lambda$ a group G_α;
- for each $\alpha \leq \beta$ a homomorphism $\phi_{\alpha\beta} : G_\beta \to G_\alpha$ such that we have equalities $\phi_{\alpha\gamma} = \phi_{\alpha\beta} \circ \phi_{\beta\gamma}$ for $\alpha \leq \beta \leq \gamma$.

The *inverse limit* of the system is defined as the subgroup of the direct product $\prod_{\alpha\in\Lambda} G_\alpha$ consisting of sequences (g_α) such that $\phi_{\alpha\beta}(g_\beta) = g_\alpha$ for all $\alpha \leq \beta$. It is denoted by $\varprojlim G_\alpha$; we shall not specify the inverse system in the notation when it is clear from the context. Also, we shall often loosely say that $\varprojlim G_\alpha$ is the inverse limit of the groups G_α, without special reference to the inverse system.

Plainly, this notion is not specific to the category of groups and one can define the inverse limit of sets, rings, modules, even of topological spaces in an analogous way.

We now come to the key definition.

Definition 1.3.3 A *profinite group* is defined to be an inverse limit of a system of finite groups. For a prime number p, a *pro-p group* is an inverse limit of finite p-groups.

Examples 1.3.4

1. A finite group is profinite; indeed, it is the inverse limit of the system $(G_\alpha, \phi_{\alpha\beta})$ for any directed index set Λ, with $G_\alpha = G$ and $\phi_{\alpha\beta} = \mathrm{id}_G$.
2. Given a group G, the set of its finite quotients can be turned into an inverse system as follows. Let Λ be the index set formed by the normal subgroups of finite index partially ordered by the following relation: $U_\alpha \leq U_\beta \Leftrightarrow U_\alpha \supset U_\beta$. For each pair $U_\alpha \leq U_\beta$ of normal subgroups we have a quotient map $\phi_{\alpha\beta} : G/U_\beta \to G/U_\alpha$. The inverse limit of this system is called the *profinite completion* of G, customarily denoted by $\widehat{G}$. There is a canonical homomorphism $G \to \widehat{G}$.
3. Take $G = \mathbf{Z}$ in the previous example. Then Λ is just the set $\mathbf{Z}_{>0}$, since each subgroup of finite index is generated by some positive integer m. The partial order is induced by the divisibility relation: $m|n$ iff $m\mathbf{Z} \supset n\mathbf{Z}$. The completion $\widehat{\mathbf{Z}}$ is usually called *zed hat* (or *zee hat* in the US). In fact, $\widehat{\mathbf{Z}}$ is also a ring, with multiplication induced by that of the $\mathbf{Z}/m\mathbf{Z}$.
4. In the previous example, taking only powers of some prime p in place of m we get a subsystem of the inverse system considered there; it is more convenient to index it by the exponent of p. With this convention the

partial order becomes the usual (total) order of $\mathbf{Z}_{>0}$. The inverse limit is $\mathbf{Z}_p$, *the additive group of p-adic integers*. This is a commutative pro-p-group. The Chinese Remainder Theorem implies that the direct product of the groups $\mathbf{Z}_p$ for all primes p is isomorphic to $\widehat{\mathbf{Z}}$. Again the $\mathbf{Z}_p$ carry a natural ring structure as well, and this is in fact a ring isomorphism.

Now we come to the main example, that of Galois groups.

Proposition 1.3.5 *Let $K|k$ be a Galois extension of fields. The Galois groups of finite Galois subextensions of $K|k$ together with the homomorphisms ϕ_{ML} : $\mathrm{Gal}(M|k) \to \mathrm{Gal}(L|k)$ form an inverse system whose inverse limit is isomorphic to $\mathrm{Gal}(K|k)$. In particular, $\mathrm{Gal}(K|k)$ is a profinite group.*

Proof Only the isomorphism statement needs a proof. For this, define a group homomorphism $\phi : \mathrm{Gal}(K|k) \to \prod \mathrm{Gal}(L|k)$ (where the product is over all finite Galois subextensions $L|k$) by sending a k-automorphism σ of K to the direct product of its restrictions to the various subfields L indexing the product. That $\sigma(L) \subset L$ for all such L follows from Proposition 1.2.4. The map ϕ is injective, since if an automorphism σ does not fix an element α of k_s, then its restriction to a finite Galois subextension containing $k(\alpha)$ is nontrivial (as we have already remarked, such an extension always exists). On the other hand, the main theorem of Galois theory assures that the image of ϕ is contained in $\varprojlim \mathrm{Gal}(L|k)$. It is actually all of $\varprojlim \mathrm{Gal}(L|k)$, which is seen as follows: take an element (σ_L) of $\varprojlim \mathrm{Gal}(L|k)$ and define a k-automorphism σ of K by putting $\sigma(\alpha) = \sigma_L(\alpha)$ with some finite Galois L containing $k(\alpha)$. The fact that σ is well-defined follows from the fact that by hypothesis the σ_L form a compatible system of automorphisms; finally, σ maps to $(\sigma_L) \in \varprojlim \mathrm{Gal}(L|k)$ by construction. □

Corollary 1.3.6 *Projection to the components of the inverse limit of the proposition yields natural surjections $\mathrm{Gal}(K|k) \to \mathrm{Gal}(L|k)$ for all finite Galois subextensions $L|k$ contained in K.*

Examples 1.3.7

1. Let $\mathbf{F}$ be a finite field of order q, and $\mathbf{F}_s$ a separable closure of $\mathbf{F}$. It is well known that for each integer $n > 0$ the extension $\mathbf{F}_s|\mathbf{F}$ has a unique subextension $\mathbf{F}_n|\mathbf{F}$ with $[\mathbf{F}_n : \mathbf{F}] = n$. Moreover, the extension $\mathbf{F}_n|\mathbf{F}$ is Galois with group $\mathrm{Gal}(\mathbf{F}_n|\mathbf{F}) \cong \mathbf{Z}/n\mathbf{Z}$, a generator being given by the *Frobenius automorphism* $\alpha \mapsto \alpha^q$. Via this isomorphism the natural projections $\mathrm{Gal}(\mathbf{F}_{mn}|\mathbf{F}) \to \mathrm{Gal}(\mathbf{F}_n|\mathbf{F})$ correspond to the projections $\mathbf{Z}/mn\mathbf{Z} \to \mathbf{Z}/n\mathbf{Z}$ (see [48], Chapter V, §5 or [106], §§43, 57). It follows

that $\mathrm{Gal}(\mathbf{F}_s|\mathbf{F}) \cong \widehat{\mathbf{Z}}$, the element 1 on the right-hand side corresponding to the Frobenius automorphism on the left.

2. Consider now the infinite extension $\mathbf{Q}(\mu_{p^\infty})$ obtained by adjoining to $\mathbf{Q}$ all p-power roots of unity for a fixed prime p. We have seen in Example 1.2.9 (2) that this is a Galois extension. It is the union of the chain of finite subextensions $\mathbf{Q}(\mu_p) \subset \mathbf{Q}(\mu_{p^2}) \subset \mathbf{Q}(\mu_{p^3}) \subset \dots$, where μ_{p^r} is the group of p^r-th roots of unity. As mentioned in Example 1.2.9 (1), one has $\mathrm{Gal}(\mathbf{Q}(\mu_{p^r})|\mathbf{Q}) \cong (\mathbf{Z}/p^r\mathbf{Z})^\times$. It follows that $\mathrm{Gal}(\mathbf{Q}(\mu_{p^\infty})|\mathbf{Q})$ is $\varprojlim(\mathbf{Z}/p^r\mathbf{Z})^\times = \mathbf{Z}_p^\times$, the group of units in the ring $\mathbf{Z}_p$. This group is known to be isomorphic to $\mathbf{Z}/(p-1)\mathbf{Z} \times \mathbf{Z}_p$ for $p > 2$, and to $\mathbf{Z}/2\mathbf{Z} \times \mathbf{Z}_2$ for $p = 2$ (see e.g. [90], Chapter II, Theorem 2).

   Similarly, one obtains that the Galois group of the extension $\mathbf{Q}(\mu)|\mathbf{Q}$ is isomorphic to $\widehat{\mathbf{Z}}^\times$, where $\mathbf{Q}(\mu)$ is the extension of $\mathbf{Q}$ generated by all roots of unity.

Profinite groups are endowed with a natural topology as follows: if G is an inverse limit of a system of finite groups $(G_\alpha, \phi_{\alpha\beta})$, endow the G_α with the discrete topology, their product with the product topology and the subgroup $G \subset \prod G_\alpha$ with the subspace topology. It immediately follows from this construction that the natural projection maps $G \to G_\alpha$ are continuous and their kernels form a basis of open neighbourhoods of 1 in G (for the last statement, note that the image of each element $g \neq 1$ of G must have nontrivial image in some G_α, by definition of the inverse limit).

To state other topological properties, we need a lemma.

Lemma 1.3.8 *Let $(G_\alpha, \phi_{\alpha\beta})$ be an inverse system of groups equipped with the discrete topology. The inverse limit $\varprojlim G_\alpha$ is a closed topological subgroup of the product $\prod G_\alpha$.*

Proof Take an element $g = (g_\alpha) \in \prod G_\alpha$. If $g \notin \varprojlim G_\alpha$, we have to show that it has an open neighbourhood which does not meet $\varprojlim G_\alpha$. By assumption for some α and β we must have $\phi_{\alpha\beta}(g_\beta) \neq g_\alpha$. Now take the subset of $\prod G_\alpha$ consisting of all elements with α-th component g_α and β-th component g_β. It is a suitable choice, being open (by the discreteness of the G_α and by the definition of topological product) and containing g but avoiding $\varprojlim G_\alpha$. □

Corollary 1.3.9 *A profinite group is compact and totally disconnected (i.e. the only connected subsets are the one-element subsets). Moreover, the open subgroups are precisely the closed subgroups of finite index.*

Proof Recall that finite discrete groups are compact, and so is a product of compact groups, by Tikhonov's theorem ([66], Theorem 37.3). Compactness of the inverse limit then follows from the lemma, as closed subspaces of compact spaces are compact. Complete disconnectedness follows from the construction. For the second statement, note that each open subgroup U is closed since its complement is a disjoint union of cosets gU which are themselves open (the map $U \mapsto gU$ being a homeomorphism in a topological group); by compactness of G, these must be finite in number. Conversely, a closed subgroup of finite index is open, being the complement of the finite disjoint union of its cosets which are also closed. □

Remarks 1.3.10

1. In fact, one may characterize profinite groups as being those topological groups which are compact and totally disconnected. See e.g. [96], §1, Theorem 2 for a proof.
2. One may ask whether all subgroups of finite index in a profinite group are actually open. This is false already for the absolute Galois group of $\mathbf{Q}$ (see Exercise 4). However, it has been conjectured for a long time whether the property holds for those profinite groups that are *topologically finitely generated* (i.e. contain a dense finitely generated subgroup). This conjecture was recently proven by Nikolov and Segal [71].

We may now state and prove the main theorem of Galois theory for possibly infinite extensions. Observe first that if L is a subextension of a Galois extension $K|k$, then K is also a Galois extension of L and $\mathrm{Gal}(K|L)$ is naturally identified with a subgroup of $\mathrm{Gal}(K|k)$.

Theorem 1.3.11 (Krull) *Let L be a subextension of the Galois extension $K|k$. Then $\mathrm{Gal}(K|L)$ is a closed subgroup of $\mathrm{Gal}(K|k)$. Moreover, the maps*

$$L \mapsto H := \mathrm{Gal}(K|L) \quad \text{and} \quad H \mapsto L := K^H$$

yield an inclusion-reversing bijection between subfields $K \supset L \supset k$ and closed subgroups $H \subset G$. A subextension $L|k$ is Galois over k if and only if $\mathrm{Gal}(K|L)$ is normal in $\mathrm{Gal}(K|k)$; in this case there is a natural isomorphism $\mathrm{Gal}(L|k) \cong \mathrm{Gal}(K|k)/\mathrm{Gal}(K|L)$.

Proof Take first a finite separable extension $L|k$ contained in K. Using Lemma 1.3.1 we may embed it in a finite Galois extension $M|k$ contained in K. Then $\mathrm{Gal}(M|k)$ is one of the standard finite quotients of $\mathrm{Gal}(K|k)$, and it contains $\mathrm{Gal}(M|L)$ as a subgroup. Let U_L be the inverse image of $\mathrm{Gal}(M|L)$ by the natural projection $\mathrm{Gal}(K|k) \to \mathrm{Gal}(M|k)$. Since the projection is continuous and $\mathrm{Gal}(M|k)$ has the discrete topology, U_L is open. We claim that $U_L = \mathrm{Gal}(K|L)$. Indeed, we have $U_L \subset \mathrm{Gal}(K|L)$, for each element of U_L fixes L; on the other hand, the image of $\mathrm{Gal}(K|L)$ by the projection

$\mathrm{Gal}(K|k) \to \mathrm{Gal}(M|k)$ is contained in $\mathrm{Gal}(M|L)$, whence the reverse inclusion. Now if $L|k$ is an arbitrary subextension of $K|k$, write it as a union of finite subextensions $L_\alpha|k$. By what we have just proven, each $\mathrm{Gal}(K|L_\alpha)$ is an open subgroup of $\mathrm{Gal}(K|k)$, hence it is also closed by Corollary 1.3.9. Their intersection is precisely $\mathrm{Gal}(K|L)$ which is thus a closed subgroup; its fixed field is exactly L, for K is Galois over L.

Conversely, given a closed subgroup $H \subset G$, it fixes some extension $L|k$ and is thus contained in $\mathrm{Gal}(K|L)$. To show equality, let σ be an element of $\mathrm{Gal}(K|L)$, and pick a fundamental open neighbourhood U_M of the identity in $\mathrm{Gal}(K|L)$, corresponding to a Galois extension $M|L$. Now $H \subset \mathrm{Gal}(K|L)$ surjects onto $\mathrm{Gal}(M|L)$ by the natural projection; indeed, otherwise its image in $\mathrm{Gal}(M|L)$ would fix a subfield of M strictly larger than L according to finite Galois theory, which would contradict our assumption that each element of $M \setminus L$ is moved by some element of H. In particular, some element of H must map to the same element in $\mathrm{Gal}(M|L)$ as σ. Hence H contains an element of the coset σU_M and, as U_M was chosen arbitrarily, this implies that σ is in the closure of H in $\mathrm{Gal}(K|L)$. But H is closed by assumption, whence the claim. The assertion about finite extensions follows from the above in view of Corollary 1.3.9.

Finally, the relation between Galois subextensions and normal subgroups is proven exactly as in the finite case. □

Remark 1.3.12 To see that the Galois theory of infinite extensions is really different from the finite case, we must exhibit non-closed subgroups in the Galois group. We have already seen such a subgroup in the last two examples of 1.3.4: the absolute Galois group of a finite field is isomorphic to $\hat{\mathbf{Z}}$, which in turn contains $\mathbf{Z}$ as a nontrivial dense (hence non-closed) subgroup; there are in fact many copies of $\mathbf{Z}$ embeddded in $\hat{\mathbf{Z}}$.

The original example of Dedekind was very similar: he worked with the extension $\mathbf{Q}(\mu_{p^\infty})|\mathbf{Q}$. However, he did not determine the Galois group itself (profinite groups were not yet discovered at the time); he just showed the existence of a non-closed subgroup. His proof was generalized in Krull [47] to establish the existence of non-closed subgroups in the Galois group of any infinite extension as follows. First one shows that given a nontrivial Galois extension $K_2|K_1$, each automorphism of K_1 may be extended to an automorphism of K_2 in at least two ways. From this one infers by taking an infinite chain of nontrivial Galois subextensions of an infinite Galois extension $L|k$ that $\mathrm{Gal}(L|k)$ is uncountable. By the same argument, all infinite closed subgroups of $\mathrm{Gal}(L|k)$ are uncountable, hence countable subgroups in an infinite Galois group are never closed.

Remark 1.3.13 The absolute Galois group is a rather fine invariant for fields of finite type. In 1995 Florian Pop proved the following remarkable theorem: *Let K, L be two infinite fields that are finitely generated over their prime field. Fix separable closures K_s, L_s of K and L, respectively, and assume there exists a continuous isomorphism $\Phi : \mathrm{Gal}(K_s|K) \xrightarrow{\sim} \mathrm{Gal}(L_s|L)$ of profinite groups. Then there exist purely inseparable extensions $K'|K$, $L'|L$ with $K' \cong L'$. Moreover, there is an isomorphism $\phi : L'L_s \xrightarrow{\sim} K'K_s$ of separable closures such that $\Phi(g) = \phi^{-1} \circ g \circ \phi$ for all $g \in \mathrm{Gal}(K'K_s|K')$.*

Here recall that an algebraic extension is called purely inseparable if all of its separable subextensions are trivial. Of course, in characteristic 0 such extensions are trivial, and one has $K = K'$, $L = L'$. In fact, already the case $L = K$ of Pop's theorem is interesting: it shows that every continuous automorphism of the absolute Galois group comes from a field automorphism. The first non-trivial case of this theorem, that of finite Galois extensions of $\mathbf{Q}$, was proven by J. Neukirch [68]. Already this special case is quite surprising: for instance, it shows that if p and q are different primes, then the absolute Galois groups of $\mathbf{Q}(\sqrt{p})$ and $\mathbf{Q}(\sqrt{q})$ cannot be isomorphic. For more on this fascinating topic, see [77] and [101].

1.4 Interlude on category theory

In the next section we shall give another formulation of Galois theory which roughly states that 'up to isomorphism it is the same to give a finite separable extension of k and a finite set equipped with a continuous transitive $\mathrm{Gal}(k_s|k)$-action'. In order to be able to formulate the 'up to isomorphism it is the same' part of the above statement rigorously, it is convenient to recall some basic notions from category theory. These notions will be of constant use in what follows.

Definition 1.4.1 A *category* consists of *objects* as well as *morphisms* between pairs of objects; given two objects A, B of a category $\mathcal{C}$, the morphisms from A to B form a set, denoted by $\mathrm{Hom}(A, B)$. (Notice that in contrast to this we do not impose that the objects of the category form a set.) These are subject to the following constraints.

1. For each object A the set $\mathrm{Hom}(A, A)$ contains a distinguished element id_A, the identity morphism of A.
2. Given two morphisms $\phi \in \mathrm{Hom}(B, C)$ and $\psi \in \mathrm{Hom}(A, B)$, there exists a canonical morphism $\phi \circ \psi \in \mathrm{Hom}(A, C)$, the composition of ϕ and ψ. The composition of morphisms should satisfy two natural axioms:

- Given $\phi \in \mathrm{Hom}(A, B)$, one has $\phi \circ \mathrm{id}_A = \mathrm{id}_B \circ \phi = \phi$.
- (Associativity rule) For morphisms $\lambda \in \mathrm{Hom}(A, B)$, $\psi \in \mathrm{Hom}(B, C)$, $\phi \in \mathrm{Hom}(C, D)$ one has $(\phi \circ \psi) \circ \lambda = \phi \circ (\psi \circ \lambda)$.

Some more definitions: a morphism $\phi \in \mathrm{Hom}(A, B)$ is an *isomorphism* if there exists $\psi \in \mathrm{Hom}(B, A)$ with $\psi \circ \phi = \mathrm{id}_A, \phi \circ \psi = \mathrm{id}_B$; we denote the set of isomorphisms between A and B by $\mathrm{Isom}(A, B)$. If the objects themselves form a set, one can associate an oriented graph to the category by taking objects as vertices and defining an oriented edge between two objects corresponding to each morphism. With this picture in mind, it is easy to conceive what the *opposite category* $\mathcal{C}^{op}$ of a category $\mathcal{C}$ is: it is 'the category with the same objects and arrows reversed'; i.e. for each pair of objects (A, B) of $\mathcal{C}$, there is a canonical bijection between the sets $\mathrm{Hom}(A, B)$ of $\mathcal{C}$ and $\mathrm{Hom}(B, A)$ of $\mathcal{C}^{op}$ preserving the identity morphisms and composition.

The *product* of two categories $\mathcal{C}_1$ and $\mathcal{C}_2$ is the category $\mathcal{C}_1 \times \mathcal{C}_2$ whose objects are pairs (C_1, C_2) with $C_i \in \mathcal{C}_i$ and whose morphisms are pairs (ϕ_1, ϕ_2) of morphisms in the $\mathcal{C}_i$. One defines arbitrary finite products of categories in a similar way.

A *subcategory* of a category $\mathcal{C}$ is just a category $\mathcal{D}$ consisting of some objects and some morphisms of $\mathcal{C}$; it is a *full* subcategory if given two objects in $\mathcal{D}$, $\mathrm{Hom}_{\mathcal{D}}(A, B) = \mathrm{Hom}_{\mathcal{C}}(A, B)$, i.e. *all* $\mathcal{C}$-morphisms between A and B are morphisms in $\mathcal{D}$.

Examples 1.4.2 Some categories we shall frequently encounter will be the category Sets of sets (with morphisms the set-theoretic maps), the category Ab of abelian groups (with group homomorphisms) or the category Top of topological spaces (with continuous maps). Both Ab and Top are naturally subcategories of Sets but they are not full subcategories; on the other hand, Ab is a full subcategory of the category of all groups.

Now comes the second basic definition of category theory.

Definition 1.4.3 A *(covariant) functor* F between two categories $\mathcal{C}_1$ and $\mathcal{C}_2$ consists of a rule $A \mapsto F(A)$ on objects and a map on sets of morphisms $\mathrm{Hom}(A, B) \to \mathrm{Hom}(F(A), F(B))$ which sends identity morphisms to identity morphisms and preserves composition. A *contravariant functor* from $\mathcal{C}_1$ to $\mathcal{C}_2$ is a functor from $\mathcal{C}_1$ to $\mathcal{C}_2^{op}$.

Examples 1.4.4 Here are some examples of functors.

1. The identity functor is the functor $\mathrm{id}_{\mathcal{C}}$ on any category $\mathcal{C}$ which leaves all objects and morphisms fixed.
2. Other basic examples of functors are obtained by fixing an object A of a category $\mathcal{C}$ and considering the covariant functor $\mathrm{Hom}(A, \ )$ (resp.

the contravariant functor Hom(, A)) from $\mathcal{C}$ to the category Sets which sends an object B the set $\mathrm{Hom}(A, B)$ (resp. $\mathrm{Hom}(B, A)$) and a morphism $\phi : B \to C$ to the set-theoretic map $\mathrm{Hom}(A, B) \to \mathrm{Hom}(A, C)$ (resp. $\mathrm{Hom}(C, A) \to \mathrm{Hom}(B, A)$) induced by composing with ϕ.

3. An example of a functor whose definition is not purely formal is given by the set-valued functor on the category Top that sends a topological space to its set of connected components. Here to see that this is really a functor one has to use the fact that a continuous map between topological spaces sends connected components to connected components.

Definition 1.4.5 If F and G are two functors with same domain $\mathcal{C}_1$ and target $\mathcal{C}_2$, a *morphism of functors* Φ between F and G is a collection of morphisms $\Phi_A : F(A) \to G(A)$ in $\mathcal{C}_2$ for each object $A \in \mathcal{C}_1$ such that for every morphism $\phi : A \to B$ in $\mathcal{C}_1$ the diagram

$$\begin{array}{ccc} F(A) & \xrightarrow{\Phi_A} & G(A) \\ {\scriptstyle F(\phi)}\downarrow & & \downarrow{\scriptstyle G(\phi)} \\ F(B) & \xrightarrow{\Phi_B} & G(B) \end{array}$$

commutes. The morphism Φ is an isomorphism if each Φ_A is an isomorphism; in this case we shall write $F \cong G$.

Remark 1.4.6 In the literature the terminology 'natural transformation' is frequently used instead of 'morphism of functors'. We prefer the latter name, as it reflects the fact that given two categories $\mathcal{C}_1$ and $\mathcal{C}_2$ one can define a new category called the *functor category* of the pair $(\mathcal{C}_1, \mathcal{C}_2)$ whose objects are functors from $\mathcal{C}_1$ to $\mathcal{C}_2$ and whose morphisms are morphisms of functors. Here the composition rule for some Φ and Ψ is induced by the composition of the morphisms Φ_A and Ψ_A for each object A in $\mathcal{C}_1$.

We can now give one of the notions which will be ubiquitous in what follows.

Definition 1.4.7 Two categories $\mathcal{C}_1$ and $\mathcal{C}_2$ are *equivalent* if there exist two functors $F : \mathcal{C}_1 \to \mathcal{C}_2$ and $G : \mathcal{C}_2 \to \mathcal{C}_1$, and two isomorphisms of functors $\Phi : F \circ G \xrightarrow{\sim} \mathrm{id}_{\mathcal{C}_2}$ and $\Psi : G \circ F \xrightarrow{\sim} \mathrm{id}_{\mathcal{C}_1}$. In this situation we say that the functor G is a *quasi-inverse* for F (and F is a quasi-inverse for G).

If we can actually find F and G with $F \circ G = \mathrm{id}_{\mathcal{C}_2}$ and $G \circ F = \mathrm{id}_{\mathcal{C}_1}$, we say that $\mathcal{C}_1$ and $\mathcal{C}_2$ are *isomorphic*. Finally, we say that $\mathcal{C}_1$ and $\mathcal{C}_2$ are *anti-equivalent* (resp. *anti-isomorphic*) if $\mathcal{C}_1$ is equivalent (resp. isomorphic) to $\mathcal{C}_2^{op}$.

One sees that equivalence of categories has all properties that equivalence relations on sets have, i.e. it is reflexive, symmetric and transitive. Also, the seemingly asymmetric definition of anti-equivalence is readily seen to be symmetric.

In practice, when one has to establish an equivalence of categories it often turns out that the construction of one functor is easy but that of the quasi-inverse is rather cumbersome. The following general lemma enables us to make do with the construction of only one functor in concrete situations. Before stating it, we introduce some terminology.

Definition 1.4.8 A functor $F : \mathcal{C}_1 \to \mathcal{C}_2$ is *faithful* if for any two objects A and B of $\mathcal{C}_1$ the map of sets $F_{AB} : \mathrm{Hom}(A, B) \to \mathrm{Hom}(F(A), F(B))$ induced by F is injective; it is *fully faithful* if all the maps F_{AB} are bijective.

The functor F is is *essentially surjective* if every object of $\mathcal{C}_2$ is isomorphic to some object of the form $F(A)$.

Lemma 1.4.9 *Two categories $\mathcal{C}_1$ and $\mathcal{C}_2$ are equivalent if and only if there exists a functor $F : \mathcal{C}_1 \to \mathcal{C}_2$ which is fully faithful and essentially surjective.*

There is an analogous characterization of anti-equivalent categories with fully faithful and essentially surjective contravariant functors (defined in the obvious way).

Proof For the proof of the 'if' part fix for each object V of $\mathcal{C}_2$ an isomorphism $i_V : F(A) \xrightarrow{\sim} V$ with some object A of $\mathcal{C}_1$. Such an isomorphism exists by the second condition. Define a functor $G : \mathcal{C}_2 \to \mathcal{C}_1$ by sending each object V in $\mathcal{C}_2$ to the A fixed above, and each morphism $\phi : V \to W$ to $G(\phi) = F_{AB}^{-1}(i_W^{-1} \circ \phi \circ i_V)$ for $\phi \in \mathrm{Hom}(V, W)$, where $B = G(W)$ and F_{AB} is the bijection appearing in the definition of fully faithfulness. The maps $i_V : F(G(V)) \xrightarrow{\sim} V$ induce an isomorphism $\Phi : F \circ G \xrightarrow{\sim} \mathrm{id}_{\mathcal{C}_2}$ by construction. To construct an isomorphism $\Psi : G \circ F \xrightarrow{\sim} \mathrm{id}_{\mathcal{C}_1}$, we first need functorial maps $\Psi_A : G(F(A)) \to A$ for each A in $\mathcal{C}_1$. By fully faithfulness of F it is enough to construct maps $F(\Psi_A) : F(G(F(A)) \to F(A)$, and we may take as $F(\Psi_A)$ the unique preimage of $\mathrm{id}_{F(A)}$ by $\Phi_{F(A)}$. A similar construction yields a map $A \to G(F(A))$ that is an inverse to Ψ_A. As the construction is functorial in A, we obtain an isomorphism of functors.

For the 'only if' part assume there exist a functor $G : \mathcal{C}_2 \to \mathcal{C}_1$ and isomorphisms of functors $\Phi : F \circ G \xrightarrow{\sim} \mathrm{id}_{\mathcal{C}_2}$ and $\Psi : G \circ F \xrightarrow{\sim} \mathrm{id}_{\mathcal{C}_1}$. Essential surjectivity is immediate: given an object C of $\mathcal{C}_2$, it is isomorphic to $F(G(C))$ via Φ. For fully faithfulness fix any two objects A, B of $\mathcal{C}_1$ and consider the sequence of maps

$$\mathrm{Hom}(A, B) \to \mathrm{Hom}(F(A), F(B)) \to \mathrm{Hom}(G(F(A)), G(F(B))) \to \mathrm{Hom}(A, B)$$

induced respectively by F_{AB}, $G_{F(A),F(B)}$ and Ψ. Their composite is the identity, which implies that F_{AB} is a bijection. □

Note that in the above proof the construction of G depended on the axiom of choice: we had to pick for each V an isomorphism i_V. Different choices define different G's but the categories are equivalent with any choice. This suggests that the notion of equivalence of categories means that 'up to isomorphism the categories have the same objects and morphisms' but this does not mean at all that there are bijections between objects and morphisms. In order to stress this point we discuss an example that expresses a basic fact from linear algebra in the language of category theory.

Example 1.4.10 Consider a field k and the category Vecf_k of finite dimensional k-vector spaces (with linear maps as morphisms). We show that this category is equivalent to a category $\mathcal{C}$ that we define as follows. The objects of $\mathcal{C}$ are to be the non-negative integers, and the set of morphisms between two integers $n, m > 0$ is to be the set of all n by m matrices with entries in k; here the identity morphisms are given by the identity matrices and composition by multiplication of matrices. There are also canonical morphisms $0 \to n$ and $n \to 0$ for each $n \geq 0$. (Notice that this is an example of a category where morphisms are not set-theoretic maps.)

To show the asserted equivalence, we introduce an auxiliary subcategory $\mathcal{C}'$ and show it is equivalent to both categories. This category $\mathcal{C}'$ is to be the full subcategory of Vecf_k spanned by the standard vector spaces k^n. The equivalence of $\mathcal{C}'$ and Vecf_k is immediate from the criterion of the previous lemma applied to the inclusion functor of $\mathcal{C}'$ to Vecf_k: the first condition is tautological as we took $\mathcal{C}'$ to be a full subcategory and the second holds because any finite dimensional vector space is isomorphic to some k^n.

We now show the equivalence of $\mathcal{C}'$ and $\mathcal{C}$. Define a functor $F : \mathcal{C}' \to \mathcal{C}$ by sending k^n to n and a morphism $k^n \to k^m$ to its matrix with respect to the standard bases. The fact that F is indeed a functor hides a nontrivial result of linear algebra, namely that the matrix of the composition of two linear maps ϕ and ψ is the product of the matrix of ϕ with that of ψ. One constructs an inverse functor G by reversing this procedure. In this case $F \circ G$ and $G \circ F$ are actually *equal* to the appropriate identity functors.

We close this brief overview with a very important notion due to Grothendieck. It will not be used until the next chapter.

Definition 1.4.11 Let $\mathcal{C}$ be a category. A functor F from $\mathcal{C}$ to the category Sets is *representable* if there is an object $C \in \mathcal{C}$ and an isomorphism of functors $F \cong \mathrm{Hom}(C, \ )$.

Recall that the latter functor sends an object A to the set of morphisms from C to A. There is also an analogous notion for contravariant functors. The object C is called the *representing object.*

The following well-known lemma is of pivotal importance. Observe that if C and D are objects of $\mathcal{C}$, every morphism $D \to C$ induces a morphism of functors $\mathrm{Hom}(C, \ ) \to \mathrm{Hom}(D, \ )$ via composition.

Lemma 1.4.12 (Yoneda Lemma) *If F and G are functors $\mathcal{C} \to \mathrm{Sets}$ represented by objects C and D, respectively, every morphism $\Phi : F \to G$ of functors is induced by a unique morphism $D \to C$ as above.*

Proof The morphism $\Phi_C : F(C) \to G(C)$ can be rewritten as a map $\mathrm{Hom}(C, C) \to \mathrm{Hom}(D, C)$ using the representability of the functors. The image of the identity morphism $\mathrm{id}_C \in \mathrm{Hom}(C, C)$ by Φ_C then identifies with a morphism $\rho : D \to C$; we claim this is the one inducing Φ. Indeed, for an object A each element of $F(A) \cong \mathrm{Hom}(C, A)$ identifies with a morphism $\phi : C \to A$. Observe that ϕ as an element of $F(A)$ is none but the image of $\mathrm{id}_C \in \mathrm{Hom}(C, C) \cong F(C)$ via $F(\phi)$. As Φ is a morphism of functors, we get $\Phi_A(\phi) = G(\phi)(\rho)$, which under the isomorphism $G(A) \cong \mathrm{Hom}(D, A)$ corresponds exactly to $\phi \circ \rho$. □

Corollary 1.4.13 *The representing object of a representable functor F is unique up to unique isomorphism.*

Proof Assume C and D both represent F, and apply the Yoneda Lemma to the identity map of F. □

We shall encounter several interesting examples of representable functors from Chapter 2 onwards.

1.5 Finite étale algebras

We now return to Galois theory and give a second variant of the main theorem which is often referred to as 'Grothendieck's formulation of Galois theory'.

We start again from a base field k, of which we fix separable and algebraic closures $k_s \subset \bar{k}$. We use the shorthand $\mathrm{Gal}\,(k)$ for $\mathrm{Gal}\,(k_s|k)$. Let L be a finite separable extension of k; here we do *not* consider L as a subextension of k_s. We know that L has only finitely many k-algebra homomorphisms into $\bar{k}$ (the number of these is equal to $[L : k]$ by Lemma 1.1.6); actually the images of these homomorphisms are contained in k_s. So we may consider the finite set $\mathrm{Hom}_k(L, k_s)$ which is endowed by a natural left action of $\mathrm{Gal}\,(k)$ given by $(g, \phi) \mapsto g \circ \phi$ for $g \in \mathrm{Gal}\,(k)$, $\phi \in \mathrm{Hom}_k(L, k_s)$.

The first property we shall show about this action is its continuity. Recall that the action of a topological group G on a topological space X is said to be continuous if the map $m : G \times X \to X$ given by $(g, x) \mapsto gx$ is continuous. In our case X is discrete, and the property is equivalent to the openness of the stabilizer G_x of each point $x \in X$. Indeed, the preimage of x (which is an open subset of X) in $G \times X$ is $U_x = \{(g, y) \in G \times X : gy = x\}$, which is the disjoint union of the sets $\{(g, y) \in G \times \{y\} : gy = x\}$ for fixed $y \in X$. Each of these is either empty or homeomorphic to G_x via the map $g \mapsto (gh, y)$, where $h \in G$ is an element with $hy = x$. Thus the openness of G_x implies that of U_x, i.e. the continuity of m. On the other hand, G_x is the preimage of x by the composite map $G \xrightarrow{i_x} G \times X \xrightarrow{m} X$, where $i_x(g) = (g, x)$, so the continuity of m implies the openness of G_x.

Lemma 1.5.1 *The above left action of* Gal (k) *on* $\mathrm{Hom}_k(L, k_s)$ *is continuous and transitive, hence* $Hom_k(L, k_s)$ *as a* Gal (k)*-set is isomorphic to the left coset space of some open subgroup in* Gal (k)*. For L Galois over k this coset space is in fact a quotient by an open normal subgroup.*

Proof The stabilizer U of an element ϕ consists of the elements of Gal (k) fixing $\phi(L)$. Hence by Theorem 1.3.11 U is open in Gal (k), or in other words Gal (k) acts continuously on $\mathrm{Hom}_k(L, k_s)$. If L is generated by a primitive element α with minimal polynomial f, each $\phi \in \mathrm{Hom}_k(L, k_s)$ is given by mapping α to a root of f in k_s. Since Gal (k) permutes these roots transitively, the Gal (k)-action on $\mathrm{Hom}_k(L, k_s)$ is transitive. The above argument also shows that the map $g \circ \phi \mapsto gU$ induces an isomorphism of $\mathrm{Hom}_k(L, k_s)$ with the left coset space $U \backslash \mathrm{Gal}\,(k)$. For U normal we obtain the quotient $\mathrm{Gal}\,(k)/U$; by Theorem 1.3.11 this case arises if and only if L is Galois over k. □

If M is another finite separable extension of k, each k-homomorphism $\phi : L \to M$ induces a map $\mathrm{Hom}_k(M, k_s) \to \mathrm{Hom}_k(L, k_s)$ by composition with ϕ. This map is Gal (k)-equivariant, so we have obtained a *contravariant functor* from the category of finite separable extensions of k to the category of finite sets with continuous transitive left Gal (k)-action.

Theorem 1.5.2 *Let k be a field with fixed separable closure k_s. The contravariant functor mapping a finite separable extension $L|k$ to the finite* Gal (k)*-set* $\mathrm{Hom}_k(L, k_s)$ *gives an anti-equivalence between the category of finite separable extensions of k and the category of finite sets with continuous and transitive left* Gal (k)*-action. Here Galois extensions give rise to* Gal (k)*-sets isomorphic to some finite quotient of* Gal (k).

Proof We check that $\mathrm{Hom}_k(\ , k_s)$ satisfies the conditions of Lemma 1.4.9. We begin by essential surjectivity. To show that any continuous transitive left Gal (k)-set S is isomorphic to some $\mathrm{Hom}_k(L, k_s)$, pick a point $s \in S$. The stabilizer of s is an open subgroup U_s of Gal (k) which fixes a finite separable extension L of k. Now define a map of Gal (k)-sets $\mathrm{Hom}_k(L, k_s) \to S$ by the rule $g \circ i \mapsto gs$, where i is the natural inclusion $L \to k_s$ and g is any element of

G. This map is well-defined since the stabilizer of i is exactly U_s and is readily seen to be an isomorphism; in fact, both Gal (k)-sets become isomorphic to the left coset space $U_s \backslash \mathrm{Gal}(k)$ by the maps sending i (resp. s) to U_s.

For fully faithfulness we have to show that given two finite separable extensions L, M of k, the set of k-homomorphisms $L \to M$ corresponds bijectively to the set of Gal (k)-maps $\mathrm{Hom}_k(M, k_s) \to \mathrm{Hom}_k(L, k_s)$. Since both $\mathrm{Hom}_k(M, k_s)$ and $\mathrm{Hom}_k(L, k_s)$ are transitive Gal (k)-sets, a Gal (k)-map f between them is determined by the image of a fixed $\phi \in \mathrm{Hom}_k(M, k_s)$. As f is Gal (k)-equivariant, the elements of the stabilizer U of ϕ fix $f(\phi)$ as well, whence an inclusion $U \subset V$, where V is the stabilizer of $f(\phi)$. By what we have just seen, taking the fixed subfields of U and V respectively, we get an inclusion of subfields of k_s which is none but the extension $f(\phi)(L) \subset \phi(M)$. Denoting by $\psi : \phi(M) \to M$ the map inverse to ϕ we readily see that $\psi \circ f(\phi)$ is the unique element of $\mathrm{Hom}_k(L, M)$ inducing f.

The last statement follows from Lemma 1.5.1. □

If we wish to extend the previous anti-equivalence to Gal (k)-sets with not necessarily transitive action, the natural replacement for finite separable extensions of k is the following.

Definition 1.5.3 A finite dimensional k-algebra A is *étale* (over k) if it is isomorphic to a finite direct product of separable extensions of k.

As above, the Gal (k)-action on k_s induces a left action on the set of k-algebra homomorphisms $\mathrm{Hom}_k(A, k_s)$.

Theorem 1.5.4 (Main Theorem of Galois Theory – Grothendieck's version) *Let k be a field. The functor mapping a finite étale k-algebra A to the finite set* $\mathrm{Hom}_k(A, k_s)$ *gives an anti-equivalence between the category of finite étale k-algebras and the category of finite sets with continuous left* Gal (k)*-action. Here separable field extensions give rise to sets with transitive* Gal (k)*-action and Galois extensions to* Gal (k)*-sets isomorphic to finite quotients of* Gal (k).

Proof This follows from the previous theorem in view of the remark that given a decomposition $A = \prod L_i$ into a product of fields and an element $\phi \in \mathrm{Hom}_k(A, k_s)$, the map ϕ induces the injection of exactly one L_i in k_s. Indeed, if $\phi(L_i) \neq 0$, then being a field, L_i injects in k_s, and on the other hand, a product $L_i \times L_j$ cannot inject in k_s since k_s has no zero-divisors. Thus $\mathrm{Hom}_k(A, k_s)$ decomposes into the disjoint union of the $\mathrm{Hom}_k(L_i, k_s)$; this is in fact its decomposition into Gal (k)-orbits. For a similar reason, given another étale k-algebra $A' = \prod L'_j$, a morphism $A \to A'$ identifies with a collection of morphisms $L_i \to L'_j$, one for each i, and these in turn correspond bijectively to morphisms of the corresponding Gal (k)-sets by the previous theorem. □

Remark 1.5.5 The theorem generalizes immediately to an arbitrary Galois extension $K|k$: if one restricts attention to finite étale k-algebras that are products of subfields of K, the functor $A \mapsto \mathrm{Hom}_k(A, K)$ induces an equivalence with the category of finite continuous left $\mathrm{Gal}(K|k)$-sets.

To conclude this section we give another characterization of finite étale k-algebras which ties in with more classical treatments. Recall that a commutative ring is *reduced* if it has no nonzero nilpotent elements.

Proposition 1.5.6 *Let A be a finite dimensional commutative k-algebra. Then the following are equivalent:*

1. *A is étale.*
2. *$A \otimes_k \bar{k}$ is isomorphic to a finite product of copies of $\bar{k}$;*
3. *$A \otimes_k \bar{k}$ is reduced.*

In the literature, finite dimensional k-algebras satisfying the third condition of the proposition are often called *separable* k-algebras. The proposition thus provides a structure theorem for these.

Note that it may well happen that A has no nilpotents but $A \otimes_k \bar{k}$ does. A typical example is the following: let k be an imperfect field of characteristic $p > 0$, and set $A := k[x]/(x^p - a)$ for an $a \in k$ that is not a p-th power in k. Then A is a degree p field extension of k, so it has no nilpotents. But $\overline{A} := A \otimes_k \bar{k} \cong \bar{k}[x]/(x^p - a)$. Choosing $\alpha \in \bar{k}$ with $\alpha^p = a$ and denoting by $\bar{x}$ the image of x in $\overline{A}$ we have $(\bar{x} - \alpha)^p = \bar{x}^p - \alpha^p = 0$ in $\overline{A}$.

For the proof we need the following lemma which is the commutative version of the Wedderburn–Artin theorem.

Lemma 1.5.7 *A finite dimensional commutative algebra over a field F is isomorphic to a direct product of finite field extensions of F if and only if it is reduced.*

The proof is taken from Fröhlich–Taylor [25].

Proof The 'only if' part is obvious. To prove the 'if' part, by decomposing a finite dimensional F-algebra A into a finite direct product of indecomposable F-algebras we may assume that A is indecomposable itself. Notice that under this restriction A can have no idempotent elements other than 0 and 1; indeed, if $e \neq 0, 1$ were an idempotent then $A \cong Ae \times A(1-e)$ would be a nontrivial direct product decomposition since $e(1-e) = e - e^2 = 0$ by assumption. The lemma will follow if we show that every nonzero element $x \in A$ is invertible and thus A is a field. Since A is finite dimensional over F, the descending chain of ideals $(x) \supset (x^2) \supset \dots (x^n) \supset \dots$ must stabilize and thus for some m we must have $x^n = x^{n+1}y$ with an appropriate y. By iterating this formula we get $x^n = x^{n+i}y^i$ for all positive integers i, in particular $x^n = x^{2n}y^n$. Thus $x^n y^n = (x^n y^n)^2$, i. e. $x^n y^n$ is an idempotent. By what has been said above there are two cases. If $x^n y^n = 0$, then $x^n = (x^n)(x^n y^n) = 0$, which is a contradiction since $x \neq 0$ by assumption and A is reduced. Otherwise, $x^n y^n = 1$ and thus x is invertible. □

Remark 1.5.8 The lemma already implies that a finite dimensional commutative algebra over a *perfect* field is étale if and only if it is reduced.

Proof of Proposition 1.5.6 The derivation of the third condition from the second is immediate; actually, the lemma applied to $A \otimes_k \bar{k}$ shows that they are equivalent. We therefore only have to prove the equivalence of the first two conditions. To see that (1) implies (2) we may restrict to finite separable extensions L of k. We then have $L = k[x]/(f)$ with some polynomial f which decomposes as a product of n distinct factors $(x - \alpha_i)$ in $\bar{k}$. We conclude by the chain of isomorphisms

$$L \otimes_k \bar{k} \cong \bar{k}[x]/(f) = \bar{k}[x]/(x - \alpha_1) \cdots (x - \alpha_n) \cong \prod_{i=1}^{n} \bar{k}[x]/(x - \alpha_i) \cong \prod_{i=1}^{n} \bar{k},$$

the middle isomorphism holding by the Chinese Remainder Theorem (see e.g. [48], Chapter II, Theorem 2.1).

Now to derive (1) from (2), let $\overline{A}$ be the quotient of A by the ideal formed by its nilpotent elements. The lemma implies that $\overline{A}$ is a sum of finite extension fields of k. Since $\bar{k}$ is reduced, each k-algebra homomorphism $A \to \bar{k}$ factors through $\overline{A}$ and hence through one of its decomposition factors L. By Lemma 1.1.6, the number of k-algebra homomorphisms $L \to \bar{k}$ can equal at most the degree of L over k, with equality if and only if $L|k$ is separable, whence $\mathrm{Hom}_k(A, \bar{k})$ has at most $\dim_k(A)$ elements with equality if and only if $A = \overline{A}$ and A is étale. To see that equality indeed holds, observe that we have a canonical bijection of finite sets

$$\mathrm{Hom}_k(A, \bar{k}) \cong \mathrm{Hom}_{\bar{k}}(A \otimes_k \bar{k}, \bar{k}).$$

[To see this, observe that given a k-algebra homomorphism $A \to \bar{k}$, tensoring by $\bar{k}$ and composing by the multiplication map gives a $\bar{k}$-homomorphism $A \otimes_k \bar{k} \to \bar{k} \otimes_k \bar{k} \to \bar{k}$; on the other hand the natural inclusion $k \to \bar{k}$ induces a k-homomorphism $A \cong A \otimes_k k \to A \otimes_k \bar{k}$ which composed by homomorphisms $A \otimes_k \bar{k} \to \bar{k}$ gives a map from the set on the right-hand side to that on the left which is clearly inverse to the previous construction.] The assumption now implies that the set on the right-hand side has $\dim_{\bar{k}}(A \otimes_k \bar{k})$ elements. But $\dim_{\bar{k}}(A \otimes_k \bar{k}) = \dim_k A$, whence the claim. □

Exercises

1. Show that an inverse limit of nonempty finite discrete sets (for any directed index set) is nonempty. [*Hint:* For such an inverse system $\{X_\alpha : \alpha \in \Lambda\}$ consider the subsets $X_{\lambda\mu} \subset \prod X_\alpha$ consisting of the sequences (x_α) satisfying $\phi_{\lambda\mu}(x_\mu) = x_\lambda$ for a fixed pair $\lambda \leq \mu$ and use the compactness of the topological product of the X_α. We shall go through this argument in a more general case in Lemma 3.4.12.]
2. Let G be a profinite group, p a prime number. A *pro-p-Sylow subgroup* of G is a pro-p-group whose image in each finite quotient of G is of index prime to p. Show that pro-p-Sylow subgroups exist and they are conjugate in G. [*Hint:* Apply the previous exercise to the inverse system formed by the sets of p-Sylow subgroups in each finite quotient of G.]
3. Let k be a perfect field, p a prime number. Show that there exists an algebraic extension $k^{(p)}|k$ such that each finite subextension is of degree prime to p, and $k^{(p)}$ has no nontrivial finite extensions of degree prime to p. Is such an extension unique inside a fixed algebraic closure? [*Hint:* Use the previous exercise.]
4. Consider the compositum E of all quadratic extensions of $\mathbf{Q}$ inside a fixed algebraic closure $\overline{\mathbf{Q}}$.
   (a) Show that $\mathrm{Gal}\,(E|\mathbf{Q})$ is uncountable and has uncountably many subgroups of index 2.
   (b) Deduce that there are uncountably many subgroups of index 2 in $\mathrm{Gal}\,(\overline{\mathbf{Q}}|\mathbf{Q})$ that are not open.
5. Let G be a profinite group acting via field automorphisms on a field K. Assume that the action is continuous when K carries the discrete topology, and that each nontrivial element in G acts nontrivially on K. Show that $G \cong \mathrm{Gal}\,(K|k)$, where $k = K^G$.
6. (Leptin, Waterhouse) Show that every profinite group G arises as the Galois group of some Galois extension $K|k$. [*Hint:* For each open normal subgroup $N \subset G$ fix a system of left coset representatives $1 = \sigma_1^N, \ldots, \sigma_m^N$. Let F be a perfect field, and $K|F$ the purely transcendental extension obtained by adjoining an indeterminate x_i^N for each σ_i^N. Make G act on F trivially, and on K via $\sigma(x_i^N) = x_j^N$, where x_j^N corresponds to σ_j^N with $\sigma_j^N N = \sigma(\sigma_i^N N)$. Verify that this action satisfies the criterion of the previous exercise.]
   [*Remark:* The statement does not hold if one requires K to be a separable closure of k. For instance, Artin and Schreier showed in [3] that among the nontrivial finite groups only $\mathbf{Z}/2\mathbf{Z}$ can arise as an absolute Galois group.]
7. Let k be a field, and A a finite étale k-algebra equipped with an action of a finite group G via k-algebra automorphisms; we call such algebras *G-algebras*. We moreover say that A is *Galois* with group G if $\dim_k(A)$ equals the order of G and $A^G = k$.
   (a) Consider the G-algebra structure on $A \otimes_k \bar{k}$ given by $g(a \otimes \alpha) = g(a) \otimes \alpha$. Prove that A is Galois with group G if and only if $A \otimes_k \bar{k}$ is isomorphic to the group algebra $\bar{k}[G]$ as a G-algebra.
   (b) Making G act on $\mathrm{Hom}_k(A, k_s)$ via $\phi \mapsto \phi \circ g$, show that in the correspondence of Theorem 1.5.4 Galois algebras with group G correspond to finite continuous $\mathrm{Gal}\,(k)$-sets with simply transitive G-action.

8. Let k be a field of characteristic different from 2, and S a continuous left Gal (k)-set with n elements. Consider the subset $\Sigma(S) \subset S^n$ consisting of n-tuples $(s_1, \dots, s_n)$ with $x_i \neq x_j$ for $i \neq j$. It inherits a continuous left action of Gal (k) from the product action on S^n, and it also has a natural action by the symmetric group S_n via permutation of the components. Denote by $\Delta(S)$ the quotient of $\Sigma(S)$ by the action of the alternating group $A_n \subset S_n$. It is a 2-element continuous left Gal (k)-set.
   (a) Show that the finite étale k-algebra corresponding to $\Delta(S)$ via Theorem 1.5.4 is isomorphic to $k \times k$ if Gal (k) acts on $\Delta(S)$ by even permutations, and is a degree 2 field extension of k otherwise.
   (b) If A is the finite étale k-algebra corresponding to S via Theorem 1.5.4, denote the k-algebra of (a) by $\Delta(A)$. Show that if $A \cong k[x]/(f)$ for a polynomial f without multiple roots, then $\Delta(A) \cong k[x]/(x^2 - d(f))$, where $d(f) \in k$ is the Vandermonde determinant formed from the roots of f.

   [*Remark:* The k-algebra $\Delta(A)$ is called the *discriminant* of the finite étale k-algebra A. For a description of $\Delta(A)$ in the general case, see [45], Proposition 18.24.]

2

Fundamental groups in topology

In the last section we saw that when studying extensions of some field it is plausible to conceive the base field as a point and a finite separable extension (or, more generally, a finite étale algebra) as a finite discrete set of points mapping to this base point. Galois theory then equips the situation with a continuous action of the absolute Galois group which leaves the base point fixed. It is natural to try to extend this situation by taking as a base not just a point but a more general topological space. The role of field extensions would then be played by certain continuous surjections, called covers, whose fibres are finite (or, even more generally, arbitrary discrete) spaces. We shall see in this chapter that under some restrictions on the base space one can develop a topological analogue of the Galois theory of fields, the part of the absolute Galois group being taken by the fundamental group of the base space.

In the second half of the chapter we give a reinterpretation of the main theorem of Galois theory for covers in terms of locally constant sheaves. Esoteric as these objects may seem to the novice, they stem from reformulating in a modern language very classical considerations from analysis, such as the study of local solutions of holomorphic differential equations. In fact, the whole concept of the fundamental group arose from Riemann's study of the monodromy representation for hypergeometric differential equations, a topic we shall briefly discuss at the end of the chapter. Our exposition therefore traces history backwards, but hopefully reflects the intimate connection between differential equations and the fundamental group.

2.1 Covers

We start with the basic definitions.

Definition 2.1.1 Let X be a topological space. A *space over* X is a topological space Y together with a continuous map $p : Y \to X$. A morphism between two spaces $p_i : Y_i \to X$ $(i = 1, 2)$ over X is given by a continuous map $f : Y_1 \to Y_2$ making the diagram

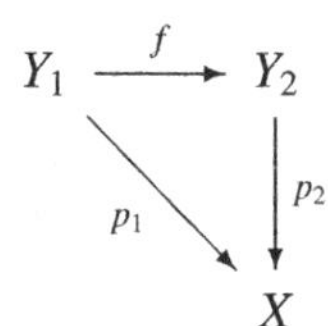

commute.

A *cover* of X is a space Y over X where the projection $p : Y \to X$ is subject to the following condition: each point of X has an open neighbourhood V for which $p^{-1}(V)$ decomposes as a disjoint union of open subsets U_i of Y such that the restriction of p to each U_i induces a homeomorphism of U_i with V.

We define a morphism between two covers of a space X to be a morphism of spaces over X.

In the literature the terms 'covering space' and 'covering' are also used for what we call a cover; we shall stick to the above terminology. Note an easy consequence of the definition: if $p : Y \to X$ is a cover, the map p is always surjective.

Example 2.1.2 Take a nonempty discrete topological space I and form the topological product $X \times I$. The first projection $X \times I \to X$ turns $X \times I$ into a space over X which is immediately seen to be a cover. It is called the *trivial cover.*

Trivial covers may at first seem very special but as the next proposition shows, every cover is locally a trivial cover.

Proposition 2.1.3 *A space Y over X is a cover if and only if each point of X has an open neighbourhood V such that the restriction of the projection $p : Y \to X$ to $p^{-1}(V)$ is isomorphic (as a space over V) to a trivial cover.*

Proof The 'if' part follows from the previous example and the 'only if' part is easily seen as follows: given a decomposition $p^{-1}(V) \cong \coprod_{i \in I} U_i$ for some index set I as in the definition of covers, mapping $u_i \in U_i$ to the pair $(p(u_i), i)$ defines a homeomorphism of $\coprod_{i \in I} U_i$ onto $V \times I$, where I is endowed with the discrete topology. By construction this is an isomorphism of covers of V. □

In the notation of the previous proof, the set I is the fibre of p over the points of V. The proof shows that the points of X over which the fibre of p equals I form an open subset of X. Thus making I vary yields a decomposition of X into a disjoint union of open subsets. In particular:

Corollary 2.1.4 *If X is connected, the fibres of p are all homeomorphic to the same discrete space I.*

Notice that this does not mean at all that the cover is trivial. Indeed, let us give an example of a nontrivial cover with a connected base.

Example 2.1.5 Consider a rectangle $XYZW$ and divide the sides XY and ZW into two equal segments by the points P and Q. Identifying the sides XY and ZW with opposite orientations we get a Möbius strip on which the image of the segment PQ becomes a closed curve C homeomorphic to a circle. The natural projection of the boundary B of the Möbius strip onto C coming from the perpendicular projection of the sides XW and YZ of the rectangle onto the segment PQ makes B a space over C. It is actually a cover since locally it is a product of a segment by a two-point space, i.e. a trivial cover of the segment. However, the cover itself is nontrivial since B is not homeomorphic to a disjoint union of two circles.

Other important examples arise from group actions on topological spaces. To obtain covers we need a technical restriction.

Definition 2.1.6 Let G be a group acting continuously from the left on a topological space Y. The action of G is *even* if each point $y \in Y$ has some open neighbourhood U such that the open sets gU are pairwise disjoint for all $g \in G$.

This terminology is that of Fulton [26]. Older texts use the much more awkward term 'properly discontinuous'. Now recall that if a group G acts from the left on a topological space Y, one may form the quotient space $G\backslash Y$ whose underlying set is by definition the set of orbits under the action of G and the topology is the finest one that makes the projection $Y \to G\backslash Y$ continuous.

Lemma 2.1.7 *If G is a group acting evenly on a connected space Y, the projection $p_G : Y \to G\backslash Y$ turns Y into a cover of $G\backslash Y$.*

Proof The map p_G is surjective, and moreover each $x \in G\backslash Y$ has an open neighbourhood of the form $V = p_G(U)$ with a U as in Definition 2.1.6. This V is readily seen to satisfy the condition of Definition 2.1.1. □

Example 2.1.8 With this tool at hand, one can give lots of examples of covers.

1. Let $\mathbf{Z}$ act on $\mathbf{R}$ by translations (which means that the automorphism defined by $n \in \mathbf{Z}$ is the map $x \mapsto x + n$). We obtain a cover $\mathbf{R} \to \mathbf{R}/\mathbf{Z}$, where $\mathbf{R}/\mathbf{Z}$ is immediately seen to be homeomorphic to a circle.
2. The previous example can be generalized to arbitrary dimension: take any basis $\{x_1, \ldots, x_n\}$ of the vector space $\mathbf{R}^n$ and make $\mathbf{Z}^n$ act on $\mathbf{R}^n$ so that the i-th direct factor of $\mathbf{Z}^n$ acts by translation by x_i. This action is clearly even and turns $\mathbf{R}^n$ into a cover of what is called a *linear torus*; for $n = 2$, this is the usual torus. The subgroup Λ of $\mathbf{R}^n$ generated by the x_i is usually called a *lattice*; thus linear tori are quotients of $\mathbf{R}^n$ by lattices.
3. For an integer $n > 1$ denote by μ_n the group of n-th roots of unity. Multiplying by elements of μ_n defines an even action on $\mathbf{C}^* := \mathbf{C} \setminus \{0\}$, whence a cover $p_n : \mathbf{C}^* \to \mathbf{C}^*/\mu_n$. In fact, the map $z \mapsto z^n$ defines a

natural homeomorphism of $\mathbf{C}^*/\mu_n$ onto $\mathbf{C}^*$ (even an isomorphism of topological groups) and via this homeomorphism p_n becomes identified with the cover $\mathbf{C}^* \to \mathbf{C}^*$ given by $z \mapsto z^n$. Note that this map does *not* extend to a cover $\mathbf{C} \to \mathbf{C}$; this phenomenon will be studied further in Chapter 3.

2.2 Galois covers

Henceforth we fix a base space X which will be assumed *locally connected* (i.e. each point has a basis of neighbourhoods consisting of connected open subsets). Given a cover $p : Y \to X$, its *automorphisms* are to be automorphisms of Y as a space over X, i.e. topological automorphisms compatible with the projection p. They form a group with respect to composition that we shall denote by $\mathrm{Aut}(Y|X)$. By convention all automorphisms will be assumed to act *from the left.* Note that for each point $x \in X$ the group $\mathrm{Aut}(Y|X)$ maps the fibre $p^{-1}(x)$ onto itself, so $p^{-1}(x)$ is equipped with a natural action of $\mathrm{Aut}(Y|X)$.

First we prove a necessary and sufficient condition for a topological automorphism of Y to be an element of $\mathrm{Aut}(Y|X)$. We begin with:

Lemma 2.2.1 *An automorphism ϕ of a connected cover $p : Y \to X$ having a fixed point must be trivial.*

Instead of proving the lemma we establish a more general statement which will also be needed later. The lemma follows from it by taking $Z = Y$, $f = \mathrm{id}$ and $g = \phi$.

Proposition 2.2.2 *Let $p : Y \to X$ be a cover, Z a connected topological space, $f, g : Z \to Y$ two continuous maps satisfying $p \circ f = p \circ g$. If there is a point $z \in Z$ with $f(z) = g(z)$, then $f = g$.*

Proof Suppose $z \in Z$ is as above, $y = f(z) = g(z)$. Take some connected open neighbourhood V of $p(y)$ satisfying the condition in the definition of a cover (such a V exists since X is locally connected) and let $U_i \cong V$ be the component of $p^{-1}(V)$ containing y. By continuity f and g must both map some open neighbourhood W of z into U_i. Since $p \circ f = p \circ g$ and p maps U_i homeomorphically onto V, f and g must agree on W. The same type of reasoning shows that if $f(z') \neq g(z')$ for some point $z' \in Y$, f and g must map a whole open neighbourhood of z' to different components of $p^{-1}(V)$. Thus the set $\{z \in Z : f(z) = g(z)\}$ is nonempty, open and closed in Z, so by connectedness it is the whole of Z. □

Here is a first application of Lemma 2.2.1.

Proposition 2.2.3 *If $p : Y \to X$ is a connected cover, the action of* $\mathrm{Aut}(Y|X)$ *on Y is even.*

Proof Let y be a point of Y, and set $x = p(y)$. Let V be a connected open neighbourhood of x such that $p^{-1}(V)$ is a disjoint union of open sets U_i as in Definition 2.1.1. One of these, say U_i, contains y. We contend that U_i satisfies the condition of Definition 2.1.6. Indeed, a nontrivial $\phi \in \mathrm{Aut}(Y|X)$ maps U_i isomorphically onto some U_j, by definition of a cover automorphism. Since Y is connected, Lemma 2.2.1 applies and shows that for $\phi \neq \mathrm{id}_Y$ we must have $i \neq j$. □

Conversely, we have:

Proposition 2.2.4 *If G is a group acting evenly on a connected space Y, the automorphism group of the cover $p_G : Y \to G\backslash Y$ is precisely G.*

Proof Notice first that we may naturally view G as a subgroup of $\mathrm{Aut}(Y|(G\backslash Y))$. Now given an element ϕ in the latter group, look at its action on an arbitrary point $y \in Y$. Since the fibres of p_G are precisely the orbits of G we may find $g \in G$ with $\phi(y) = gy$. Applying Lemma 2.2.1 to the automorphism $\phi \circ g^{-1}$ we get $g = \phi$. □

Now given a connected cover $p : Y \to X$, we may form the quotient of Y by the action of $\mathrm{Aut}(Y|X)$. It is immediate from the definition of cover automorphisms that the projection p factors as a composite of continuous maps

$$Y \to \mathrm{Aut}(Y|X)\backslash Y \xrightarrow{\overline{p}} X$$

where the first map is the natural projection.

Definition 2.2.5 A cover $p : Y \to X$ is said to be *Galois* if Y is connected and the induced map $\overline{p}$ above is a homeomorphism.

Remark 2.2.6 Note the similarity of the above definition with that of a Galois extension of fields. This analogy is further confirmed by remarking that the cover p_G in Proposition 2.2.4 is Galois.

Proposition 2.2.7 *A connected cover $p : Y \to X$ is Galois if and only if* $\mathrm{Aut}(Y|X)$ *acts transitively on each fibre of p.*

Proof Indeed, the underlying set of $\mathrm{Aut}(Y|X)\backslash Y$ is by definition the set of orbits of Y under the action of $\mathrm{Aut}(Y|X)$, and so the map $\overline{p}$ is one-to-one precisely when each such orbit is equal to a whole fibre of p, i.e. when $\mathrm{Aut}(Y|X)$ acts transitively on each fibre. □

Remark 2.2.8 In fact, for a connected cover $p : Y \to X$ to be Galois it suffices for $\mathrm{Aut}(Y|X)$ to act transitively on *one fibre*. Indeed, in this case $\mathrm{Aut}(Y|X)\backslash Y$ is a connected cover of X where one of the fibres consists of a single element; it is thus isomorphic to X by Remark 2.1.4.

Example 2.2.9 Consider the linear torus $X = \mathbf{R}^n/\Lambda$ with lattice $\Lambda \cong \mathbf{Z}^n$, and let $m > 1$ be an integer. The multiplication-by-m map of $\mathbf{R}^n$ maps Λ into itself and hence induces a map $X \to X$. It is a Galois cover with group $(\mathbf{Z}/m\mathbf{Z})^n$.

Now we can state the topological analogue of Theorem 1.2.5.

Theorem 2.2.10 *Let $p : Y \to X$ be a Galois cover. For each subgroup H of $G = \mathrm{Aut}(Y|X)$ the projection p induces a natural map $\overline{p}_H : H\backslash Y \to X$ which turns $H\backslash Y$ into a cover of X.*

Conversely, if $Z \to X$ is a connected cover fitting into a commutative diagram

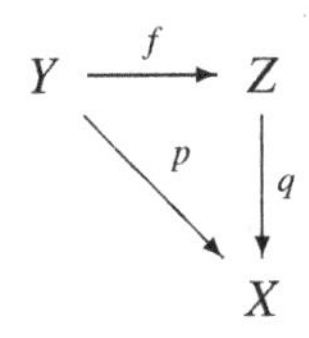

then $f : Y \to Z$ is a Galois cover and actually $Z \cong H\backslash Y$ for the subgroup $H = \mathrm{Aut}(Y|Z)$ of G. The maps $H \mapsto H\backslash Y$, $Z \mapsto \mathrm{Aut}(Y|Z)$ induce a bijection between subgroups of G and intermediate covers Z as above. The cover $q : Z \to X$ is Galois if and only if H is a normal subgroup of G, in which case $\mathrm{Aut}(Z|X) \cong G/H$.

Before starting the proof we need a general lemma on covers.

Lemma 2.2.11 *Assume given a connected cover $q : Z \to X$ and a continuous map $f : Y \to Z$. If the composite $q \circ f : Y \to X$ is a cover, then so is $f : Y \to Z$.*

Proof Let z be a point of Z, $x = q(z)$ and V a connected open neighbourhood of x satisfying the property of Definition 2.1.1 for both $p = q \circ f$ and q, giving rise to decompositions $p^{-1}(V) = \coprod U_i$ and $q^{-1}(V) = \coprod V_j$. Here for each U_i its image $f(U_i)$ is a connected subset of Z mapping onto V by q, hence there is some j with $f(U_i) \subset V_j$. But this is in fact a homeomorphism since both sides get mapped homeomorphically onto V by q. This implies in particular that $f(Y)$ is open in Z.

Now to prove the lemma we first show that f is surjective. For this it is enough to see by connectedness of Z and openness of $f(Y)$ that the complement of $f(Y)$ in Z is open. If z is a point of $Z \setminus f(Y)$ and V is a neighbourhood of

$x = q(z)$ as above, the whole component V_j of $q^{-1}(V)$ containing z must be disjoint from $f(Y)$. Indeed, otherwise by the previous argument the whole of V_j would be contained in $f(Y)$, which is a contradiction. This settles the openness of $Z \setminus f(Y)$, and to conclude the proof of the lemma it remains to notice that the preimage of the above V_j by f is a disjoint union of some U_i. □

Proof of Theorem 2.2.10 Since $H \subset \mathrm{Aut}(Y|X)$, the projection p factors as a composite $Y \stackrel{p_H}{\to} H\backslash Y \stackrel{\overline{p}_H}{\to} X$. Here $\overline{p}_H$ is continuous because p is continuous and p_H is a local homeomorphism by Lemma 2.1.7. By Proposition 2.1.3, over sufficiently small subsets V of X we have $p^{-1}(V) \cong V \times F$ with a discrete set F equipped with an H-action. The open set $\overline{p}_H^{-1}(V) \subset H\backslash Y$ will then be isomorphic to a product of V by the discrete set of H-orbits of F, so by applying Proposition 2.1.3 again we conclude that $\overline{p}_H : H\backslash Y \to X$ is a cover.

For the converse, apply the previous lemma to see that $f : Y \to Z$ is a cover. Then $H = \mathrm{Aut}(Y|Z)$ is a subgroup of G, so to show that the cover is Galois it suffices by Proposition 2.2.7 to check that H acts transitively on each fibre of f. To do so, take a point $z \in Z$ and let y_1 and y_2 be two points of $f^{-1}(z)$. They are both contained in the fibre $p^{-1}(q(z))$, so since $p : Y \to X$ is Galois, we have $y_1 = \phi(y_2)$ with some $\phi \in G$. We are done if we show $\phi \in H$, which is equivalent to saying that the subset $S = \{y \in Y : f(y) = f(\phi(y))\}$ is equal to the whole of Y. But this follows from Proposition 2.2.2, applied to our current Y, Z and f as well as $g = f \circ \phi$.

It is immediate that the two constructions above are inverse to each other, so only the last statement remains, and it is proven similarly as the corresponding statement in the Galois theory of fields (see the proof of Theorem 1.2.5). One implication is easy: if H is normal in G, then G/H acts naturally on $Z = H\backslash Y$, and this action preserves the projection q. So we obtain a group homomorphism $G/H \to \mathrm{Aut}(Z|X)$ which is readily seen to be injective. But $(G/H)\backslash Z \cong G\backslash Y \cong X$, so $G/H \cong \mathrm{Aut}(Z|X)$ and $q : Z \to X$ is Galois. For the converse assume that $Z \to X$ is a Galois cover. We first show that each element ϕ of $G = \mathrm{Aut}(Y|X)$ induces an automorphism of Z over X. In other words, we need an automorphism $\psi : Z \to Z$ which can be inserted into the commutative diagram

$$\begin{array}{ccc} Y & \xrightarrow{\phi} & Y \\ {\scriptstyle f}\downarrow & & \downarrow{\scriptstyle f} \\ Z & & Z \\ {\scriptstyle q}\downarrow & & \downarrow{\scriptstyle q} \\ X & \xrightarrow{\mathrm{id}} & X \end{array}$$

For this take a point $y \in Y$ with image $x = (q \circ f)(y)$ in X. By commutativity of the diagram $f(y)$ and $f(\phi(y))$ are in the same fibre $q^{-1}(x)$ of q. Since $\mathrm{Aut}(Z|X)$ acts transitively on the fibres of q, there is an automorphism $\psi \in \mathrm{Aut}(Z|X)$ with the property that $\psi(f(y)) = f(\phi(y))$. In fact, ψ is the unique element of $\mathrm{Aut}(Z|X)$ with this property, for if $\lambda \in \mathrm{Aut}(Z|X)$ is another one, Lemma 2.2.1 implies that $\psi \circ \lambda^{-1}$ is the identity. We contend that ψ is the map we are looking for, i.e. the maps $\psi \circ f$ and $f \circ \phi$ are the same. Indeed, both are continuous maps from the connected space Y to Z that coincide at the point y, and moreover their compositions with q are equal, so the assertion follows from Proposition 2.2.2. The map $\phi \mapsto \psi$ is in fact a homomorphism $G \to \mathrm{Aut}(Z|X)$. Its kernel is none but $H = \mathrm{Aut}(Y|Z)$, which is thus a normal subgroup in G. □

2.3 The monodromy action

Our next goal is to prove an analogue of Theorem 1.5.4 for covers. The role of the absolute Galois group will be played by the fundamental group of the base space, about which we quickly recall the basic facts.

Let X be a topological space. A *path* in X is a continuous map $f : [0, 1] \to X$, where $[0, 1]$ is the closed unit interval. The *endpoints* of the path are the points $f(0)$ and $f(1)$; if they coincide, the path is called a *closed path* or a *loop*. Two paths $f, g : [0, 1] \to X$ are called *homotopic* if $f(0) = g(0)$, $f(1) = g(1)$ and there is a continuous map $h : [0, 1] \times [0, 1] \to X$ with $h(0, y) = f(y)$ and $h(1, y) = g(y)$ for all $y \in [0, 1]$. It is an easy exercise to check that homotopy of paths is an equivalence relation.

Now given two paths $f, g : [0, 1] \to X$ with $f(0) = g(1)$, define their composition $f \bullet g : [0, 1] \to X$ by setting $(f \bullet g)(x) = g(2x)$ for $0 \le x \le 1/2$ and $(f \bullet g)(x) = f(2x - 1)$ for $1/2 \le x \le 1$. It is again an easy exercise to verify that this operation passes to the quotient modulo homotopy equivalence, i.e. if f_1, f_2 are homotopic paths with $f_1(1) = f_2(1) = g(0)$ then so are $f_1 \bullet g$ and $f_2 \bullet g$, and similarly for the homotopy class of g.

Remark 2.3.1 The above convention for composition of paths is the one used by Deligne in his fundamental works [12] and [13]. It differs from the convention of many textbooks: most authors define the composition by first going through f and then through g. However, several reasons speak for our convention. One is that it is parallel to the usual convention for composition of functions. For another particularly pregnant one, see Remark 2.6.3 below.

Composition of paths thus induces a multiplication map on the set $\pi_1(X, x)$ of homotopy classes of closed paths with endpoint equal to some fixed $x \in X$.

In fact, $\pi_1(X, x)$ equipped with this operation is a group: the unit element is the class of the constant path $[0, 1] \to \{x\}$ and the inverse of a class given by a path $f : [0, 1] \to X$ is the class of the path f^{-1} obtained by composing f with the map $[0, 1] \to [0, 1]$, $x \mapsto 1 - x$. It is called the *fundamental group of X with base point x*. If X is *path-connected*, i.e. any two points x and y may be joined by a path f, then $\pi_1(X, x)$ is non-canonically isomorphic to $\pi_1(X, y)$ via $g \mapsto f \bullet g \bullet f^{-1}$, hence the isomorphism class of the fundamental group does not depend on the base point. A path-connected space is always connected; it is *simply connected* if its fundamental group is trivial.

We now show that given a cover $p : Y \to X$, the fibre $p^{-1}(x)$ over a point $x \in X$ carries a natural action by the group $\pi_1(X, x)$. This will be a consequence of the following lemma on 'lifting paths and homotopies'.

Lemma 2.3.2 *Let $p : Y \to X$ be a cover, y a point of Y and $x = p(y)$.*

1. *Given a path $f : [0, 1] \to X$ with $f(0) = x$, there is a unique path $\tilde{f} : [0, 1] \to Y$ with $\tilde{f}(0) = y$ and $p \circ \tilde{f} = f$.*
2. *Assume moreover given a second path $g : [0, 1] \to X$ homotopic to f. Then the unique $\tilde{g} : [0, 1] \to Y$ with $\tilde{g}(0) = y$ and $p \circ \tilde{g} = g$ has the same endpoint as $\tilde{f}$, i.e. we have $\tilde{f}(1) = \tilde{g}(1)$.*

Actually, the proof will show that in the second situation the liftings $\tilde{f}$ and $\tilde{g}$ are homotopic, but this will not be needed later.

Proof For the first statement, note first that uniqueness follows from Proposition 2.2.2 applied with X, Y and Z replaced by our current X, $[0, 1]$ and Y. Existence is immediate in the case of a trivial cover. To reduce the general case to this, for each $x \in f([0, 1])$ choose some open neighbourhood V_x satisfying the condition in the definition of a cover. The sets $f^{-1}(V_x)$ form an open covering of the interval $[0, 1]$ from which we may extract a finite subcovering since the interval is compact. We may then choose a subdivision $0 = t_0 \leq t_1 \leq \cdots \leq t_n = 1$ of $[0, 1]$ such that each closed interval $[t_i, t_{i+1}]$ is contained in some $f^{-1}(V_x)$, hence the cover is trivial over each $f([t_i, t_{i+1}])$. We can now construct $\tilde{f}$ inductively: given a lifting $\tilde{f}_i$ of the path f restricted to $[t_0, t_i]$ (the case $i = 0$ being trivial), we may construct a lifting of the restriction of f to $[t_i, t_{i+1}]$ starting from $\tilde{f}_i(t_i)$; piecing this together with f_i gives f_{i+1}.

For statement (2) we first show that given a homotopy $h : [0, 1] \times [0, 1] \to X$ with $h(0, t) = f(t)$ and $h(1, t) = g(t)$, there is a lifting $\tilde{h} : [0, 1] \times [0, 1] \to Y$ with $p \circ \tilde{h} = h$, $\tilde{h}(0, t) = \tilde{f}(t)$ and $\tilde{h}(1, t) = \tilde{g}(t)$. The construction is similar to that for f: first choose a sufficiently fine finite subdivision of $[0, 1] \times [0, 1]$ into small subsquares S_{ij} so that over each $h(S_{ij})$ the cover is trivial. That this may be done is assured by a well-known fact from the topology of compact metric spaces called Lebesgue's lemma (see e.g. Munkres [66], Lemma 27.5; we have used a trivial case of it above). Then proceed by piecing together liftings over each subsquare, moving 'serpent-wise' from the point $(0, 0)$ (for which we put $\widetilde{h}(0, 0) = y$) towards the point $(1, 1)$. Note that by uniqueness of path lifting it is sufficient to find a local lifting which coincides with the previous one at the left corner of the side where two squares meet; they will then coincide over the whole of the common side. Again by uniqueness of path lifting we get successively that the path $t \mapsto \widetilde{h}(0, t)$ is $\widetilde{f}$ (since both are liftings of f starting from y), the path $s \mapsto \widetilde{h}(s, 0)$ is the constant path $[0, 1] \to \{y\}$ and that $t \mapsto \widetilde{h}(1, t)$ is none but $\widetilde{g}$. Finally, $s \mapsto \widetilde{h}(s, 1)$ is a path joining $\widetilde{f}(1)$ and $\widetilde{g}(1)$ which lifts the constant path $[0, 1] \to \{f(1)\}$; by uniqueness it must coincide with the constant path $[0, 1] \to \{\widetilde{f}(1)\}$, whence $\widetilde{f}(1) = \widetilde{g}(1)$. □

We can now construct the promised left action of $\pi_1(X, x)$ on the fibre $p^{-1}(x)$.

Construction 2.3.3 Given $y \in p^{-1}(x)$ and $\alpha \in \pi_1(X, x)$ represented by a path $f : [0, 1] \to X$ with $f(0) = f(1) = x$, we define $\alpha y := \widetilde{f}(1)$, where $\widetilde{f}$ is the unique lifting $\widetilde{f}$ to Y with $\widetilde{f}(0) = y$ given by part (1) of the lemma above. By part (2) of the lemma αy does not depend on the choice of f, and it lies in $p^{-1}(x)$ by construction. This is indeed a left action of $\pi_1(X, x)$ on $p^{-1}(x)$: $(\alpha \bullet \beta)y = \alpha(\beta y)$ for $\alpha, \beta \in \pi_1(X, x)$. It is called the *monodromy* action on the fibre $p^{-1}(x)$.

The monodromy action is the analogue of the Galois action on homomorphisms encountered in Theorem 1.5.4. We now state a category equivalence analogous to the algebraic case as follows. Fix a space X and a point $x \in X$. First we define a functor Fib_x from the category of covers of X to the category of sets equipped with a left $\pi_1(X, x)$-action by sending a cover $p : Y \to X$ the fibre $p^{-1}(x)$. This is indeed a functor since a morphism $f : Y \to Z$ of covers respects the fibres over x by definition, and sends the unique lifting of a closed path through x starting with a point in $y \in Y$ to the unique lifting in Z starting with $f(y)$, by uniqueness of the lifting.

Theorem 2.3.4 *Let X be a connected and locally simply connected topological space, and $x \in X$ a base point. The functor* Fib_x *induces an equivalence of the category of covers of X with the category of left $\pi_1(X, x)$-sets. Connected covers correspond to $\pi_1(X, x)$-sets with transitive action and Galois covers to coset spaces of normal subgroups.*

Here local simply connectedness means that each point has a basis of simply connected open neighbourhoods.

The proof of this classification result relies on two crucial facts. The first one uses the notion of representable functors introduced in Definition 1.4.11.

Theorem 2.3.5 *For a connected and locally simply connected topological space X and a base point $x \in X$ the functor* Fib_x *is representable by a cover* $\widetilde{X}_x \to X$.

The cover $\widetilde{X}_x$ depends on the choice of the base point x, and comes equipped with a canonical point in the fibre $\pi^{-1}(x)$ called the *universal element*. Let us spell this out in detail. By definition, cover maps from $\pi : \widetilde{X}_x \to X$ to a fixed cover $p : Y \to X$ correspond bijectively (and in a functorial way) to points of the fibre $p^{-1}(x) \subset Y$. In particular, since $\widetilde{X}_x$ itself is a cover of X via π, we have a canonical isomorphism $\mathrm{Fib}_x(\widetilde{X}_x) \cong \mathrm{Hom}_X(\widetilde{X}_x, \widetilde{X}_x)$, where Hom_X denotes the set of maps of spaces over X. Via this isomorphism the identity map of $\widetilde{X}_x$ corresponds to a canonical element $\widetilde{x}$ in the fibre $\pi^{-1}(x)$; this is the universal element. Now for an arbitrary cover $p : Y \to X$ and element $y \in \pi^{-1}(x)$ the cover map $\pi_y : \widetilde{X}_x \to Y$ corresponding to y via the isomorphism $\mathrm{Fib}_x(Y) \cong \mathrm{Hom}_X(\widetilde{X}_x, Y)$ maps $\widetilde{x}$ to y by commutativity of the diagram

$$\begin{array}{ccc} \mathrm{Hom}_X(\widetilde{X}_x, \widetilde{X}_x) & \xrightarrow{\cong} & \mathrm{Fib}_x(\widetilde{X}_x) \\ \downarrow & & \downarrow \\ \mathrm{Hom}_X(\widetilde{X}_x, Y) & \xrightarrow{\cong} & \mathrm{Fib}_x(Y) \end{array}$$

where the vertical maps are induced by π_y.

We next recover the monodromy action. Notice that given an automorphism $\phi : \widetilde{X}_x \to \widetilde{X}_x$ of $\widetilde{X}_x$ as a cover of X, composition by ϕ induces a bijection of the set $\mathrm{Hom}_X(\widetilde{X}_x, Y)$ onto itself for each cover Y. In this way we obtain a *right* action on $\mathrm{Hom}_X(\widetilde{X}_x, Y) \cong \mathrm{Fib}_x(Y)$ from the *left* action of $\mathrm{Aut}(\widetilde{X}_x|X)$ on $\widetilde{X}_x$. We would like to compare it with the monodromy action, which is a left action. To this end we introduce the following notion:

Definition 2.3.6 For a group G the *opposite group* G^{op} is the group with the same underlying set as G but with multiplication defined by $(x, y) \mapsto yx$.

Note that $(G^{op})^{op} = G$, and moreover G is canonically isomorphic to G^{op} via the map $g \mapsto g^{-1}$. This being said, the above right action of $\mathrm{Aut}(\widetilde{X}_x|X)$ on $\widetilde{X}_x$ becomes a left action of $\mathrm{Aut}(\widetilde{X}_x|X)^{op}$.

Theorem 2.3.7 *The cover $\widetilde{X}_x$ is a connected Galois cover of X, with automorphism group isomorphic to $\pi_1(X, x)$. Moreover, for each cover $Y \to X$ the left action of* $\mathrm{Aut}(\widetilde{X}_x|X)^{op}$ *on* $\mathrm{Fib}_x(Y)$ *given by the previous construction is exactly the monodromy action of $\pi_1(X, x)$.*

We postpone the proof of the above two theorems to the next section, and prove Theorem 2.3.4 assuming their validity.

Proof of Theorem 2.3.4 The proof is strictly parallel to that of Theorem 1.5.4: we check that the functor satisfies the conditions of Lemma 1.4.9. For full faithfulness we have to show that given two covers $p : Y \to X$ and $q : Z \to X$, each map $\phi : \mathrm{Fib}_x(Y) \to \mathrm{Fib}_x(Z)$ of $\pi_1(X, x)$-sets comes from a unique map $Y \to Z$ of covers of X. For this we may assume Y, Z are connected and consider the map $\pi_y : \widetilde{X}_x \to Y$ corresponding to a fixed $y \in \mathrm{Fib}_x(Y)$. By Theorem 2.2.10 the map π_y realizes Y as the quotient of $\widetilde{X}_x$ by the stabilizer $U_y = \mathrm{Aut}(\widetilde{X}_x|Y)$ of y; let $\psi_y : Y \xrightarrow{\sim} U_y\backslash\widetilde{X}_x$ be the inverse map. Since U_y injects into the stabilizer of $\phi(y)$ via ϕ, the natural map $\pi_z : \widetilde{X}_x \to Z$ corresponding to $\phi(y)$ induces a map $U_y\backslash\widetilde{X}_x \to Z$ by passing to the quotient; composing it with ψ_y gives the required map $Y \to Z$. For essential surjectivity we have to show that each left $\pi_1(X, x)$-set S is isomorphic to the fibre of some cover of X. For S transitive we may take the quotient of $\widetilde{X}_x$ by the action of the stabilizer of some point; in the general case we decompose S into its $\pi_1(X, x)$-orbits and take the disjoint union of the corresponding covers. □

Remark 2.3.8 If we compare the above theorems with Theorem 1.5.4, we see that the cover $\widetilde{X}_x$ plays the role of a separable closure k_s; the choice of x corresponds to the choice of the separable closure. The fundamental group is the counterpart of the absolute Galois group. The functor inducing the equivalence is $A \mapsto \mathrm{Hom}(A, k_s)$ in the case of fields (it is contravariant), and $Y \mapsto \mathrm{Hom}_X(\widetilde{X}_x, Y) \cong \mathrm{Fib}_x(Y)$ in the topological case.

We now state a corollary of Theorem 2.3.4 that is even closer to Theorem 1.5.4 in its formulation and will be invoked in subsequent chapters. First a definition: call a cover $Y \to X$ *finite* if it has finite fibres; for connected X these have the same cardinality, called the *degree* of X.

Corollary 2.3.9 *For X and x as in the Theorem 2.3.4, the functor* Fib_x *induces an equivalence of the category of* finite *covers of X with the category of finite continuous left $\widehat{\pi_1(X, x)}$-sets. Connected covers correspond to finite $\widehat{\pi_1(X, x)}$-sets with transitive action and Galois covers to coset spaces of open normal subgroups.*

Here $\widehat{\pi_1(X, x)}$ denotes the *profinite completion* of $\pi_1(X, x)$ (Example 1.3.4 (2)). We need a well-known lemma from group theory.

Lemma 2.3.10 *In a group G each subgroup H of finite index contains a normal subgroup N of finite index.*

Proof Consider the natural left representation ρ_H of G on the left coset space of H, and take $N := \ker(\rho_H)$. It is of finite index in G as $[G : H]$ is finite, and it is contained in H as it fixes H considered as a coset. □

Proof of Corollary 2.3.9 For a finite connected cover $p : Y \to X$ the action of $\pi_1(X, x)$ on $p^{-1}(x)$ factors through a finite quotient, so we obtain an action of $\widehat{\pi_1(X, x)}$ as well. The stabilizer of each point $y \in Y$ is a subgroup of finite index, and hence contains a normal subgroup of finite index by the lemma. Therefore the stabilizer of y under the action of $\widehat{\pi_1(X, x)}$ is an open subgroup in the profinite topology (being a union of cosets of an open normal subgroup), which means that the action is continuous. Conversely, a continuous action of $\widehat{\pi_1(X, x)}$ on a finite set factors through a finite quotient which is also a quotient of $\pi_1(X, x)$, and as such gives rise to a finite cover $Y \to X$. □

2.4 The universal cover

In this section we prove Theorems 2.3.4 and 2.3.7. We begin with the construction of the space $\widetilde{X}_x$.

Construction 2.4.1 We construct the space $\widetilde{X}_x$ as follows. The points of $\widetilde{X}_x$ are to be homotopy classes of paths starting from x. To define the projection π, we pick for each point $\widetilde{y} \in \widetilde{X}_x$ a path $f : [0, 1] \to X$ with $f(0) = x$ representing $\widetilde{y}$, and put $\pi(\widetilde{y}) = f(1) = y$. This gives a well-defined map since homotopic paths have the same endpoints by definition. We next define the topology on $\widetilde{X}_x$ by taking as a basis of open neighbourhoods of a point $\widetilde{y}$ the following sets $\widetilde{U}$: we start from a simply connected neighbourhood U of y, and if $f : [0, 1] \to X$ is a path representing $\widetilde{y}$, we define $\widetilde{U}_{\widetilde{y}}$ to be the set of homotopy classes of paths obtained by composing the homotopy class of f with the homotopy class of some path $g : [0, 1] \to X$ with $g(0) = y$ and $g([0, 1]) \subset U$. Notice that since U is assumed to be simply connected, two such g having the same endpoints have the same homotopy class. Thus in more picturesque terms, $\widetilde{U}$ is obtained by 'continuing homotopy classes of paths arriving at y to other points of U'. This indeed gives a basis of open neighbourhoods of $\widetilde{y}$, for given two neighbourhoods $\widetilde{U}_{\widetilde{y}}$ and $\widetilde{V}_{\widetilde{y}}$, their intersection $\widetilde{U}_{\widetilde{y}} \cap \widetilde{V}_{\widetilde{y}}$ contains $\widetilde{W}_{\widetilde{y}}$ with some simply connected neighbourhood W of y contained in $U \cap V$; one also sees immediately that π is continuous with respect to this topology. The inverse image by π of a simply connected neighbourhood of a point y will be the disjoint union of the open sets $\widetilde{U}_{\widetilde{y}}$ for all inverse images $\widetilde{y}$ of y, so we have obtained a cover of X. Finally note that there is a 'universal element' $\widetilde{x}$ of the fibre $\pi^{-1}(x)$ corresponding to the homotopy class of the constant path.

Proof of Theorem 2.3.5 We show that the cover $\widetilde{X}_x \to X$ constructed above represents the functor Fib_x. This means that for a cover $p : Y \to X$ each point y of the fibre $p^{-1}(x)$ should correspond in a canonical and functorial manner to a morphism $\pi_y : \widetilde{X}_x \to Y$ over X. We define π_y as follows: given a point $\widetilde{x}' \in \widetilde{X}_x$ represented by a path $f : [0, 1] \to X$, we send it to $\widetilde{f}(1)$, where $\widetilde{f} : [0, 1] \to Y$ is the unique path lifting f with $\widetilde{f}(0) = y$ whose existence is guaranteed by Lemma 2.3.2 (1). Part (2) of the lemma implies that this map is well defined, and there is no difficulty in checking that it is indeed a map of covers. The map $y \mapsto \pi_y$ is a bijection between $p^{-1}(x)$ and the set $\mathrm{Hom}_X(\widetilde{X}_x, Y)$, an inverse being given by sending a morphism ϕ to the image $\phi(\widetilde{x})$ of the universal element $\widetilde{x}$. Finally, the above bijection is functorial: given a morphism $Y \to Y'$ of covers of X mapping y to some $y' \in Y'$, the induced map $\mathrm{Hom}_X(\widetilde{X}_x, Y) \to \mathrm{Hom}_X(\widetilde{X}_x, Y')$ maps π_y to $\pi_{y'}$, since these are the maps sending $\widetilde{x}$ to y and y', respectively. □

The proof of Theorem 2.3.7 will be in several steps. We begin with:

Lemma 2.4.2 *The space $\widetilde{X}_x$ is connected.*

Proof It is enough to see that $\widetilde{X}_x$ is path-connected, for which we show that there is a path in $\widetilde{X}_x$ connecting the universal point $\widetilde{x}$ to any other point $\widetilde{x}'$. Indeed, let $f : [0, 1] \to X$ be a path representing $\widetilde{x}'$. The multiplication map $m : [0, 1] \times [0, 1] \to [0, 1]$, $(s, t) \mapsto st$ is continuous, hence so is $f \circ m$ and the restriction of $f \circ m$ to each subset of the form $\{s\} \times [0, 1]$; such a restriction defines a path f_s from $\widetilde{x}$ to $f(s)$, with f_0 the constant path $[0, 1] \to \{x\}$ and $f_1 = f$. The definition of the topology of $\widetilde{X}_x$ implies that the map sending $s \in [0, 1]$ to the homotopy class of f_s is continuous and thus defines the path we need; in fact, it is the unique lifting of f to $\widetilde{X}_x$ beginning at $\widetilde{x}$. Alternatively, one may start by taking this unique lifting and check by going through the construction that its endpoint is indeed $\widetilde{x}'$. □

Next we prove:

Proposition 2.4.3 *The cover $\pi : \widetilde{X}_x \to X$ is Galois.*

For the proof we need some auxiliary statements.

Lemma 2.4.4 *A cover of a simply connected and locally path-connected space is trivial.*

Here 'locally path-connected' means that each point has a basis of path-connected open neighbourhoods.

Proof It is enough to show that given a *connected* cover $p : Y \to X$ of a space X as in the lemma, the map p is injective. For this, note first that since X is locally path-connected and p is a local homeomorphism, Y has a covering by path-connected open subsets. The connectedness assumption on Y then implies that it must be path-connected as well. Now consider two points y_0, y_1 in Y with $p(y_0) = p(y_1)$. By path-connectedness of Y there is a path $\widetilde{f} : [0, 1] \to Y$ with $\widetilde{f}(0) = y_0$ and $\widetilde{f}(1) = y_1$. The path $\widetilde{f}$ must be the unique lifting starting from y_0 of the path $f = p \circ \widetilde{f}$ which is a closed path around $x = p(y_0) = p(y_1)$. Since X is simply connected, this path is homotopic to the constant path $[0, 1] \to \{x\}$ of which the constant path $[0, 1] \to \{y_0\}$ provides the unique lifting to Y starting from y_0. By Lemma 2.3.2 (2) this is only possible if $y_0 = y_1$. □

Corollary 2.4.5 *Let X be a locally simply connected space. Given two covers $p : Y \to X$ and $q : Z \to Y$, their composite $q \circ p : Z \to X$ is again a cover of X.*

Proof Given a point of X, choose a simply connected neighbourhood U. According to the proposition, the restriction of p to $p^{-1}(U)$ gives a trivial cover of U. Repeating this argument for q over each of the connected components of $p^{-1}(U)$ (which are simply connected themselves, being isomorphic to U) we get that the restriction of $p \circ q$ to $(p \circ q)^{-1}(U)$ is a trivial cover of U. □

Proof of Proposition 2.4.3 By Remark 2.2.8 it is enough to show that the group $\mathrm{Aut}(\widetilde{X}_x|X)$ acts transitively on the fibre $\pi^{-1}(x)$. For each point $\widetilde{y}$ of the fibre $\pi^{-1}(x)$ Theorem 2.3.5 gives a continuous map $\pi_{\widetilde{y}} : \widetilde{X}_x \to \widetilde{X}_x$ compatible with π and mapping the universal element $\widetilde{x}$ to $\widetilde{y}$. We show that $\pi_{\widetilde{y}}$ is an automorphism. Since $\widetilde{X}_x$ is connected, by Lemma 2.2.11 $\pi_{\widetilde{y}}$ endows $\widetilde{X}_x$ with a structure of a cover over itself; in particular it is surjective. Take an element $\widetilde{z} \in \pi_{\widetilde{y}}^{-1}(\widetilde{x})$. Since $\pi \circ \pi_{\widetilde{y}} : \widetilde{X}_x \to X$ is also a cover of X according to Corollary 2.4.5, we may apply Theorem 2.3.5 to this cover to get a continuous and surjective map $\pi_{\widetilde{z}} : \widetilde{X}_x \to \widetilde{X}_x$ with $\pi_{\widetilde{z}}(\widetilde{x}) = \widetilde{z}$ and $\pi \circ \pi_{\widetilde{y}} \circ \pi_{\widetilde{z}} = \pi$. But $\pi_{\widetilde{y}} \circ \pi_{\widetilde{z}}(\widetilde{x}) = \widetilde{x}$, hence $\pi_{\widetilde{y}} \circ \pi_{\widetilde{z}}$ is the identity map of $\widetilde{X}_x$ by Lemma 2.2.2. By surjectivity of $\pi_{\widetilde{z}}$ this implies that $\pi_{\widetilde{y}}$ is injective and we are done. □

We now turn to the second statement of Theorem 2.3.7.

Proposition 2.4.6 *There is a natural isomorphism* $\mathrm{Aut}(\widetilde{X}_x|X)^{op} \cong \pi_1(X, x)$.

Proof First observe that $\widetilde{X}_x$ is endowed with a natural right action of $\pi_1(X, x)$ defined as follows: given a point $\widetilde{x}' \in \widetilde{X}_x$ and an element $\alpha \in \pi_1(X, x)$ with respective path representatives f and f_α, we may take the composition $f \bullet f_\alpha$ and then take the homotopy class of the product. It is straightforward to check that the map $\phi_\alpha : \widetilde{X}_x \to \widetilde{X}_x$ thus obtained is continuous and compatible with π, i.e. it is a cover automorphism. As by convention $\mathrm{Aut}(\widetilde{X}_x|X)$ acts on $\widetilde{X}_x$ from the left, this defines a group homomorphism $\pi_1(X, x) \to \mathrm{Aut}(\widetilde{X}_x|X)^{op}$

that is moreover injective, since any nontrivial α moves the distinguished element $\widetilde{x}$. It remains to prove the surjectivity of this homomorphism. For this purpose, take an arbitrary $\phi \in \mathrm{Aut}(\widetilde{X}_x|X)$ and a point $\widetilde{x}' \in \widetilde{X}_x$ represented by some path $f : [0, 1] \to X$. The point $\phi(\widetilde{x}')$ is then represented by some $g : [0, 1] \to X$ with $g(1) = f(1)$. Now $f^{-1} \bullet g$ is a closed path around x in X satisfying $f \bullet (f^{-1} \bullet g) = g$; denote by α its class in $\pi_1(X, x)$. The automorphism $\phi \circ \phi_\alpha^{-1}$ fixes $\widetilde{x}'$, so it is the identity by connectedness of $\widetilde{X}_x$ and Lemma 2.2.1. This shows $\phi_\alpha = \phi$. □

Proof of Theorem 2.3.7 Everything was proven above except for the last statement concerning the monodromy action. By Theorem 2.3.5 each point y of the fibre corresponds to a morphism of covers $\pi_y : \widetilde{X}_x \to Y$ and the proof shows that π_y maps points of $\widetilde{X}_x$ represented by paths f to points of the form $\widetilde{f}(1)$ where $\widetilde{f} : [0, 1] \to Y$ is the lifting of f with $\widetilde{f}(0) = y$; in particular y, being the image of the class of the constant path $c : [0, 1] \to \{x\}$, corresponds to the constant path $[0, 1] \to \{y\}$. By the proof of the previous lemma an element of $\pi_1(X, x)$ represented by a path f_α acts on c by mapping it to the class of f_α. Hence the action of α on $p^{-1}(x)$ maps y to $\widetilde{f_\alpha}(1)$, where $\widetilde{f_\alpha} : [0, 1] \to Y$ is the canonical lifting of f_α starting from y. This is the monodromy action. □

We now examine the dependence of the fundamental group on the choice of the base point. Assume given a path-connected and locally simply connected space X and two base points $x, y \in X$. Pick a path f from x to y. There is a map $\widetilde{X}_y \to \widetilde{X}_x$ induced on homotopy classes by the map $g \mapsto g \bullet f$ (here g is a path starting from y representing a point of $\widetilde{X}_y$). It only depends on the homotopy class of f and is an isomorphism of spaces over X, the inverse coming from composition with f^{-1}.

Proposition 2.4.7 *The above construction yields a bijection between homotopy classes of paths joining x to y and isomorphisms $\widetilde{X}_y \xrightarrow{\sim} \widetilde{X}_x$ in the category of covers of X.*

Proof An isomorphism $\lambda : \widetilde{X}_y \to \widetilde{X}_x$ takes the distinguished element $\widetilde{y} \in \widetilde{X}_y$ of the fibre over y to an element $\lambda(\widetilde{y}) \in \widetilde{X}_x$. It must also lie above y, and hence is the homotopy class of a path from x to y. The reader will check that it induces the isomorphism λ in the manner described above. □

Remark 2.4.8 A cover isomorphism $\lambda : \widetilde{X}_y \to \widetilde{X}_x$ induces a group isomorphism $\mathrm{Aut}(\widetilde{X}_y|X) \xrightarrow{\sim} \mathrm{Aut}(\widetilde{X}_x|X)$ by the map $\phi \mapsto \lambda \circ \phi \circ \lambda^{-1}$. Via the isomorphism of Proposition 2.4.6 it corresponds to an isomorphism $\lambda^{op} : \pi_1(X, y) \xrightarrow{\sim} \pi_1(X, x)$. If λ is induced by the homotopy class of a path f in the above construction, then λ^{op} corresponds at the level of homotopy classes of paths to the map $g \mapsto f^{-1} \bullet g \bullet f$. We thus recover the familiar dependence of the fundamental group on the base point. Changing the path f to f_1 changes λ^{op} to its composite with the inner automorphism of $\pi_1(X, x)$ induced by the class of $f_1^{-1} \bullet f$, so λ^{op} is uniquely determined up to an inner automorphism of $\pi_1(X, x)$.

Most textbooks neglect the role of the base point explained above and call any cover isomorphic to some $\widetilde{X}_x$ a *universal cover* of X. The next proposition shows that one can easily detect universal covers in practice.

Proposition 2.4.9 *Let X be a path-connected and locally simply connected space. A cover $\widetilde{X} \to X$ is universal if and only if it is simply connected.*

For the proof we need a lemma.

Lemma 2.4.10 *Consider a space X as in the proposition, a base point $x \in X$ and a connected cover $p : Y \to X$. The cover map $\widetilde{X}_x \to X$ factors through Y, and $\widetilde{X}_x$ represents* Fib_y *for each $y \in p^{-1}(x)$.*

Proof Since $\widetilde{X}_x$ represents Fib_x, the point y corresponds to a canonical map $\pi_y : \widetilde{X}_x \to Y$ which turns $\widetilde{X}_x$ into a cover of Y by virtue of Lemma 2.2.11. Our task is to show that this cover represents the functor Fib_y. So take a cover $q : Z \to Y$ and pick a point $z \in q^{-1}(y)$. Since by Corollary 2.4.5 the composition $p \circ q$ turns Z into a cover of X with $z \in (p \circ q)^{-1}(x)$, the point z corresponds to a morphism $\pi_z : \widetilde{X}_x \to Z$ of covers *of* X mapping the universal point $\widetilde{x}$ of $\widetilde{X}_x$ to z. It is now enough to see that π_z is also a morphism of covers of Y, i.e. $\pi_y = q \circ \pi_z$. But $p \circ \pi_y = p \circ q \circ \pi_z$ by construction, and moreover both π_y and $q \circ \pi_z$ map the universal point $\widetilde{x}$ to y, so the assertion follows from Lemma 2.2.2. □

Proof of Proposition 2.4.9 To prove simply connectedness of $\widetilde{X}_x$ for some $x \in X$, apply the lemma to see that $\widetilde{X}_x$ as a trivial cover of itself represents the fibre functor $\mathrm{Fib}_{\widetilde{x}}$ for the universal element $\widetilde{x}$. Then it follows from Theorem 2.3.7 that $\pi_1(\widetilde{X}_x, \widetilde{x}) \cong \mathrm{Aut}(\widetilde{X}_x|\widetilde{X}_x)^{op} = \{1\}$. Conversely, if X' is a simply connected cover of X, then $\widetilde{X}' \cong \pi_1(X', x')\backslash\widetilde{X}_x \cong \widetilde{X}_x$ with some point $x' \in X'$, by Lemma 2.4.10 and Theorem 2.3.7. □

Example 2.4.11 Since $\mathbf{R}^n$ is simply connected for any n, we see that the first two examples in 2.1.8 actually give universal covers of the circle and of linear tori, respectively. On the other hand, the third example there does not give a universal cover since $\mathbf{C}^*$ is not simply connected. However, the complex plane $\mathbf{C}$, being a two-dimensional $\mathbf{R}$-vector space, is simply connected and the exponential map $\mathbf{C} \to \mathbf{C}^*$, $z \mapsto \exp(z)$ is readily seen to be a cover. Hence $\mathbf{C}$ is the universal cover of $\mathbf{C}^*$.

Having determined the universal covers, we can compute the fundamental groups as well. The first two examples were quotients by group actions, so we obtain that the fundamental group of a circle is isomorphic to $\mathbf{Z}$, and that of a linear torus of dimension n to $\mathbf{Z}^n$. In the case of $\mathbf{C}^*$ the fundamental group is again $\mathbf{Z}$, because the exponential map is periodic with respect to $2\pi i$.

Example 2.4.12 Let us spell out in more detail an example similar to that of $\mathbf{C}^*$; it will serve in the next chapter. Let $\dot{D}$ be the punctured complex disc $\{z \in \mathbf{C} : z \neq 0,\ |z| < 1\}$. As in the previous example, the exponential map $z \mapsto \exp(z)$ restricted to the left half plane $L = \{z \in \mathbf{C} : \operatorname{Re} z < 0\}$ furnishes a universal cover of $\dot{D}$. The automorphism groups of fibres are isomorphic to $\mathbf{Z}$, the action of $n \in \mathbf{Z}$ being given via translation by $2n\pi i$. Thus if we let $\mathbf{Z}$ act on L via translation by multiplies of $2\pi i$, the disc $\dot{D}$ becomes the quotient of L by this action. But then by Theorem 2.2.10 each connected cover of $\dot{D}$ is isomorphic to a quotient of L by a subgroup of $\mathbf{Z}$. These subgroups are 0 and the subgroups $k\mathbf{Z}$ for integers $k \geq 1$; the corresponding covers of $\dot{D}$ are L and $\dot{D}$ itself via the map $z \mapsto z^k$. If we choose a base point $x \in \dot{D}$, we obtain an isomorphism $\pi_1(\dot{D}, x) \cong \mathbf{Z}$. A path representative of a generator is given by a circle around 0 going through x, oriented counterclockwise.

The previous considerations showed that one way to eliminate the role of the base point is to work up to non-canonical isomorphism. A better way is to consider all possible base points at the same time, as we explain next.

We first need to recall the notion of the *fibre product* $Y \times_X Z$ of two spaces $p : Y \to X$ and $q : Z \to X$ over X. By definition it is the subspace of $Y \times Z$ consisting of points (y, z) satisfying $p(y) = q(z)$. It is equipped with natural projections $q_Y : Y \times_X Z \to Y$ and $p_Z : Y \times_X Z \to Z$ making the diagram

$$\begin{array}{ccc} Y \times_X Z & \xrightarrow{q_Y} & Y \\ {\scriptstyle p_Z}\downarrow & & \downarrow{\scriptstyle p} \\ Z & \xrightarrow{q} & X \end{array}$$

commute. In fact, it satisfies a universal property: it represents the set-valued functor on the category of spaces over X that maps a space $S \to X$ to the set of pairs of morphisms $(\phi : S \to Y,\ \psi : S \to Z)$ over X satisfying $p \circ \phi = q \circ \psi$. In case $p : Y \to X$ is a cover, then so is $p_Z : Y \times_X Z \to Z$ and moreover the fibre $p_Z^{-1}(z)$ over $z \in Z$ is canonically isomorphic to $p^{-1}(q(z))$. Indeed, using Proposition 2.1.3 it suffices to check these properties in the case when $Y \to X$ is a trivial cover, which is straightforward. The cover $p_Z : Y \times_X Z \to Z$ is called the *pullback* of $p : Y \to X$ along q and is also denoted by $q^*Y \to Z$. Now we come to:

Construction 2.4.13 Let again X be a connected and locally simply connected space. We construct a space $\widetilde{X}$ over $X \times X$ as follows. For each pair of points $(x, y) \in X \times X$ we consider the set $\widetilde{X}_{x,y}$ of homotopy classes of paths from x to y and we define $\widetilde{X}$ to be the disjoint union of the $\widetilde{X}_{x,y}$ for all pairs (x, y). The projection $\widetilde{X} \to X \times X$ is induced by mapping a path to its endpoints and the topology on $\widetilde{X}$ is defined similarly as in Construction 2.4.1: one continues homotopy classes of paths into small neighbourhoods of *both* of their endpoints. The fact that $\widetilde{X}$ is a cover of $X \times X$ is again checked as in 2.4.1. Notice that the cover $\widetilde{X}_x$ constructed in 2.4.1 is none but the pullback of $\widetilde{X} \to X \times X$ via the inclusion map $\{x\} \times X \to X \times X$. Thus $\widetilde{X}$ can be thought of as a continuous family of the $\widetilde{X}_x$; it is called the *path space* or the *fundamental groupoid* of X. This is the promised base point free construction. The name 'fundamental groupoid' is explained in Exercise 7.

Remarks 2.4.14

1. Fix a pair of points $(x, y) \in X \times X$. Pulling back the cover $\widetilde{X} \to X \times X$ via the inclusion map $\{(x, y)\} \to X \times X$ we get back the space $\widetilde{X}_{x,y}$ viewed as a cover of the one point space $\{(x, y)\}$. Notice that it carries a natural right action $\widetilde{X}_{x,y} \times \pi_1(X, y) \to \widetilde{X}_{x,y}$ by the fundamental group $\pi_1(X, y)$ coming from composition of paths. This action is in fact simply transitive: given two paths $f, g : [0, 1] \to X$ with $f(0) = g(0) = x$ and $f(1) = g(1) = y$, we may map f to g by composing with the closed path $g \bullet f^{-1}$ around y. A topological space equipped with a continuous simply transitive action of a topological group G is sometimes called a *G-torsor*. In our case we thus obtain discrete $\pi_1(X, y)$-torsors $\widetilde{X}_{x,y}$ and we may view $\widetilde{X}_x$ as a continuous family of $\pi_1(X, y)$-torsors. The space $\widetilde{X}$ is then a continuous 2-parameter family of torsors.
2. An interesting cover of X is given by the pullback of $\widetilde{X} \to X \times X$ via the diagonal map $\Delta : X \to X \times X$ sending $x \in X$ to $(x, x) \in X \times X$. The fibre of the resulting cover $\widetilde{X}_\Delta$ over a point $x \in X$ is precisely $\pi_1(X, x)$; in particular, it carries a group structure. Thus $\widetilde{X}_\Delta$ encodes the fundamental groups of X with respect to varying base points. For some of its properties see Exercise 6.

2.5 Locally constant sheaves and their classification

In this section we shall reformulate Theorem 2.3.4 in terms of locally constant sheaves. The first step is to introduce presheaves and sheaves.

Definition 2.5.1 Let X be a topological space. A *presheaf of sets* $\mathcal{F}$ on X is a rule that associates with each nonempty open subset $U \subset X$ a set $\mathcal{F}(U)$ and each inclusion of open sets $V \subset U$ a map $\rho_{UV} : \mathcal{F}(U) \to \mathcal{F}(V)$, the maps ρ_{UU} being identity maps and the identity $\rho_{UW} = \rho_{VW} \circ \rho_{UV}$ holding for a tower of inclusions $W \subset V \subset U$. Elements of $\mathcal{F}(U)$ are called *sections* of $\mathcal{F}$ over U.

Remarks 2.5.2

1. Similarly, one defines a *presheaf of groups* (resp. abelian groups, or rings, etc.) by requiring the $\mathcal{F}(U)$ to be groups (resp. abelian groups, rings, etc.) and the ρ_{UV} to be homomorphisms.
2. Here is a more fancy formulation of the definition. Let us associate a category X_{Top} with our space X by taking as objects the nonempty open subsets $U \subset X$, and by defining $\text{Hom}(V, U)$ to be the one-element set consisting of the natural inclusion $V \to U$ whenever $V \subset U$, and to be empty otherwise. Then a presheaf of sets is just a set-valued contravariant functor on the category X_{Top}.

With this interpretation we see immediately that presheaves of sets (abelian groups, etc.) on a fixed topological space X form a category: as morphisms one takes morphisms of contravariant functors. Recall that by definition this means that a morphism of presheaves $\Phi : \mathcal{F} \to \mathcal{G}$ is a collection of maps (or homomorphisms) $\Phi_U : \mathcal{F}(U) \to \mathcal{G}(U)$ such that for each inclusion $V \subset U$ the diagram

$$\begin{array}{ccc} \mathcal{F}(V) & \xrightarrow{\Phi_V} & \mathcal{G}(V) \\ \rho^{\mathcal{F}}_{UV} \big\downarrow & & \big\downarrow \rho^{\mathcal{G}}_{UV} \\ \mathcal{F}(U) & \xrightarrow{\Phi_U} & \mathcal{G}(U) \end{array}$$

commutes.

Example 2.5.3 The basic example to bear in mind is that of continuous real-valued functions defined locally on open subsets of X; in this case, the maps ρ_{UV} are given by restriction of functions to some open subset.

Motivated by the example, given an inclusion $V \subset U$, we shall also use the more suggestive notation $s|_V$ instead of $\rho_{UV}(s)$ for a section $s \in \mathcal{F}(U)$.

The presheaf in the above example has a particular property, namely that continuous functions may be patched together over open sets. More precisely, given two open subsets U_1 and U_2 and continuous functions $f_i : U_i \to \mathbf{R}$ for $i = 1, 2$ with the property that $f_1(x) = f_2(x)$ for all $x \in U_1 \cap U_2$, we may unambiguously define a continuous function $f : U_1 \cup U_2 \to \mathbf{R}$ by setting $f(x) = f_i(x)$ if $x \in U_i$. We now axiomatize this property and state it as a definition.

Definition 2.5.4 A presheaf $\mathcal{F}$ (of sets, abelian groups, etc.) is a *sheaf* if it satisfies the following two axioms:

1. Given a nonempty open set U and a covering $\{U_i : i \in I\}$ of U by nonempty open sets, if two sections $s, t \in \mathcal{F}(U)$ satisfy $s|_{U_i} = t|_{U_i}$ for all $i \in I$, then $s = t$.
2. For an open covering $\{U_i : i \in I\}$ of U as above, given a system of sections $\{s_i \in \mathcal{F}(U_i) : i \in I\}$ with the property that $s_i|_{U_i \cap U_j} = s_j|_{U_i \cap U_j}$ whenever $U_i \cap U_j \neq \emptyset$, there exists a section $s \in \mathcal{F}(U)$ such that $s|_{U_i} = s_i$ for all $i \in I$. By the previous property such an s is unique.

The category of sheaves of sets (or abelian groups, etc.) on a space X is defined as the full subcategory of the corresponding presheaf category. This means that a morphism of sheaves is just a morphism of the underlying presheaves.

Examples 2.5.5

1. If D is a connected open subset of $\mathbf{C}$, we define the *sheaf of holomorphic functions* on D to be the sheaf of rings $\mathcal{O}$ whose sections over some open subset $U \subset D$ are the complex functions holomorphic on U. This construction carries over to any complex manifold. One defines the sheaf of analytic functions on some real analytic manifold, or the sheaf of C^∞ functions on a C^∞ manifold in a similar way.
2. Let S be a topological space (or abelian group, etc.) and X another topological space. Define a sheaf $\mathcal{F}_S$ on X by taking $\mathcal{F}_S(U)$ to be the set (abelian group, etc.) of continuous functions $U \to S$ for all nonempty open $U \subset X$. As in the case of real-valued functions, this is a sheaf.
3. In the previous example assume moreover that S is *discrete.* In this case $\mathcal{F}_S$ is called the *constant sheaf* on X with value S. The name comes from the fact that over a *connected* open subset U the sections of $\mathcal{F}_S$ are just constant functions, i.e. $\mathcal{F}_S(U) = S$.
4. Here is a more eccentric example. Fix an abelian group A and a point x of a given topological space X. Define a presheaf $\mathcal{F}^x$ of abelian groups on X by the rule $\mathcal{F}^x(U) = A$ if $x \in U$ and $\mathcal{F}^x(U) = 0$ otherwise, the restriction morphisms being the obvious ones. This presheaf is easily seen to be a sheaf, called the *skyscraper sheaf* with value A over x.

Given an open subset U of a topological space X, there is an obvious notion of the *restriction* $\mathcal{F}|_U$ of a presheaf $\mathcal{F}$ from X to U: one simply considers the sections of $\mathcal{F}$ only over those open sets that are contained in U. This remark enables us to define locally constant sheaves.

Definition 2.5.6 A sheaf $\mathcal{F}$ on a topological space is *locally constant* if each point of X has an open neighbourhood U such that the restriction of $\mathcal{F}$ to U is isomorphic (in the category of sheaves on U) to a constant sheaf.

In fact, as we shall instantly show, these are very familiar objects. Assume henceforth that *all spaces are locally connected.* First a definition that will ultimately explain the use of the terminology 'section' for sheaves.

Definition 2.5.7 Let $p : Y \to X$ be a space over X, and $U \subset X$ an open set. A *section* of p over U is a continuous map $s : U \to Y$ such that $p \circ s = \mathrm{id}_U$.

Given a space $p : Y \to X$ over X, define a presheaf $\mathcal{F}_Y$ of sets on X as follows: for an open set $U \subset X$ take $\mathcal{F}_Y(U)$ as the set of sections of p over U, and for an inclusion $V \subset U$ define the restriction map $\mathcal{F}_Y(U) \to \mathcal{F}_Y(V)$ by restricting sections to V.

Proposition 2.5.8 *The presheaf $\mathcal{F}_Y$ just defined is a sheaf. If moreover $p : Y \to X$ is a cover, then $\mathcal{F}_Y$ is locally constant. It is constant if and only if the cover is trivial.*

We call the sheaf $\mathcal{F}_Y$ the *sheaf of local sections* of the space $p : Y \to X$ over X.

Proof The sheaf axioms follow from the fact that sections over U are continuous functions $U \to Y$ and therefore satisfy the patching properties. Now assume $p : Y \to X$ is a cover. Given a point $x \in X$, take a connected open neighbourhood V of x over which the cover is trivial, i.e. isomorphic to $V \times F$ where F is the fibre over x. The image of a section $V \to Y$ is a connected open subset mapped isomorphically onto V by p, hence it must be one of the connected components of $p^{-1}(V)$. Thus sections over V correspond bijectively to points of the fibre F and the restriction of $\mathcal{F}_Y$ to V is isomorphic to the constant sheaf defined by F. We get the constant sheaf if and only if we may take V to be the whole connected component of X containing x in the above argument. □

Thus for instance Example 2.1.5 which shows a simple example of a non-trivial cover also gives, via the preceding proposition, a simple example of a locally constant but non-constant sheaf.

Given a morphism $\phi : Y \to Z$ of covers of X, it induces a natural morphism $\mathcal{F}_Y \to \mathcal{F}_Z$ of locally constant sheaves by mapping a local section $s : U \to Y$ of $p : Y \to X$ to $\phi \circ s$. To see that we indeed obtain a local section of $q : Z \to X$ over U in this way, it is enough to show by the sheaf axioms that there is a covering of U by open subsets U_i such that $(\phi \circ s)|_{U_i}$ is a local section over U_i for all i. But this holds if we choose each U_i to be so small that both covers are trivial over U_i. Thus the rule $Y \mapsto \mathcal{F}_Y$ is a *functor.*

Theorem 2.5.9 *The above functor induces an equivalence between the category of covers of X and that of locally constant sheaves on X.*

For the proof of the theorem we construct a functor in the reverse direction; the construction will in fact work for an arbitrary presheaf on X. We first need a definition:

Definition 2.5.10 Let $\mathcal{F}$ be a presheaf of sets on a topological space X, and let x be a point of X. The *stalk* $\mathcal{F}_x$ of $\mathcal{F}$ at X is defined as the disjoint union of the sets $\mathcal{F}(U)$ for all open neighbourhoods U of x in X modulo the following equivalence relation: $s \in \mathcal{F}(U)$ and $t \in \mathcal{F}(V)$ are equivalent if there exists an open neighbourhood $W \subset U \cap V$ of x with $s|_W = t_W$.

In case $\mathcal{F}$ is a sheaf of abelian groups (or rings, etc.), the stalk $\mathcal{F}_x$ carries a natural structure of abelian group (ring, etc.) For example, if $\mathcal{F}$ is the sheaf of continuous real-valued functions on X, then $\mathcal{F}_x$ is the ring of 'germs of continuous functions' at x.

Remark 2.5.11 The above definition is a special case of a more general construction. Namely, a *(filtered) direct system* of sets $(S_\alpha, \phi_{\alpha\beta})$ consists of sets S_α indexed by a directed partially ordered set Λ together with maps $\phi_{\alpha\beta} : S_\alpha \to S_\beta$ for all $\alpha \leq \beta$. The *direct limit* of the system is defined as the quotient of the disjoint union of the S_α modulo the equivalence relation in which $s_\alpha \in S_\alpha$ and $s_\beta \in S_\beta$ are equivalent if $\phi_{\alpha\gamma}(s_\alpha) = \phi_{\beta\gamma}(s_\beta)$ for some $\gamma \geq \alpha, \beta$. In our case Λ consists of the open neighbourhoods of x with the partial order in which $U \leq V$ means $V \subset U$. This partially ordered set is directed since for any U, V we have $U, V \leq U \cap V$. The sets $\mathcal{F}(U)$ form a direct system indexed by Λ whose direct limit is $\mathcal{F}_x$. Again, one can define direct limits of systems of abelian groups (rings, etc.) in an analogous manner.

Notice that a morphism of presheaves $\mathcal{F} \to \mathcal{G}$ induces a natural morphism $\mathcal{F}_x \to \mathcal{G}_x$ on stalks. Hence taking the stalk of a presheaf at some point defines a functor from the category of presheaves of sets (abelian groups, etc.) on X to the category of sets (abelian groups, etc.)

Construction 2.5.12 Having the notion of stalks at hand, we now associate a space $p_\mathcal{F} : X_\mathcal{F} \to X$ over X with a presheaf of sets $\mathcal{F}$ in such a way that moreover $p_\mathcal{F}^{-1}(x) = \mathcal{F}_x$ for all $x \in X$. As a set, $X_\mathcal{F}$ is to be the disjoint union of the stalks $\mathcal{F}_x$ for all $x \in X$. The projection $p_\mathcal{F}$ is induced by the constant maps $\mathcal{F}_x \to \{x\}$. To define the topology on $X_\mathcal{F}$, note first that given an open set $U \subset X$, a section $s \in \mathcal{F}(U)$ gives rise to a map $i_s : U \to X_\mathcal{F}$ sending $x \in U$ to the image of s in $\mathcal{F}_x$. Now the topology on $X_\mathcal{F}$ is to be the coarsest one in which the sets $i_s(U)$ are open for all U and s; one checks that the projection $p_\mathcal{F}$ and the maps i_s are continuous for this topology. In case $\mathcal{F}$ is locally constant, the space $X_\mathcal{F}$ is a cover of X. Indeed, if U is a connected open subset of X such that $\mathcal{F}|_U$ is isomorphic to the constant sheaf defined by a set F, then we have $\mathcal{F}_x = F$ for all $x \in U$ and hence $p_{\mathcal{F}}^{-1}(U)$ is isomorphic to $U \times F$, where F carries the discrete topology.

A morphism $\phi : \mathcal{F} \to \mathcal{G}$ of presheaves induces maps $\mathcal{F}_x \to \mathcal{G}_x$ for each $x \in X$, whence a map of sets $\Phi : X_{\mathcal{F}} \to X_{\mathcal{G}}$ compatible with the projections onto X.

Lemma 2.5.13 *The map Φ is a morphism of spaces over X.*

Proof To see that Φ is continuous, pick an open subset $U \subset X$, a section $t \in \mathcal{G}(U)$ and a point $x \in X$. The basic open set $i_t(U)$ meets the stalk $\mathcal{G}_x$ in the image t_x of t in $\mathcal{G}_x$. Each preimage $s_x \in \Phi^{-1}(t_x)$ lies in $\mathcal{F}_x$ and comes from a section $s \in \mathcal{F}(V)$, with a $V \subset U$ containing x that can be chosen so small that $\phi(s) = t|_V$. Then s_x is contained in the basic open set $i_s(V) \subset X_{\mathcal{F}}$ whose image by Φ is exactly $i_t(V)$. Thus $\Phi^{-1}(i_s(U))$ is open in $X_{\mathcal{F}}$. □

By the lemma, the rule $\mathcal{F} \to X_{\mathcal{F}}$ is a functor from the category of sheaves on X to that of spaces over X. On the full subcategory of locally constant sheaves it takes values in the category of covers of X, and the stalk $\mathcal{F}_x$ at a point x equals the fibre of $X_{\mathcal{F}}$ over x.

Proof of Theorem 2.5.9 We have to show that given a locally constant sheaf $\mathcal{F}$ we have $\mathcal{F}_{X_{\mathcal{F}}} \cong \mathcal{F}$ functorially in $\mathcal{F}$ and conversely, given a cover $Y \to X$ we have $X_{\mathcal{F}_Y} \cong Y$ functorially in Y. In any case we have a natural morphism of sheaves $\mathcal{F} \to \mathcal{F}_{X_{\mathcal{F}}}$ sending a section $s \in \mathcal{F}(U)$ to the local section $i_s : U \to X_{\mathcal{F}}$ it defines, and a morphism of covers $Y \to X_{\mathcal{F}_Y}$ sending a point $y \in Y$ in the fibre Y_x over a point $x \in X$ to the corresponding point of the fibre $\mathcal{F}_{Y,x} = Y_x$ of $X_{\mathcal{F}_Y}$ over x. To show that these maps are isomorphisms it is enough to show that their restrictions over a suitable open covering of X are. Now choose the open covering $\{U_i : i \in I\}$ so that $\mathcal{F}|_{U_i}$ is constant for each i. Replacing U_i by X we may thus assume $\mathcal{F}$ is a constant sheaf with values in a set F. But then $X_{\mathcal{F}} \cong X \times F$, and conversely the sheaf of local sections of the trivial cover $X \times F \to X$ is the constant sheaf defined by F. This finishes the proof. □

Now that we have proven the theorem, we may combine it with Theorem 2.2.10 to obtain:

Theorem 2.5.14 *Let X be a connected and locally simply connected topological space, and let x be a point in X. The category of locally constant sheaves of sets on X is equivalent to the category of sets endowed with a left action of $\pi_1(X, x)$. This equivalence is induced by the functor mapping a sheaf $\mathcal{F}$ to its stalk $\mathcal{F}_x$ at x.*

Finally, we consider sheaves with values in sets with additional structure.

Theorem 2.5.15 *Let X and x be as above, and let R be a commutative ring. The category of locally constant sheaves of R-modules on X is equivalent to the category of left modules over the group ring $R[\pi_1(X, x)]$.*

Proof The stalk $\mathcal{F}_x$ is an R-module by construction, and it is also equipped with a left action by $\pi_1(X, x)$ *as a set*. To show that it is an $R[\pi_1(X, x)]$-module we have to show that the action of $\pi_1(X, x)$ is compatible with the R-module structure. For this let $\mathcal{F} \times \mathcal{F}$ be the direct product sheaf defined by $(\mathcal{F} \times \mathcal{F})(U) = \mathcal{F}(U) \times \mathcal{F}(U)$ over all open $U \subset X$; its stalk over a point x is just $\mathcal{F}_x \times \mathcal{F}_x$. The addition law on $\mathcal{F}$ is a morphism of sheaves $\mathcal{F} \times \mathcal{F} \to \mathcal{F}$ given over an open set U by the formula $(s_1, s_2) \mapsto s_1 + s_2$; the morphism $\mathcal{F}_x \times \mathcal{F}_x \to \mathcal{F}_x$ induced on the stalk at x is the addition law on $\mathcal{F}_x$. But this latter map is a map of $\pi_1(X, x)$-sets, which means precisely that $\sigma(s_1 + s_2) = \sigma s_1 + \sigma s_2$ for all $s_1, s_2 \in \mathcal{F}_x$ and $\sigma \in \pi_1(X, x)$. One verifies the compatibility with multiplication by elements of R in a similar way. The rest of the proof is a straightforward modification of that given for sheaves of sets; we leave the details to the reader. □

2.6 Local systems

In this section we investigate a most interesting special case of the preceding construction, which is also the one that historically first arose.

Definition 2.6.1 A *complex local system* on X is a locally constant sheaf of finite dimensional complex vector spaces. If X is connected, all stalks must have the same dimension, which is called the *dimension* of the local system.

With this definition we can state the following corollary of Theorem 2.5.15:

Corollary 2.6.2 *Let X be a connected and locally simply connected topological space, and x a point in X. The category of complex local systems on X is equivalent to the category of finite dimensional left representations of $\pi_1(X, x)$.*

Thus to give a local system on X is the same as giving a homomorphism $\pi_1(X, x) \to \mathrm{GL}(n, \mathbf{C})$ for some n. This representation is called the *monodromy representation* of the local system.

Remark 2.6.3 This is a point where the reader may appreciate the advantage of the convention we have chosen for the multiplication rule in $\pi_1(X, x)$ in the previous chapter. With our convention, the monodromy representation is indeed a homomorphism for the usual multiplication rule of matrices in $\mathrm{GL}(n, \mathbf{C})$. Had we defined the composition of paths 'the other way round', we would be forced to use here an unorthodox matrix multiplication of columns by rows instead of rows by columns.

The following example shows where to find local systems 'in nature'. It uses the straightforward notion of a *subsheaf* of a sheaf $\mathcal{F}$: it is a sheaf whose sections over each open set U form a subset (subgroup, subspace etc.) of $\mathcal{F}(U)$.

Example 2.6.4 Let $D \subset \mathbf{C}$ be a connected open subset. Consider over D a homogeneous n-th order linear differential equation

$$y^{(n)} + a_1 y^{(n-1)} + \cdots + a_{n-1} y' + a_n y = 0 \tag{2.1}$$

where the a_i are holomorphic functions on D. We look at local holomorphic solutions of the equation over each open subset $U \subset D$. As $\mathbf{C}$-linear combinations of local solutions of (2.1) over U are again local solutions, they form a $\mathbf{C}$-vector space $\mathcal{S}(U)$. Moreover, by a classical theorem due to Cauchy (see [23], Theorem 11.2 or any basic text on differential equations) each point of D has an open neighbourhood U where the $\mathbf{C}$-vector space $\mathcal{S}(U)$ has a finite basis $x_1, \ldots, x_n$. The second sheaf axiom then shows that the local solutions of the equation (2.1) form a subsheaf $\mathcal{S} \subset \mathcal{O}^n$; this is a sheaf of complex vector spaces. Since the restrictions of the above x_i to smaller open sets still form a basis for the solutions, we conclude that the sheaf $\mathcal{S}$ is a complex local system of dimension n.

A similar observation can be made when one considers solutions of a system of n linear differential equations in n variables, given in matrix form by $y' = Ay$, where A is an $n \times n$ matrix of holomorphic functions and $y = (y_1, \ldots, y_n)$ is an n-tuple of variables. In fact, by a classical trick the solutions of (2.1) are the same as the solutions of the linear system in the n variables $y_1 = y, y_2 = y', \ldots, y_n = y^{(n-1)}$ given by $y_i' = y_{i+1}$ for $1 \leq i \leq n-1$ and $y_n' = -a_1 y_n - \cdots - a_n y_1$.

Remark 2.6.5 According to Corollary 2.6.2, the local system $\mathcal{S}$ of the above example is uniquely determined by an n-dimensional left representation of $\pi_1(X, x)$, where x is a point of D (notice that D is locally simply connected). Let us describe this representation explicitly. Take a closed path $f : [0, 1] \to D$ representing an element $\gamma \in \pi_1(X, x)$ and take an element $s \in \mathcal{S}_x$ which is, in classical terminology, a germ of a (vector-valued) holomorphic function satisfying the equation (2.1). Now s is naturally a point of the fibre over x of the cover $p_{\mathcal{S}} : D_{\mathcal{S}} \to D$ associated with $\mathcal{S}$ by Theorem 2.5.9. By definition of the monodromy action, the element γ acts on s as follows: s is mapped to the element $\widetilde{f}(1)$ of the fibre $p_{\mathcal{S}}^{-1}(x) = \mathcal{S}_x$, where $\widetilde{f}$ is the unique lifting of f to $D_{\mathcal{S}}$. By looking at the construction of the unique lifting in the proof of Lemma 2.3.2, one can make this even more explicit as follows. There exist open subsets $U_1, \dots, U_k$ of D such that the $f^{-1}(U_1), \dots, f^{-1}(U_k)$ give an open covering of $[0, 1]$ and $\mathcal{S}$ is constant over each U_i. There are moreover sections $s_i \in \mathcal{S}(U_i)$ such that the restrictions of s_i and s_{i+1} to $U_i \cap U_{i+1}$ coincide for all $1 \le i \le k-1$, and such that s_1 (resp. s_k) maps to s (resp. γs) in $\mathcal{S}_x$. Classically this is expressed by saying that γs is the analytic continuation of the holomorphic function germ s along the path f representing γ.

Notice that the existence and the uniqueness of γs are guaranteed by the fact that $\mathcal{S}$ is a locally constant sheaf and hence $D_{\mathcal{S}} \to D$ is a cover. Had we worked with the bigger sheaf $\mathcal{O}^n$ instead of $\mathcal{S}$, the analytic continuation of an arbitrary germ may not have been possible.

Example 2.6.6 Let us work out the simplest nontrivial case of the above theory in detail. Take as D an open disc in the complex plane centred around 0, of radius $1 < R \le \infty$, with the point 0 removed. We choose 1 as base point for the fundamental group of D. We study the local system of solutions of the first order differential equation

$$y' = fy \tag{2.2}$$

where f is a holomorphic function on D that extends meromorphically into 0. It is well known that the solutions to (2.2) in some neighbourhood of a point $x \in D$ are constant multiples of functions of the form $\exp \circ F$ where F is a primitive of f. Thus the solution sheaf is a locally constant sheaf of one-dimensional complex vector spaces. The reason why it is locally constant but not constant is that, as we learn from complex analysis, the primitive F exists locally but not globally on the whole of D. A well-defined primitive F_- of f exists, for instance, over $U_- = D \setminus (0, -iR)$ and another primitive F_+ over $U_+ = D \setminus (0, iR)$. The intersection $U_- \cap U_+$ splits in two connected components $C_- \subset \{z : \mathrm{Re}(z) < 0\}$ and $C_+ \subset \{z : \mathrm{Re}(z) > 0\}$. As F_+ and F_- may differ only by a constant on each component, we are allowed to choose them in such a way that $F_- = F_+$ on C_-. The local system of solutions to (2.2) is isomorphic over U_- to the constant sheaf defined by the one-dimensional subspace of $\mathcal{O}(U_-)$ generated by $\exp \circ F_-$, and similarly for U_+.

Now we compute the monodromy representation $\pi_1(D, 1) \to GL(1, \mathbf{C})$ of this local system. We have seen in Example 2.4.12 that $\pi_1(D, 1) \cong \mathbf{Z}$. Explicitly, a generator γ is given by the class of the path $g : [0, 1] \to D$, $t \mapsto e^{2\pi i t}$ which 'goes counterclockwise around the unit circle'. A one-dimensional representation of $\pi_1(D, 1)$ is determined by the image m of γ in $GL(1, \mathbf{C}) \cong \mathbf{C}^*$. By the recipe of the previous remark, in our case m can be described as follows: given a holomorphic function germ ϕ defined in a neighbourhood of 1 and satisfying (2.2) with $y = \phi$, the analytic continuation of ϕ along the path g representing γ is precisely $m\phi$. But we may take for ϕ the function $\exp \circ F_-$;

when continuing it analytically along g, we obtain $\exp \circ F_+$ since we have to switch from F_- to F_+ somewhere on C_-. Thus

$$m = \exp(F_-(1))(\exp(F_+(1)))^{-1} = \exp(F_-(1) - F_-(-1) + F_+(-1) - F_+(1))$$
$$= \exp\left(\int_\gamma f\right) = \exp(2\pi i \operatorname{Res}_0(f))$$

by the Residue Theorem (see e.g. Rudin [80], Theorem 10.42), where $\operatorname{Res}_0(f)$ denotes the residue of f at 0. So we have expressed m in terms of the function f occurring in the equation (2.2).

Remark 2.6.7 One sees from the above example that for any one-dimensional monodromy representation on the punctured open disc D we may find a differential equation of type (2.2) whose local system has the given monodromy; if m is the image of γ, one may take, for example, $y'(z) = \mu z^{-1} y(z)$ with $\mu \in \mathbf{C}$ satisfying $\exp(2\pi i \mu) = m$. This equation has the additional virtue that the coefficient μz^{-1} has only a simple pole at 0.

We can generalize this to higher dimension. An n-dimensional representation of the punctured open disc D of radius R is again determined by the image of γ which is a matrix $M \in \mathrm{GL}_n(\mathbf{C})$. To find a system of n differential equations $y' = Ay$ with the above monodromy representation and with coefficients holomorphic on D except for a simple pole at the origin, we may take $A = Bz^{-1}$, where $B \in \mathrm{GL}_n(\mathbf{C})$ is a matrix with $\exp(2\pi i B) = M$. Here $\exp(2\pi i B)$ is defined by the absolutely convergent power series

$$\exp(2\pi i B) := \sum_{i=0}^{\infty} \frac{(2\pi i B)^i}{i!}.$$

Given $M \in \mathrm{GL}_n(\mathbf{C})$, one constructs B with $\exp(2\pi i B) = M$ using the Jordan decomposition of M (see [23], 11.8 for details).

2.7 The Riemann–Hilbert correspondence

Our discussion of complex local systems would not be complete without the introduction of connections. We shall only define them in the classical situation where the basis is a connected open subset $D \subset \mathbf{C}$, but everything holds in much greater generality; see the remarks below.

To begin with, consider the ring sheaf $\mathcal{O}$ of holomorphic functions on our open domain $D \subset \mathbf{C}$. A *sheaf of $\mathcal{O}$-modules* is a sheaf of abelian groups $\mathcal{F}$ on X such that for each open $U \subset X$ the group $\mathcal{F}(U)$ is equipped with an

$\mathcal{O}(U)$-module structure $\mathcal{O}(U) \times \mathcal{F}(U) \to \mathcal{F}(U)$ making the diagram

$$\begin{array}{ccc} \mathcal{O}(U) \times \mathcal{F}(U) & \longrightarrow & \mathcal{F}(U) \\ \downarrow & & \downarrow \\ \mathcal{O}(V) \times \mathcal{F}(V) & \longrightarrow & \mathcal{F}(V) \end{array}$$

commute for each inclusion of open sets $V \subset U$. A morphism of $\mathcal{O}$-modules is a morphism of sheaves of abelian groups compatible with the $\mathcal{O}$-module structure just described.

We say that $\mathcal{F}$ is *locally free* if every point of D has an open neighbourhood $V \subset D$ such that $\mathcal{F}|_V \cong \mathcal{O}^n|_V$ for some $n > 0$, where $\mathcal{O}^n$ denotes the n-fold direct sum of $\mathcal{O}$. The integer n is called the *rank* of $\mathcal{F}$. We say that $\mathcal{F}$ is *free* of rank n if there is actually an isomorphism $\mathcal{F} \cong \mathcal{O}^n$ on the whole of D.

Remark 2.7.1 Locally free sheaves on D are related to holomorphic vector bundles on D. The latter are defined as complex manifolds $\mathcal{E}$ equipped with a holomorphic surjection $p : \mathcal{E} \to D$ (see Remark 3.1.2 below for these concepts) such that $p^{-1}(z)$ has the structure of a complex vector space for each $z \in D$, and moreover each $z \in D$ has an open neighbourhood $V \subset D$ for which there is an isomorphism $p^{-1}(V) \cong V \times \mathbf{C}^n$ of complex manifolds over V inducing vector space isomorphisms on the fibres. Given a holomorphic vector bundle $p : \mathcal{E} \to D$, its *sheaf of holomorphic sections* $\mathcal{H}_\mathcal{E}$ is given over $U \subset D$ by the holomorphic maps $s : U \to \mathcal{E}$ satisfying $p \circ s = \mathrm{id_U}$. The restriction of $\mathcal{H}_\mathcal{E}$ to an open subset $V \subset D$ where $p^{-1}(V) \cong V \times \mathbf{C}^n$ is isomorphic to $\mathcal{O}^n|_V$, therefore $\mathcal{H}_\mathcal{E}$ is a locally free sheaf. This construction induces an equivalence between the category of holomorphic vector bundles on D and that of locally free sheaves on D, the functor in the reverse direction being defined by sending $\mathcal{F}|_V$ to $V \times \mathbf{C}^n$, and then patching the resulting complex manifolds together. For the above reason locally free sheaves are often called vector bundles in the literature.

We shall also need the sheaf Ω^1_D of holomorphic 1-forms on D. It may be defined as follows. Given a holomorphic function f on some $U \subset D$, consider the function $df : U \to \mathbf{C}$ given by $z \mapsto f'(z)$. A section of Ω^1_D over U is then defined as a complex function $\omega_U : U \to \mathbf{C}$ such that each $z \in U$ has an open neighbourhood $V \subset U$ mapped isomorphically onto an open disc around 0 in $\mathbf{C}$ by a holomorphic function $g \in \mathcal{O}(V)$ such that $\omega_U|_V = f\,dg$ with some $f \in \mathcal{O}(V)$. This is a sheaf of $\mathcal{O}$-modules on D.

The above definition is a bit clumsy, but it extends well to more general situations. Note that $dz : D \to C$ is a global section of Ω^1_D (because $d(z - a) = dz$ for all $a \in D$), and so is df for an arbitrary $f \in \mathcal{O}(D)$ (because $df = f'dz$). In particular, we may identify the function $df/dz : D \to \mathbf{C}$ with $f' \in \mathcal{O}(V)$.

Now we come to the main definition.

Definition 2.7.2 A *holomorphic connection* on D is a pair $(\mathcal{E}, \nabla)$, where $\mathcal{E}$ is a locally free sheaf on D, and $\nabla : \mathcal{E} \to \mathcal{E} \otimes_{\mathcal{O}} \Omega^1_D$ is a morphism of sheaves of $\mathbf{C}$-vector spaces satisfying the 'Leibniz rule'

$$\nabla(fs) = df \otimes s + f\nabla(s) \tag{2.3}$$

for all $U \subset D$, $f \in \mathcal{O}(U)$ and $s \in \mathcal{E}(U)$. We call ∇ the *connection map*.

Here the tensor product $\mathcal{E} \otimes_{\mathcal{O}} \Omega^1_D$ is defined in the obvious way, by the rule $U \mapsto \mathcal{E}(U) \otimes_{\mathcal{O}(U)} \Omega^1_D(U)$. Holomorphic connections form a category: a morphism $(\mathcal{E}, \nabla) \to (\mathcal{E}', \nabla')$ is a morphism of $\mathcal{O}$-modules $\phi : \mathcal{E} \to \mathcal{E}'$ making the diagram

$$\begin{array}{ccc} \mathcal{E} & \xrightarrow{\nabla} & \mathcal{E} \otimes_{\mathcal{O}} \Omega^1_D \\ \phi \downarrow & & \downarrow \phi\otimes\mathrm{id} \\ \mathcal{E}' & \xrightarrow{\nabla'} & \mathcal{E}' \otimes_{\mathcal{O}} \Omega^1_D \end{array}$$

commute.

Example 2.7.3 Assume that $\mathcal{E} \cong \mathcal{O}^n$ is a free $\mathcal{O}$-module. In this case we may identify sections $s \in \mathcal{E}(U)$ with n-tuples $(f_1, \dots, f_n)$ of holomorphic functions on U. We can define an obvious connection map $d : \mathcal{O}^n \to (\Omega^1_D)^{\oplus n}$ by setting $d(f_1, \dots, f_n) = (df_1, \dots, df_n)$; the formula (2.3) follows from the usual Leibniz rule for differentiation.

Now given any other connection map ∇ on $\mathcal{O}^n$, formula (2.3) implies that the map $\nabla - d : \mathcal{O}^n \to (\Omega^1_D)^{\oplus n}$ satisfies $(\nabla - d)(fs) = f(\nabla - d)(s)$ for all $f \in \mathcal{O}(D)$ and $s \in \mathcal{E}(D)$. It follows that $\nabla - d$ is given by an $n \times n$ matrix Ω of holomorphic 1-forms, called the *connection matrix* of ∇. In other words,

$$\nabla((f_1, \dots, f_n)) = (df_1, \dots, df_r) + \Omega(f_1, \dots, f_n).$$

By the remarks preceding Definition 2.7.2, we can actually identify each entry ω_{ij} of Ω with a 1-form of the shape $f_{ij}\,dz$, so 'division of Ω by dz' yields a matrix $[f_{ij}]$ with entries in $\mathcal{O}(D)$. Hence setting $f = (f_1, \dots, f_n)$ and $A = -[f_{ij}]$ we see that $\nabla(f) = 0$ if and only if f satisfies the system of differential equations $y' = Ay$.

In general, a section $s \in \mathcal{E}(U)$ of a connection $(\mathcal{E}, \nabla)$ is called *horizontal* if it satisfies $\nabla(s) = 0$. Horizontal sections form a subsheaf of $\mathbf{C}$-vector spaces $\mathcal{E}^\nabla \subset \mathcal{E}$.

Lemma 2.7.4 *The sheaf $\mathcal{E}^\nabla$ is a local system of dimension equal to the rank of $\mathcal{E}$.*

Proof According to a general fact on non-compact Riemann surfaces ([23], Theorem 30.4), every locally free sheaf on D is actually free. Therefore by the above example the sections of $\mathcal{E}^\nabla$ over $U \subset D$ correspond to local solutions of a linear system $y' = Ay$ of holomorphic differential equations on D. We conclude by Cauchy's theorem reinterpreted in Example 2.6.4. □

We can now prove a proposition that belongs to the family of statements that usually go by the name of 'Riemann–Hilbert correspondence' in the literature.

Proposition 2.7.5 *The functor $(\mathcal{E}, \nabla) \mapsto \mathcal{E}^\nabla$ induces an equivalence between the category of holomorphic connections on D and that of complex local systems on D.*

Proof To construct a functor in the reverse direction, take a local system $\mathcal{L}$ on D. The rule $U \mapsto \mathcal{L}(U) \otimes_{\mathbf{C}} \mathcal{O}(U)$ defines a locally free sheaf $\mathcal{E}_\mathcal{L}$ on D. We define a connection map $\nabla_\mathcal{L}$ on $\mathcal{E}_\mathcal{L}$ as follows. Given an open subset U where $\mathcal{L}|_U \cong \mathbf{C}^n$, fix a $\mathbf{C}$-basis $s_1, \dots, s_n$ of $\mathcal{L}(U)$. Then each section of $\mathcal{E}_\mathcal{L}(U)$ can be uniquely written as a sum $\sum s_i \otimes f_i$ with some $f_i \in \mathcal{O}(U)$. Now define $\nabla_\mathcal{L}|_U$ by setting $\nabla_\mathcal{L}(\sum s_i \otimes f_i) := \sum s_i \otimes df_i$. As two bases of $\mathcal{L}(U)$ differ by a matrix whose entries are in $\mathbf{C}$ and hence are annihilated by the differential d, the definition does not depend on the choice of the s_i. Therefore the $\nabla_\mathcal{L}|_U$ defined over the various U patch together to a map $\nabla_\mathcal{L}$ defined over the whole of D.

To check that the functors $(\mathcal{E}, \nabla) \mapsto \mathcal{E}^\nabla$ and $\mathcal{L} \mapsto (\mathcal{E}_\mathcal{L}, \nabla_\mathcal{L})$ induce an equivalence of categories, one may argue over open subsets where $\mathcal{E}$ (resp. $\mathcal{L}$) are trivial, and there it follows from the construction. □

Remark 2.7.6 The Riemann–Hilbert correspondence holds for arbitrary Riemann surfaces, and extends to higher dimensional complex manifolds as well. In the higher dimensional case, however, one has to impose a further condition on the connection $(\mathcal{E}, \nabla)$ which is automatic in dimension 1: namely that it is *flat* (or *integrable*). This means the following. Write Ω^1_X for the sheaf of holomorphic 1-forms on the complex manifold X, and Ω^2_X for its second exterior power. For each connection $(\mathcal{E}, \nabla)$ one may define a map $\nabla_1 : \mathcal{E} \otimes_\mathcal{O} \Omega^1_X \to \mathcal{E} \otimes_\mathcal{O} \Omega^2_X$ by setting $\nabla_1(s \otimes \omega) = \nabla(s) \wedge \omega + s \otimes d\omega$. Here $d : \Omega^1_X \to \Omega^2_X$ is the usual differential given by $d(f dz_i) = df \wedge dz_i$, where $dz_1, \dots, dz_n$ are free generators of Ω^1_X in the neighbourhood of a point. One then says that the connection is flat if the composite map $\nabla_1 \circ \nabla$ is trivial. The Riemann–Hilbert correspondence for complex manifolds asserts that the category of holomorphic flat connections on a complex manifold X is equivalent to that of complex local systems on X. For a proof, see e.g. [110], Proposition 9.11.

Combined with Corollary 2.6.2, Proposition 2.7.5 implies that every finite dimensional representation of $\pi_1(D, x)$ for some base point x is the monodromy representation of some linear system of holomorphic differential equations. Assume now that D is a complement of finitely many points in $\mathbf{C}$. Then one may consider a more subtle question, namely realizing representations of the fundamental group as monodromy representations of linear systems satisfying additional restrictions on the local behaviour of the coefficients around the missing points.

Traditionally, this problem is studied over the projective line $\mathbf{P}^1(\mathbf{C})$. The precise definition of $\mathbf{P}^1(\mathbf{C})$ as a Riemann surface will be given in Example 3.1.3 (2), but it can be easily introduced in an ad hoc way: it is just the complex plane with a point ∞ added at infinity. A system of complex open neighbourhoods of ∞ is given by the complements of closed discs around 0, and a function f is holomorphic around ∞ if $z \mapsto f(z^{-1})$ is holomorphic around 0. Thus we may extend the notion of sheaves of $\mathcal{O}$-modules to $\mathbf{P}^1(\mathbf{C})$. One difference should be noted, however: whereas over a domain $D \subset \mathbf{C}$ all locally free sheaves are free, over $\mathbf{P}^1(\mathbf{C})$ this is not so any more.

Consider a finite set $S = \{x_1, \dots, x_m\}$ of points of $\mathbf{P}^1(\mathbf{C})$, and set henceforth $D := \mathbf{P}^1(\mathbf{C}) \setminus S$. We may safely assume $x_m = \infty$. Define an extension $\Omega^1(S)$ of Ω^1_D to an $\mathcal{O}$-module on $\mathbf{P}^1(\mathbf{C})$ as follows: the sections of $\Omega^1(S)$ are $\mathbf{P}^1(\mathbf{C})$-valued functions that restrict to sections of Ω^1_D over open subsets contained in D, and in a neighbourhood of an x_i can be written in the form $f\,dz$ with f having at most a simple pole at x_i (for $x_i = \infty$ one of course requires a representation of the form $f\,d(z^{-1})$). The sheaf $\Omega^1(S)$ is called the sheaf of 1-forms with *logarithmic poles along* S. The name is explained by writing $f = g(z - x_i)^{-1}$ with a function g holomorphic around x_i: we then obtain $f\,dz = g\,(z - x_i)^{-1} d(z - x_i)$, and we recognize the logarithmic derivative of $z - x_i$ on the right-hand side.

A *connection with logarithmic poles along* S is a pair $(\overline{\mathcal{E}}, \overline{\nabla})$, where $\overline{\mathcal{E}}$ is a locally free sheaf on $\mathbf{P}^1(\mathbf{C})$, and $\overline{\nabla} : \overline{\mathcal{E}} \to \overline{\mathcal{E}} \otimes_{\mathcal{O}} \Omega^1(S)$ is a $\mathbf{C}$-linear map satisfying (2.3). As in Example 2.7.3, in the case when $\overline{\mathcal{E}}$ is free a connection with logarithmic poles corresponds to a linear system $y' = Ay$ of differential equations where the entries of the matrix A are holomorphic functions on D that have at worst a simple pole at the x_i. Such linear systems are called *Fuchsian*.

Proposition 2.7.7 *Given a holomorphic connection* $(\mathcal{E}, \nabla)$ *on* D, *there is a connection* $(\overline{\mathcal{E}}, \overline{\nabla})$ *on* $\mathbf{P}^1(\mathbf{C})$ *with logarithmic poles along* S *satisfying* $(\overline{\mathcal{E}}, \overline{\nabla})|_D \cong (\mathcal{E}, \nabla)$.

The proof uses a general construction for patching sheaves that will also serve later.

Construction 2.7.8 Let X be a topological space, $\{U_i : i \in I\}$ an open covering of X, and $\mathcal{F}_i$ a sheaf of sets on U_i for $i \in I$. Assume further given for each pair $i \neq j$ isomorphisms

$$\theta_{ij} : \mathcal{F}_i|_{U_i \cap U_j} \xrightarrow{\sim} \mathcal{F}_j|_{U_i \cap U_j}$$

satisfying $\theta_{jk} \circ \theta_{ij} = \theta_{ik}$ over $U_i \cap U_j \cap U_k$ for every triple (i, j, k) of different indices. Then there exists a sheaf $\mathcal{F}$ on X with $\mathcal{F}|_{U_i} = \mathcal{F}_i$ for each $i \in I$; it is unique up to unique isomorphism.

To define $\mathcal{F}(U)$ over an open subset $U \subset X$, set

$$\mathcal{F}(U) := \{(s_i)_{i \in I} : s_i \in \mathcal{F}_i(U \cap U_i) \quad \text{and}$$
$$\theta_{ij}(s_i|_{U \cap U_i \cap U_j}) = s_j|_{U \cap U_i \cap U_j} \quad \text{for all} \quad i \neq j\}.$$

The $\mathcal{F}(U)$ together with the obvious restriction maps form a presheaf $\mathcal{F}$, and the sheaf axioms for the $\mathcal{F}_i$ imply that $\mathcal{F}$ is in fact a sheaf. Its restrictions over the U_i yield the $\mathcal{F}_i$ by construction. The verification of the isomorphism statement is left to the readers.

One says that $\mathcal{F}$ is obtained by *patching* (or *gluing*) the $\mathcal{F}_i$ together. Of course the lemma also holds for sheaves with additional structure (sheaves of groups, rings, etc.) In particular, patching locally free sheaves on a complex domain yields a locally free sheaf. Also, patching locally constant sheaves on a topological space results in a locally constant sheaf.

Proof of Proposition 2.7.7 Take small open discs $D_i \subset \mathbf{P}^1(\mathbf{C})$ around each x_i that do not meet, and write n for the rank of the connection $(\mathcal{E}, \nabla)$. We know from Remark 2.6.7 that for each i there exists an $n \times n$ system $y' = A_i y$ of linear differential equations having a simple pole at x_i and holomorphic elsewhere. Write $(\mathcal{E}_i, \nabla_i)$ for the connection given by $\mathcal{E}_i = \mathcal{O}|_{D_i}^n$ and $\nabla_i((f_1, \ldots, f_n)) = (df_1, \ldots, df_r) + A_i(f_1, \ldots, f_n)$. We can cover each open set $D \cap D_i = D_i \setminus \{x_i\}$ by two simply connected open subsets U_{i+} and U_{i-} as in Example 2.6.6. Over each U_{i+} and U_{i-} the locally constant sheaves $\mathcal{E}^{\nabla}$ and $\mathcal{E}_i^{\nabla_i}$ are both trivial of dimension n by Lemma 2.4.4. Consequently, we may patch the locally free sheaves $\mathcal{E} \cong \mathcal{E}^{\nabla} \otimes_{\mathbf{C}} \mathcal{O}$ and $\mathcal{E}_i \cong \mathcal{E}_i^{\nabla_i} \otimes_{\mathbf{C}} \mathcal{O}$ together using the above construction. The restrictions of $\mathcal{E}$ and $\mathcal{E}_i$ to the U_{i+} and U_{i-} are both equipped with the trivial connection map since they correspond to trivial local systems, so the connections also patch. □

Corollary 2.7.9 *There is an essentially surjective functor from the category of connections on* $\mathbf{P}^1(\mathbf{C})$ *with logarithmic poles along* S *to that of finite dimensional left representations of* $\pi_1(D, x)$ *for some base point* $x \in D$.

Proof Combine the proposition with Proposition 2.7.5 and Corollary 2.6.2. □

Remarks 2.7.10

1. In his fundamental work [12], Deligne proved a vast generalization of the above results. He considered a complex algebraic variety U, and introduced the notion of *regular connections* on the associated analytic space U^{an}. In the case when U is a Zariski open subset in a smooth projective variety X whose complement S is a divisor with normal crossings (which means that its irreducible components are smooth of codimension 1 and meet transversely), a regular connection is a holomorphic flat connection on U^{an} that extends to a meromorphic connection on X with logarithmic poles along S. The extension is then actually an *algebraic* connection, which in our very special case corresponds to the fact that the functions occurring in the construction of our connections are meromorphic over $\mathbf{P}^1(\mathbf{C})$ and hence are *rational functions*. Deligne showed that all holomorphic flat connections on U^{an} are regular, and consequently every finite dimensional representation of the fundamental group of U^{an} can be constructed in a purely algebraic way. For a nice introduction to these ideas, see [40].
2. Corollary 2.7.9 does *not* say that every representation $\rho : \pi_1(D, x) \to \mathrm{GL}_n(\mathbf{C})$ is the monodromy representation of a Fuchsian system of linear differential equations, because the sheaf $\overline{\mathcal{E}}$ we constructed above is not necessarily free. The question whether every such ρ is the monodromy representation of a Fuchsian system is usually called the *Riemann–Hilbert problem* in the literature, though it is due neither to Riemann nor to Hilbert. However, the 21st problem on Hilbert's famous list, which he did not formulate completely rigorously, may be interpreted in this way.

   The fundamental group of $D = \mathbf{P}^1(\mathbf{C}) \setminus S$ is known to have a presentation of the form $\langle \gamma_1, \ldots, \gamma_m \mid \gamma_1 \cdots \gamma_n = 1 \rangle$, where γ_i is the class of a closed path going through z that turns around x_i. Thus the representation ρ is determined by the image of the γ_i in $\mathrm{GL}_n(\mathbf{C})$, i.e. by a system of m invertible matrices $M_1, \ldots, M_m \in \mathrm{GL}_n(\mathbf{C})$ satisfying $M_1 \cdots M_m = 1$.

   It was believed for a long time that Pljemelj gave a positive answer to the question as early as 1908. However, the last step of his argument contains a gap, and only works under the additional assumption that one of the matrices M_i is diagonalizable; see [37], Theorem 18.6 for a proof of Pljemelj's result. At the end of the 1980s Bolibrukh came up with a series of counterexamples showing that the answer can be negative in general, already for $n = 3$. For expositions of the problem and of Bolibrukh's counterexamples, see the Bourbaki seminar by Beauville [4] as well as the book of Ilyashenko and Yakovenko [37].

One way to eliminate the mistake from Pljemelj's proof is to allow *apparent singularities* for the system $y' = Ay$. This means that the entries of A may have poles outside the x_i, but the associated monodromy matrix should be the identity. Pljemelj's theorem then immediately yields a positive answer: one simply adds one more x_i with the identity as monodromy matrix and applies the theorem. Thus if Hilbert meant that apparent singularities are allowed, then Pljemelj already solved the problem.

3. There is a variant of the Riemann–Hilbert problem with a single n-th order linear differential equation

$$y^{(n)} + a_1 y^{(n-1)} + \cdots + a_{n-1} y' + a_n y = 0. \tag{2.4}$$

Such an equation is called *Fuchsian* if its coefficients are holomorphic outside a finite set $S = \{x_1, \ldots, x_m\} \subset \mathbf{P}^1(\mathbf{C})$, and at each x_j the coefficient a_i has a pole of order at most i. The relation with the previous definition is as follows. Assume for simplicity that $x_j = 0$ and write ∂ for the differential operator $y \mapsto zy'$ (recall that z is the coordinate function on $\mathbf{C}$). After multiplying the equation by z^n we may rewrite it in the form

$$\partial^n y + b_1 \partial^{n-1} y + \cdots + b_{n-1} \partial y + b_n y = 0, \tag{2.5}$$

where the b_i are holomorphic at 0 as a consequence of the Fuchsian condition. By this equation the $\mathbf{C}^n$-valued function $(\partial^{n-1} y, \partial^{n-2} y, \ldots, \partial y, y)$ satisfies a matrix equation of the form

$$y' = Bz^{-1} y,$$

where B is a matrix of holomorphic functions. This is a Fuchsian system in the previous sense.

A simple counting argument shows that one cannot hope to realize each representation of $\pi_1(D, x) \to \mathrm{GL}_n(\mathbf{C})$ as the monodromy representation of an n-th order Fuchsian equation: Fuchsian equations 'depend on fewer parameters' than monodromy representations do. However, if apparent singularities are allowed, then such a realization is possible; this is already contained in Pljemelj's work.

Example 2.7.11 The case $n = 2$, $m = 3$ of the Riemann–Hilbert problem has a long and glorious history, with a groundbreaking contribution by Riemann himself. He introduced the concept of *local exponents* of an equation (2.4) around a singular point x_j. We quote the later definition of Fuchs: for $x_j = 0$ the local exponents are the roots of the *indicial equation*

$$x^n + b_1(0) x^{n-1} + \cdots + b_{n-1}(0) x + b_n(0) = 0,$$

with the b_i as in (2.5). In the case $n = 2$, $m = 3$ one may assume $\{x_1, x_2, x_3\} = \{0, 1, \infty\}$, and then gets three pairs of local exponents at 0, 1 and ∞, traditionally denoted by (α, α'), (β, β') and (γ, γ'), respectively. If one moreover imposes $\alpha + \alpha' + \beta + \beta' + \gamma + \gamma' = 1$, it turns out that there is (up to equivalence) a unique second order Fuchsian differential equation on $\mathbf{P}^1(\mathbf{C}) \setminus \{0, 1, \infty\}$ with the above local exponents, namely

$$y'' + \frac{(1+\gamma+\gamma')z + (\alpha+\alpha'-1)}{z(z-1)}\, y' + \frac{\gamma\gamma' z^2 + (\beta\beta' - \gamma\gamma' - \alpha\alpha')z + \alpha\alpha'}{z^2(z-1)^2}\, y = 0.$$

Setting $\alpha = \beta = 0$, $\gamma = a$, $\gamma' = b$, $\alpha' = 1 - c$, $\beta' = c - a - b$ yields the famous *hypergeometric differential equation*

$$y'' + \frac{(1+a+b)z - c}{z(z-1)}\, y' + \frac{ab}{z(z-1)}\, y = 0$$

studied previously by Euler, Gauss, Kummer and others; the general equation can be reduced to this form by suitable transformations.

Riemann computed (and also defined!) the monodromy representation associated with such equations. The results are quite complicated. For instance, under the additional assumption that none of the differences $\alpha - \alpha'$, $\beta - \beta'$, $\gamma - \gamma'$ are integers, one obtains for the monodromy representation ρ

$$\rho(\gamma_0) = \begin{bmatrix} \exp(2\pi i\alpha) & 0 \\ 0 & \exp(2\pi i\alpha') \end{bmatrix},$$

$$\rho(\gamma_1) = \begin{bmatrix} \frac{\tau\exp(2\pi i\beta) - \exp(2\pi i\beta')}{\tau - 1} & \frac{\exp(2\pi i\beta) - \exp(2\pi i\beta')}{\tau - 1} \\ \frac{\tau\exp(2\pi i\beta') - \tau\exp(2\pi i\beta)}{\tau - 1} & \frac{\tau\exp(2\pi i\beta') - \exp(2\pi i\beta)}{\tau - 1} \end{bmatrix},$$

where γ_0, γ_1 are the standard loops around 0 and 1, respectively, and

$$\tau = \frac{\sin((\beta' + \alpha + \gamma)\pi)\sin((\beta + \alpha' + \gamma)\pi)}{\sin((\beta + \alpha + \gamma)\pi)\sin((\beta' + \alpha' + \gamma)\pi)}.$$

These formulae are taken from Section 3.1 of Varadarajan's expository paper [108], where the degenerate cases are also discussed. See also [5] for an introduction to hypergeometric equations.

Exercises

1. Show that if a *finite* group G acts on a Hausdorff space Y such that none of its elements has a fixed point, the action is even and thus it yields a Galois cover $Y \to G\backslash Y$ for connected Y.

2. Show that a connected cover $Y \to X$ is Galois with automorphism group G if and only if the fibre product $Y \times_X Y$ is isomorphic to the trivial cover $Y \times G \to Y$ (where G carries the discrete topology).
3. Let X be a connected and locally simply connected topological space and $x \in X$ a fixed base point. Consider a connected cover $p : Y \to X$ and a base point $y \in p^{-1}(x)$.
   (a) Show that there is a natural injective homomorphism $\pi_1(Y, y) \to \pi_1(X, x)$.
   (b) Viewing $\pi_1(Y, y)^{op}$ as a subgroup of $\mathrm{Aut}(\widetilde{X}_x|X)$ via the isomorphism $\mathrm{Aut}(\widetilde{X}_x|X) \cong \pi_1(X, x)^{op}$, establish an isomorphism $\pi_1(Y, y)^{op} \backslash \widetilde{X}_x \cong Y$.
4. Let G be a connected and locally simply connected topological group with unit element e. Let $\pi : \widetilde{G}_e \to G$ be a universal cover, and $\widetilde{e}$ the universal element in the fibre $\pi^{-1}(e)$.
   (a) Equip $\widetilde{G}_e$ with a natural structure of a topological group with unit element $\widetilde{e}$ for which π becomes a homomorphism of topological groups.
   (b) Show that a group structure with the above properties is unique.
   (c) Construct an exact sequence of topological groups

$$1 \to \pi_1(G, e)^{op} \to \widetilde{G}_e \to G \to 1,$$

   the topology on $\pi_1(G, e)$ being discrete.
5. Let $p : Y \to X$ be a cover of a connected and locally simply connected topological space X. For a point $x \in X$, we have defined two canonical left actions on the fibre $p^{-1}(x)$: one is the action by $\mathrm{Aut}(Y|X)$ and the other is that by $\pi_1(X, x)^{op}$. Verify that these two actions commute, i.e. that $\alpha(\phi(y)) = \phi(\alpha y)$ for $y \in \pi^{-1}(x)$, $\phi \in \mathrm{Aut}(Y|X)$ and $\alpha \in \pi_1(X, x)$. [*Hint*: Use Theorems 2.3.5 and 2.3.7.]
6. Let $\widetilde{X}_\Delta \to X$ be the cover of the connected and locally simply connected space X introduced in Remark 2.4.14 (2).
   (a) Verify that the $\pi_1(X, x)$-space corresponding to $\widetilde{X}_\Delta$ via Theorem 2.3.4 is $\pi_1(X, x)$ acting on itself via inner automorphisms.
   (b) Let $f : [0, 1] \to X$ be a path with endpoints $x = f(0)$ and $y = f(1)$. The pullback $f^*\widetilde{X}_\Delta$ is a trivial cover of $[0, 1]$ by simply connectedness of $[0, 1]$, whence an isomorphism of fibres $\widetilde{X}_{\Delta,x} \xrightarrow{\sim} \widetilde{X}_{\Delta,y}$. Verify that under the identifications $\widetilde{X}_{\Delta,x} \cong \pi_1(X, x)$ and $\widetilde{X}_{\Delta,y} \cong \pi_1(X, y)$ this isomorphism corresponds to the isomorphism of fundamental groups $\pi_1(X, x) \xrightarrow{\sim} \pi_1(X, y)$ induced by $g \mapsto f \bullet g \bullet f^{-1}$.
7. (Deligne) Let X be a connected and locally simply connected topological space. A *groupoid cover* over X is a cover $\pi : \Pi \to X \times X$ together with the following additional data. Denote by s and t the maps $\Pi \to X$ obtained by composing π with the two projections $X \times X \to X$, and consider the fibre product $\Pi \times_X \Pi$ with respect to the map s on the left and t on the right. The groupoid structure is given by three morphisms of spaces over $X \times X$: a multiplication map $m : \Pi \times_X \Pi \to \Pi$, a unit map $e : X \to \Pi$ (where X is considered as a space over $X \times X$ by means of the diagonal map), and an inverse map $\iota : \Pi \to \Pi$ which satisfies $t \circ \iota = s$, $s \circ \iota = t$.

These are subject to the following conditions:

$$m(m \times \mathrm{id}) = m(\mathrm{id} \times m), \quad m(e \times \mathrm{id}) = m(\mathrm{id} \times e) = \mathrm{id},$$
$$m(\mathrm{id} \times \iota) = e \circ t, \quad m(\iota \times \mathrm{id}) = e \circ s.$$

A morphism of groupoid covers is a morphism of covers of $X \times X$ compatible with the above additional data.

(a) Verify that the space $\widetilde{X} \to X \times X$ of Construction 2.4.13 carries the structure of a groupoid cover over X.

(b) Fix a base point (x, x) of $X \times X$ coming from a point $x \in X$. Show that the cover $\widetilde{X} \to X \times X$ corresponds via Theorem 2.3.4 to the underlying set of $\pi_1(X, x)$ together with the left action of $\pi_1(X, x) \times \pi_1(X, x)$ given by $(\gamma_1, \gamma_2)\gamma = \gamma_1\gamma\gamma_2^{-1}$.

(c) Show that for every groupoid cover $\Pi \to X \times X$ there is a unique morphism of groupoid covers $\widetilde{X} \to \Pi$.

8. Let X be a connected and locally path-connected, but not necessarily locally simply connected topological space. Construct a profinite group G such that the category of finite covers of X becomes equivalent to the category of finite sets equipped with a continuous left G-action.
9. Let S be a set having at least two elements, and let X be a topological space. Define a presheaf $\mathcal{F}_S$ of sets on X by setting $\mathcal{F}_S(U) = S$ for all nonempty open sets $U \subset X$ and $\rho_{UV} = \mathrm{id}_S$ for all open inclusions $V \subset U$. Give a necessary and sufficient condition on X for $\mathcal{F}_S$ to be a sheaf.
10. Let X be a topological space. Show that the category of sheaves on X is equivalent to the category of those spaces $p : Y \to X$ over X where the projection p is a local homeomorphism (i.e. each point in Y has an open neighbourhood such that $p|_U$ is a homeomorphism onto its image). [*Hint:* Begin by showing that for a sheaf $\mathcal{F}$ the projection $p_{\mathcal{F}} : X_{\mathcal{F}} \to X$ is a local homeomorphism.]
11. Let $\mathcal{F}$ be a presheaf of sets. Show that there is a natural morphism of presheaves $\rho : \mathcal{F} \to \mathcal{F}_{X_{\mathcal{F}}}$, and moreover every morphism $\mathcal{F} \to \mathcal{G}$ with $\mathcal{G}$ a sheaf factors as a composite $\mathcal{F} \xrightarrow{\rho} \mathcal{F}_{X_{\mathcal{F}}} \to \mathcal{G}$. [*Remark:* For this reason $\mathcal{F}_{X_{\mathcal{F}}}$ is called the *sheaf associated with* $\mathcal{F}$.]

3

Riemann surfaces

One obtains more information on covers of a topological space when it carries additional structure, for instance when it is a complex manifold. The complex manifolds of dimension 1 are called Riemann surfaces, and they already have a rich theory. The study of their covers creates a link between the Galois theory of fields and that of covers: finite étale algebras over the field of meromorphic functions on a connected compact Riemann surface correspond up to isomorphism to *branched* covers of the Riemann surface; by definition, the latter are topological covers outside a discrete exceptional set. As we shall see, all proper holomorphic surjections of Riemann surfaces define branched covers. The dictionary between branched covers and étale algebras over the function field has purely algebraic consequences: as an application, we shall prove that every finite group occurs as the Galois group of a finite Galois extension of the rational function field $\mathbf{C}(t)$.

Parts of this chapter were inspired by the expositions in [17] and [23].

3.1 Basic concepts

Let X be a Hausdorff topological space. A *complex atlas* on X is an open covering $\mathcal{U} = \{U_i : i \in I\}$ of X together with maps $f_i : U_i \to \mathbf{C}$ mapping U_i homeomorphically onto an open subset of $\mathbf{C}$ such that for each pair $(i, j) \in I^2$ the map $f_j \circ f_i^{-1} : f_i(U_i \cap U_j) \to \mathbf{C}$ is holomorphic. The maps f_i are called *complex charts.* Two complex atlases $\mathcal{U} = \{U_i : i \in I\}$ and $\mathcal{U}' = \{U_i' : i \in I'\}$ on X are *equivalent* if their union (defined by taking all the U_i and U_i' as a covering of X together with all complex charts) is also a complex atlas. Note that the extra condition to be satisfied here is that the maps $f_j' \circ f_i^{-1}$ should be holomorphic on $f_i(U_i \cap U_j')$ for all $U_i \in \mathcal{U}$ and $U_j \in \mathcal{U}'$.

Definition 3.1.1 A *Riemann surface* is a Hausdorff space together with an equivalence class of complex atlases.

We shall often refer to the equivalence class of atlases occurring in the above definition as the *complex structure* on the Riemann surface.

Remark 3.1.2 The above definition is the case $n = 1$ of that of n-dimensional *complex manifolds:* these are Hausdorff spaces equipped with an equivalence class of n-dimensional complex atlases defined in the same way as above, except that the complex charts f_i map onto open subsets of $\mathbf{C}^n$, and the $f_j \circ f_i^{-1}$ are to be holomorphic maps from $f_i(U_i \cap U_j')$ to $\mathbf{C}^n$.

Examples 3.1.3

1. An open subset $U \subset \mathbf{C}$ is endowed with a structure of a Riemann surface by the trivial covering $\mathcal{U} = \{U\}$ and the inclusion $i : U \to \mathbf{C}$.
2. *The complex projective line.* Consider the real 2-sphere S^2, and fix antipodal points $0, \infty \in S^2$. We define two complex charts on S^2 as follows. We first map the complement of ∞ homeomorphically onto the complex plane $\mathbf{C}$ via stereographic projection; we call the resulting homeomorphism z. Then we define a homeomorphism of the complement of 0 onto $\mathbf{C}$ by mapping ∞ to 0 and using the map $1/z$ at the other points. Since the map $z \mapsto 1/z$ is holomorphic on $\mathbf{C} \setminus \{0\}$, this is a complex atlas; the resulting Riemann surface is the complex projective line $\mathbf{P}^1(\mathbf{C})$.
3. *Complex tori.* Consider $\mathbf{C}$ as a two-dimensional real vector space and let $c_1, c_2 \in \mathbf{C}$ be a basis over $\mathbf{R}$. The c_i generate a discrete subgroup Λ of $\mathbf{C}$ isomorphic to $\mathbf{Z} \times \mathbf{Z}$. The topological quotient space is homeomorphic to a torus. We define a complex atlas on T as follows. We cover $\mathbf{C}$ by sufficiently small open discs D_i such that neither of them contains two points congruent modulo Λ. The image of each D_i by the projection $p : \mathbf{C} \to T$ is an open subset of T by definition of the quotient topology, and the projection maps D_i homeomorphically onto its image. The images of the D_i thus form an open covering of T, and the complex charts f_i are to be the inverses of the projection maps $p|_{D_i} : D_i \to p(D_i)$. The coordinate changes $f_j \circ f_i^{-1}$ on $f_i(U_i \cap U_j)$ are translations by elements of Λ, so we have indeed obtained a complex atlas.
4. *Smooth affine plane curves.* Let X be the closed subset of $\mathbf{C}^2$ defined as the locus of zeros of a polynomial $f \in \mathbf{C}[x, y]$. Assume that there is no point of X where the partial derivatives $\partial_x f$ and $\partial_y f$ both vanish. We can then endow X with the structure of a Riemann surface as follows. In the neighbourhood of a point where $\partial_y f$ is nonzero define a complex chart by mapping a point to its x-coordinate; similarly, for points where $\partial_x f$ is nonzero we take the y-coordinate. By the inverse function theorem for holomorphic functions, in a sufficiently small neighbourhood the above mappings are indeed homeomorphisms. Secondly, the holomorphic version of the implicit function theorem ([28], p. 19) implies that in the neighbourhood of points where both x and y define a complex chart, the transition function from x to y is holomorphic, i.e. when $\partial_y f$ does not vanish at some point, we may express y as a holomorphic function of x and vice versa. So we have defined a complex atlas.

In the second and third examples we obtain *compact* Riemann surfaces; the other two are non-compact. In fact, one may define a compact version of the last example by considering smooth *projective* plane curves: these are to be the closed subsets of the complex projective plane $\mathbf{P}^2(\mathbf{C})$ arising as the locus of zeros of some homogeneous polynomial $F \in \mathbf{C}[x, y, z]$ such that the partial derivatives $\partial_x F, \partial_y F, \partial_z F$ have no common zero. The complex structure is defined in a similar way as above, or by means of a covering by smooth affine curves.

Definition 3.1.4 Let Y and X be Riemann surfaces. A *holomorphic* (or *analytic*) map $\phi : Y \to X$ is a continuous map such that for each pair of open subsets $U \subset X$, $V \subset Y$ satisfying $\phi(V) \subset U$ and complex charts $f : U \to \mathbf{C}$, $g : V \to \mathbf{C}$ the functions $f \circ \phi \circ g^{-1} : g(V) \to \mathbf{C}$ are holomorphic.

It follows from the definition of equivalence between atlases that the above definition does not depend on the atlases chosen. Riemann surfaces together with holomorphic maps form a category.

We define a *holomorphic function* on an open subset $U \subset X$ to be a holomorphic map $U \to \mathbf{C}$, where $\mathbf{C}$ is equipped with its usual complex structure. For instance, a complex chart $f : U \to \mathbf{C}$ is a holomorphic map.

Remark 3.1.5 The *sheaf of holomorphic functions* on a Riemann surface X is defined by associating with an open subset $U \subset X$ the ring $\mathcal{O}(U)$ of holomorphic functions on U. One can check that the complex structure on X is uniquely determined by its underlying topological space and the sheaf $\mathcal{O}$. This is the starting point of the general definition of complex analytic spaces.

3.2 Local structure of holomorphic maps

In this section we study holomorphic maps between Riemann surfaces from a topological viewpoint. Henceforth we shall tacitly assume that *the maps under consideration are non-constant on all connected components*, i.e. they do not map a whole component to a single point.

We have seen in Example 2.4.12 that the map $z \mapsto z^k$ defines a cover of $\mathbf{C}^\times$ by itself but its extension to $\mathbf{C}$ does not. The next proposition shows that locally every holomorphic map of Riemann surfaces is of this shape.

Proposition 3.2.1 *Let $\phi : Y \to X$ be a holomorphic map of Riemann surfaces, and y a point of Y with image $x = \phi(y)$ in X. There exist open neighbourhoods V_y (resp. U_x) of y (resp. x) satisfying $\phi(V_y) \subset U_x$, as well as complex charts $g_y : V_y \to \mathbf{C}$ and $f_x : U_x \to \mathbf{C}$ satisfying $f_x(x) = g_y(y) = 0$ such that* *the diagram*

$$\begin{array}{ccc} V_y & \xrightarrow{\phi} & U_x \\ {\scriptstyle g_y}\downarrow & & \downarrow{\scriptstyle f_x} \\ \mathbf{C} & \xrightarrow{z \mapsto z^{e_y}} & \mathbf{C} \end{array}$$

commutes with an appropriate positive integer e_y that does not depend on the choice of the complex charts.

Proof By performing affine linear transformations in $\mathbf{C}$ and by shrinking U_x and V_y if necessary, one may find charts g_y and f_x satisfying all conditions of the proposition except perhaps the last one. In particular, $f_x \circ \phi \circ g_y^{-1}$ is a holomorphic function in a neighbourhood of 0 which vanishes at 0. As such, it must be of the form $z \mapsto z^{e_y} H(z)$ where H is a holomorphic function with $H(0) \neq 0$. Denote by log a fixed branch of the logarithm in a neighbourhood of $H(0)$. It is then a basic fact from complex analysis that by shrinking V_y if necessary the formula $h := \exp((1/e_y) \log H))$ defines a holomorphic function h on $g_y(V_y)$ with $h^{e_y} = H$. Thus by replacing g_y by its composition with the map $z \mapsto zh(z)$ we obtain a chart that satisfies the required properties. The independence of e_y of the charts follows from the fact that changing a chart amounts to composing with an invertible holomorphic function. □

Definition 3.2.2 The integer e_y of the proposition is called the *ramification index* or *branching order* of ϕ at y. The points y with $e_y > 1$ are called *branch points*. We denote the set of branch points of ϕ by S_ϕ.

Corollary 3.2.3 *A holomorphic map between Riemann surfaces is open (i.e. it maps open sets onto open sets).*

Proof Indeed, the map $z \to z^e$ is open. □

Corollary 3.2.4 *The fibres of ϕ and the set S_ϕ are discrete closed subsets of Y.*

Proof The proposition implies that each point of Y has a punctured open neighbourhood containing no branch points where ϕ is finite-to-one. □

Now we restrict our attention to *proper* maps. Recall that a continuous map of locally compact topological spaces is proper if the preimage of each compact subset is compact. If the spaces in question are moreover Hausdorff spaces, then a proper map is closed, i.e. maps closed subsets to closed subsets. This follows from the easy fact that in a locally compact Hausdorff space a subset is closed if and only if its intersection with every compact subset is closed. Note that Riemann surfaces are locally compact Hausdorff spaces (since they are locally homeomorphic to open subsets of $\mathbf{C}$).

Examples 3.2.5

1. A continuous map $Y \to X$ of locally compact Hausdorff spaces is automatically proper if Y is compact. The main case of interest for us will be that of compact Riemann surfaces.
2. A finite cover $p : Y \to X$ of locally compact topological spaces is proper. Indeed, given a compact subset $Z \subset X$ and an open covering $\mathcal{U}$ of $p^{-1}(Z)$, we may refine $\mathcal{U}$ in a covering by open subsets $U_i \subset p^{-1}(Z)$ so that the cover is trivial over each $p(U_i)$. As p is an open map, the $p(U_i)$ form an open covering of Z, whence we may extract a finite subcovering $\mathcal{V}$. The U_i with $p(U_i) \in \mathcal{V}$ yield a finite open covering of $p^{-1}(Z)$ because the cover $p : Y \to X$ is finite.
3. In the next chapter we shall see that if X and Y are smooth complex affine plane curves and $\phi : Y \to X$ is a *finite* algebraic morphism, then ϕ is proper as a map of Riemann surfaces.

We can now state the main topological properties of proper holomorphic maps.

Proposition 3.2.6 *Let X be a connected Riemann surface, and $\phi : Y \to X$ a proper holomorphic map. The map ϕ is surjective with finite fibres, and its restriction to $Y \setminus \phi^{-1}(\phi(S_\phi))$ is a finite topological cover of $X \setminus \phi(S_\phi)$.*

Proof Finiteness of fibres follows from the previous corollary, since discrete subsets of a compact space are finite. Surjectivity of ϕ holds because $\phi(Y)$ is open in X (by Corollary 3.2.3) as well as closed (as ϕ is proper), and X is connected. For the last statement note that by Proposition 3.2.1 each of the finitely many preimages of $x \in X \setminus \phi(S_\phi)$ has an open neighbourhood mapping homeomorphically onto some open neighbourhood of x; the intersection of these is a distinguished open neighbourhood of x as in the definition of a cover. □

We call a proper surjective map of locally compact Hausdorff spaces that restricts to a finite cover outside a discrete closed subset a *finite branched cover.* Its degree is by definition the degree of the cover obtained by restriction. With this terminology the proposition says that proper holomorphic maps of Riemann surfaces give rise to finite branched covers.

We now state a theorem that will show in particular that a proper holomorphic map is determined by its topological properties. First some notation: given a connected Riemann surface X and a discrete closed subset $S \subset X$, we denote by $\mathrm{Hol}_{X,S}$ the category of Riemann surfaces Y equipped with a proper holomorphic map $Y \to X$ whose branch points all lie above points in S. A morphism in this category is a holomorphic map compatible with the projections onto X.

Theorem 3.2.7 *In the above situation mapping a Riemann surface $\phi : Y \to X$ over X to the topological cover $Y \setminus \phi^{-1}(S) \to X \setminus S$ obtained by restriction of ϕ induces an equivalence of the category* $\mathrm{Hol}_{X,S}$ *with the category of finite topological covers of $X \setminus S$.*

The proof will be in several steps. The following lemma essentially handles the case $S = \emptyset$.

Lemma 3.2.8 *Let X be a Riemann surface, and $p : Y \to X$ a connected cover of X as a topological space. The space Y can be endowed with a unique complex structure for which p becomes a holomorphic mapping.*

In fact, the proof will show that it is enough to require that p is a local homeomorphism.

Proof Each point $y \in Y$ has a neighbourhood V that projects homeomorphically onto a neighbourhood U of $p(y)$. Take a complex chart $f : U' \to \mathbf{C}$ with $U' \subset U$; $f \circ p$ will then define a complex chart in a neighbourhood of y. It is immediate that we obtain a complex atlas in this way, and uniqueness follows from the fact that for any complex structure on Y the restriction of p to V must be an analytic isomorphism. □

The following proposition shows that the functor of Theorem 3.2.7 is essentially surjective.

Proposition 3.2.9 *Assume given a connected Riemann surface X, a discrete closed set S of points of X and a finite connected cover $\phi' : Y' \to X'$, where $X' := X \setminus S$. There exists a Riemann surface Y containing Y' as an open subset and a proper holomorphic map $\phi : Y \to X$ such that $\phi|_{Y'} = \phi'$ and $Y' = Y \setminus \phi^{-1}(S)$.*

Proof Fix a point $x \in S$. By performing an affine linear transformation in $\mathbf{C}$ if necessary we find a connected open neighbourhood U_x of x avoiding the other points in S and a complex chart mapping U_x homeomorphically onto the open unit disc $D \subset \mathbf{C}$. The restriction of ϕ' to $\phi'^{-1}(U_x \setminus \{x\})$ is a finite cover, hence $\phi'^{-1}(U_x \setminus \{x\})$ decomposes as a finite disjoint union of connected components V_x^i each of which is a cover of $U_x \setminus \{x\}$. Via the isomorphism of $U_x \setminus \{x\}$ with the punctured disc $\dot{D} = D \setminus \{0\}$ each V_x^i becomes isomorphic to a finite connected cover of $\dot{D}$, hence by Example 2.4.12 it must be isomorphic to a cover $\dot{D} \to \dot{D}$ given by $z \mapsto z^k$ for some $k > 1$. Now choose 'abstract' points y_x^i for all i and x, and define Y as the disjoint union of Y' with the y_x^i. Define an extension ϕ of ϕ' to Y by mapping each y_x^i to x. For each i and x extend the holomorphic isomorphism $\rho_x^i : V_x^i \xrightarrow{\sim} \dot{D}$ to a bijection $\bar{\rho}_x^i : V_x^i \cup \{y_x^i\} \xrightarrow{\sim} D$ by sending y_x^i to 0, and define the topology on Y in such a way that this bijection becomes a homeomorphism extending ρ_x^i. Use the ρ_x^i as complex charts in the neighbourhoods of the points y_x^i. Together with the canonical complex structure on Y' defined in the previous lemma they form a complex atlas on Y. The map ϕ is holomorphic, as it is holomorphic away from the y_x^i by the lemma above, and in the neighbourhood of these looks like the map $z \mapsto z^k$. Finally, the map ϕ is proper, because ϕ' is proper by Example 3.2.5 (2), the fibres of ϕ are finite, and the compact subsets of X' differ from those of X by finitely many points. □

Proof of Theorem 3.2.7 In view of the proposition above it remains to prove that the functor of the theorem is fully faithful. This means the following: given two Riemann surfaces Y and Z equipped with proper holomorphic maps ϕ_Y and ϕ_Z onto X with all branch points above S and a morphism of covers $\rho' : Y' \to Z'$ over X' with $Y' = Y \setminus \phi_Y^{-1}(S)$ and $Z' = Z \setminus \phi_Z^{-1}(S)$, there is a unique holomorphic map $\rho : Y \to Z$ over X extending ρ'. We know from Lemma 2.2.11 that $\rho' : Y' \to Z'$ is a cover, so it is holomorphic with respect to the unique complex structure on Y' by Lemma 3.2.8. This complex structure must be compatible with that of Y, because $\phi_Y|_{Y'} = \phi_Z \circ \rho'$ is holomorphic with respect to both. It follows that for each point $y \in \phi_Y^{-1}(S)$ the map ρ' must send a sufficiently small open neighbourhood of y holomorphically isomorphic to $\dot{D}$ to a similar neighbourhood of a point $z \in \phi_Z^{-1}(S)$. Setting $\rho(y) = z$ we obtain the required holomorphic extension, by a similar argument as in the previous proof. □

By the theorem the automorphism group of $Y \to X$ as an object of $\mathrm{Hol}_{X,S}$ is the same as that of the cover $Y' \to X'$. Therefore it makes sense to call Y a *finite Galois branched cover* of X if Y' is Galois over X'. We conclude with some simple topological properties of Galois branched covers that will be needed later.

Proposition 3.2.10 *Let $\phi : Y \to X$ be a proper holomorphic map of connected Riemann surfaces that is topologically a Galois branched cover. Then the following hold.*

1. *The group* $\mathrm{Aut}(Y|X)$ *acts transitively on all fibres of ϕ.*
2. *If $y \in Y$ is a branch point with ramification index e, then so are all points in the fibre $\phi^{-1}(\phi(y))$. Moreover, the stabilizers of these points in* $\mathrm{Aut}(Y|X)$ *are conjugate cyclic subgroups of order e.*

Proof As $\mathrm{Aut}(Y|X)$ acts transitively on all fibres not containing branch points, the first statement follows from the continuity of automorphisms. Most of the second statement then results from the automorphism property. Finally, the assertion about stabilizers follows from the fact that an automorphism fixing

y must stabilize a sufficiently small open neighbourhood of y over which ϕ is isomorphic to the cover $z \mapsto z^e$ of the open disc. □

3.3 Relation with field theory

We begin with a basic definition.

Definition 3.3.1 Let X be a Riemann surface. A *meromorphic function* f on X is a holomorphic function on $X \setminus S$, where $S \subset X$ is a discrete closed subset, such that moreover for all complex charts $\phi : U \to \mathbf{C}$ the complex function $f \circ \phi^{-1} : \phi(U) \to \mathbf{C}$ is meromorphic.

Note that meromorphic functions on a Riemann surface X form a ring with respect to the usual addition and multiplication of functions; we denote this ring by $\mathcal{M}(X)$.

Lemma 3.3.2 *If X is connected, the ring $\mathcal{M}(X)$ is a field.*

Proof For a nonzero $f \in \mathcal{M}(X)$ the function $1/f$ will be seen to give an element of $\mathcal{M}(X)$ once we show that the zeros of f form a discrete closed subset. Indeed, if it is not discrete, then it has a limit point x. Composing f with any complex chart containing x we get a holomorphic function on some complex domain whose set of zeros has a limit point. By the Identity Principle of complex analysis ([80], Theorem 10.18) this implies that the function is identically 0, so f is 0 in some neighbourhood of x. Now consider the set of those points y of X for which f vanishes identically in a neighbourhood of y. This set is open by definition, but it is also closed for it contains all of its boundary points by the previous argument. Since X is connected, this implies $f = 0$, a contradiction. □

The field $\mathcal{M}(X)$ contains a subfield isomorphic to $\mathbf{C}$ given by the constant functions. Surprisingly, in the case when X is compact, it is not obvious at all that there are other functions in $\mathcal{M}(X)$ as well. We shall use a somewhat stronger form of this fact:

Theorem 3.3.3 (Riemann's Existence Theorem) *Let X be a compact Riemann surface, $x_1, \dots, x_n \in X$ a finite set of points, and $a_1, \dots, a_n$ a sequence of complex numbers. There exists a meromorphic function $f \in \mathcal{M}(X)$ such that f is holomorphic at all the x_i and $f(x_i) = a_i$ for $1 \le i \le n$.*

The proof uses nontrivial analytic techniques and cannot be given here. See e.g. [23], Corollary 14.13.

Remark 3.3.4 The theorem easily follows from the following seemingly weaker statement: *given a compact Riemann surface X and two points $x_1, x_2 \in X$, there exists a meromorphic function f on X holomorphic at the x_i with $f(x_1) \neq f(x_2)$.* Indeed, the function $f_1 := f - f(x_2)$ satisfies $f_1(x_1) \neq 0$ but $f_1(x_2) = 0$, and there is a similar function for x_2. If n points $x_1, \ldots, x_n$ are given, we obtain by induction functions f_i with $f_i(x_i) \neq 0$ but $f_i(x_j) = 0$ for $i \neq j$. The theorem then follows by taking a suitable linear combination.

Consider now a holomorphic map $\phi : Y \to X$ of Riemann surfaces which is not constant on any connected component. It induces a ring homomorphism $\phi^* : \mathcal{M}(X) \to \mathcal{M}(Y)$ via $\phi^*(f) := f \circ \phi$. In the case when X and Y are compact and X is connected, the map ϕ is proper and surjective with finite fibres. Our first goal is to show that in this case $\mathcal{M}(Y)$ becomes a finite étale algebra over $\mathcal{M}(X)$ via ϕ^*. Note that by compactness Y must be a finite disjoint union of connected compact Riemann surfaces Y_i. Since in this case $\mathcal{M}(Y) \cong \prod \mathcal{M}(Y_i)$, we only have to show that for compact and connected X and Y the field extension $\mathcal{M}(Y)|\phi^*\mathcal{M}(X)$ is finite. We prove somewhat more:

Proposition 3.3.5 *Let $\phi : Y \to X$ be a non-constant holomorphic map of compact connected Riemann surfaces which has degree d as a branched cover. The induced field extension $\mathcal{M}(Y)|\phi^*\mathcal{M}(X)$ is finite of degree d.*

The key lemma is the following.

Lemma 3.3.6 *Let $\phi : Y \to X$ be a proper holomorphic map of connected Riemann surfaces which has degree d as a branched cover. Every meromorphic function $f \in \mathcal{M}(Y)$ satisfies a (not necessarily irreducible) polynomial equation of degree d over $\mathcal{M}(X)$.*

Proof Let S be the set of branch points of ϕ. Each point $x \notin \phi(S)$ has some open neighbourhood U such that $\phi^{-1}(U)$ decomposes as a finite disjoint union of open sets $V_1, \ldots, V_d$ homeomorphic to U. Let s_i be the (holomorphic) section of ϕ mapping U onto V_i and put $f_i = f \circ s_i$. The function f_i is then a meromorphic function on U. Put

$$F = \prod (t - f_i) = t^d + a_{n-1}t^{d-1} + \cdots + a_0.$$

The coefficients a_j, being the elementary symmetric polynomials of the f_i, are meromorphic on U. Now if $x_1 \notin \phi(S)$ is another point with distinguished open neighbourhood U_1, then on $U \cap U_1$ the coefficients of the polynomial F_1 corresponding to the similar construction over U_1 must coincide with those of F since the roots of the two polynomials are the same meromorphic functions over $U \cap U_1$. Hence the a_j extend to meromorphic functions on $X \setminus \phi(S)$. To see that they extend to meromorphic functions on the whole of X, pick a point $x \in S$ and consider a coordinate chart $f_x : U_x \to \mathbf{C}$ in a neighbourhood U_x of x contained in U with $f_x(x) = 0$. Then $f_x \circ \phi$ defines a holomorphic function in some neighbourhood of each $y \in \phi^{-1}(x)$, with $(f_x \circ \phi)(y) = 0$. As f extends meromorphically to all y, we find $k > 0$ such that $(f_x \circ \phi)^k f$ is holomorphic at all $y \in \phi^{-1}(x)$; in particular, this function is bounded in a punctured neighbourhood of each y. Composing with the s_i we obtain that the functions $f_x^k f_i$, and hence also the $f_x^{kd} a_j$ are bounded on $U_x \setminus \{x\}$, so they extend to holomorphic functions on the whole of U_x by Riemann's removable singularity theorem (see e.g. [80], Theorem 10.20). This shows that the a_j are meromorphic on X and hence $F \in \mathcal{M}(X)[t]$. Finally, we have $(\phi^* F)(f) = 0$ because over U we have $(\phi^* F \circ s_i)(f \circ s_i) = F(f_i) = 0$ for each i. □

Proof of Proposition 3.3.5 We first show that we may find a function $f \in \mathcal{M}(Y)$ satisfying an *irreducible* polynomial equation of degree d over $\mathcal{M}(X)$. Take a point $x \in X$ which is not the image of a branch point, and let $y_1, \ldots, y_d$ be its inverse images in Y. By Theorem 3.3.3 we find $f \in \mathcal{M}(Y)$ that is holomorphic at all the y_i and the values $f(y_i)$ are all distinct. By the lemma f satisfies an irreducible polynomial equation $(\phi^* a_n) f^n + \cdots + (\phi^* a_0) = 0$, with $a_i \in \mathcal{M}(X)$ and $n \leq d$. If the a_i are all holomorphic at x, then the polynomial $a_n(x)t^n + \cdots + a_0(x) \in \mathbf{C}[t]$ has d distinct complex roots (namely the $f(y_i)$), whence we must have $n = d$. If by chance one of the a_i happens to have a pole in x, observe that by continuity all points x' in a sufficiently small open neighbourhood of x still have the property that they are not images of branch points, and moreover f is holomorphic and takes distinct values at all of their preimages. We may choose x' as above so that all the a_i are holomorphic at x' and apply the argument with x' instead of x.

Next observe that with f as above we have $\mathcal{M}(Y) \cong \mathcal{M}(X)(f)$. Indeed, if g is another element of $\mathcal{M}(Y)$, we have $\mathcal{M}(X)(f, g) = \mathcal{M}(X)(h)$ with some function $h \in \mathcal{M}(Y)$ according to the theorem of the primitive element. In particular we have $\mathcal{M}(X)(f) \subset \mathcal{M}(X)(h)$, but since h should also have degree at most d over $\mathcal{M}(X)$ by the lemma, this inclusion must be an equality, i.e. $g \in \mathcal{M}(X)(f)$. □

According to the proposition and the remarks preceding it, the rule $Y \mapsto \mathcal{M}(Y)$ gives a contravariant functor from the category of compact Riemann surfaces mapping holomorphically onto a fixed connected compact Riemann surface X to the category of finite étale algebras over $\mathcal{M}(X)$.

Theorem 3.3.7 *The above functor is an anti-equivalence of categories. In this anti-equivalence finite Galois branched covers of X correspond to finite Galois extensions of $\mathcal{M}(X)$ of the same degree.*

We first prove that the functor is essentially surjective.

Proposition 3.3.8 *Let X be a connected compact Riemann surface, and let A be a finite étale algebra over $\mathcal{M}(X)$. There exists a compact Riemann surface Y mapping holomorphically onto X such that $\mathcal{M}(Y)$ is isomorphic to A as an $\mathcal{M}(X)$-algebra.*

Proof Since a non-connected compact Riemann surface Y is a finite disjoint union of its connected components Y_i and $\mathcal{M}(Y) \cong \prod \mathcal{M}(Y_i)$, by decomposing A into a direct product of fields it is enough to treat the case of a finite field extension $L|\mathcal{M}(X)$. Denote by α a primitive element in L, by F the minimal polynomial of α over $\mathcal{M}(X)$, and by d the degree of F. The irreducible polynomial F is separable over $\mathcal{M}(X)$, so the ideal (F', F) of the polynomial ring $\mathcal{M}(X)[t]$ (which is a principal ideal domain) must be the whole ring. Therefore there are functions $A, B \in \mathcal{M}(X)$ satisfying $AF + BF' = 1$. Denote by F_x the complex polynomial obtained from F by evaluating its coefficients at a point $x \in X$ where all of them are holomorphic. From the above equation we infer that F_x and F_x' may have a common root only at those points $x \in X$ where one of the functions A, B has a pole. Therefore if we denote by $S \subset X$ the discrete closed set consisting of poles of the coefficients of F as well as those of A and B, we get that on $X' = X \setminus S$ all coefficients of F are holomorphic and $F_x(a) = 0$ for $a \in \mathbf{C}$ implies $F_x'(a) \neq 0$. We conclude that for $x \in X'$ the polynomial $F_x \in \mathbf{C}[t]$ has d distinct roots $a_1, \ldots, a_d$.

Now for an open subset $U \subset X'$ denote by $\mathcal{F}(U)$ the set of holomorphic functions f on U satisfying $F(f) = 0$. Together with the natural restriction maps they form a sheaf of sets $\mathcal{F}$ on X'. We contend that $\mathcal{F}$ is locally constant on X', and each stalk has cardinality d. Indeed, by the holomorphic version of the implicit function theorem (see e.g. [28], p. 19) given a point $x \in X'$ and a root a_i of the polynomial $F_x \in \mathbf{C}[T]$ for some x, the condition $F_x'(a_i) \neq 0$ implies that there is a holomorphic function f_i defined in a neighbourhood of x with $F(f_i) = 0$ and $f_i(x) = a_i$. For each of the d distinct roots of F_x we thus find d different functions f_i that are all sections of $\mathcal{F}$ in some open neighbourhood of x. In a connected open neighbourhood V the sheaf $\mathcal{F}$ cannot have more sections, since the product of the polynomials $(t - f_i)$ already gives a factorization of F in the polynomial ring $\mathcal{M}(V)[t]$. Thus over V the sheaf $\mathcal{F}$ is isomorphic to the constant sheaf given by the finite set of the $\{f_1, \ldots, f_d\}$.

Once we know that $\mathcal{F}$ is locally constant, invoking Theorem 2.5.9 yields a cover $p_{\mathcal{F}} : X'_{\mathcal{F}} \to X'$. We can then apply Proposition 3.2.9 to each of the connected components X'_j of $X'_{\mathcal{F}}$, and get compact Riemann surfaces X_j mapping holomorphically onto X. We still have to show that $j = 1$, i.e. $X_{\mathcal{F}}$ is connected. Indeed, define a function f on $X'_{\mathcal{F}}$ by putting $f(f_i) = f_i(p_{\mathcal{F}}(f_i))$. One sees by the method of proof of Proposition 3.3.5 that f extends to a meromorphic function on each X_j and by applying Proposition 3.3.5 we see that f as an element of $\mathcal{M}(X_i)$ has a minimal polynomial G over $\mathcal{M}(X)$ of degree at most d_j, where d_j is the cardinality of the fibres of the cover $X'_j \to X'$. But since manifestly $F(f) = 0$, G must divide F in the polynomial ring $\mathcal{M}(X)[t]$, whence $F = G$ by irreducibility of F and $d_j = d$. This proves that there is only one X_j, i.e. $X_\mathcal{F}$ is connected.

Finally, mapping f to α induces an inclusion of fields $\mathcal{M}(Y) \subset L$ which must be an equality by comparing degrees. □

Remark 3.3.9 One may replace the use of the implicit function theorem in the above proof by using ideas from residue calculus in complex analysis as in [23], Corollary 8.8.

Proof of Theorem 3.3.7 Essential surjectivity was proven in the previous proposition. It is enough to prove fully faithfulness when Y is connected. We first show that given a pair (Y, ϕ) as in the theorem and a generator f of the extension $\mathcal{M}(Y)|\mathcal{M}(X)$ with minimal polynomial F, the cover of X given by the restriction of ϕ to the complement of branch points and inverse images of poles of coefficients of F is canonically isomorphic to the cover $X_\mathcal{F}$ defined in the previous proof. This isomorphism is best defined on the associated sheaves of local sections: just map a local section s_i to the holomorphic function $f \circ s_i$. By Theorem 3.2.7 the isomorphism extends uniquely to an isomorphism of Y with the compactification $X'_\mathcal{F}$ of $X_\mathcal{F}$ defined in the previous proof. Thus it suffices to check fully faithfulness for holomorphic maps of Riemann surfaces of the form $X'_\mathcal{F}$. So assume given a tower of finite field extensions $M|L|\mathcal{M}(X)$, with L and M corresponding by the previous proof to canonical maps of Riemann surfaces $X'_\mathcal{F} \to X$ and $X'_\mathcal{G} \to X$, respectively. But by construction the map $X'_\mathcal{G} \to X$ factors through a holomorphic map $X'_\mathcal{G} \to X'_\mathcal{F}$ inducing the extension $M|L$, which is the one we were looking for.

It remains to prove the second statement of the theorem. Given a branched cover $Y \to X$ of degree d, we have $\mathrm{Aut}(Y|X) \cong \mathrm{Aut}(\mathcal{M}(Y)|\mathcal{M}(X))$ by the fully faithfulness just proven, and we also know from Proposition 3.3.5 that the extension $\mathcal{M}(Y)|\mathcal{M}(X)$ has degree d. The group $\mathrm{Aut}(Y|X)$ is of order at most d, with equality if and only if the finite branched cover $\phi : Y \to X$ is Galois over X. Similarly, the group $\mathrm{Aut}(\mathcal{M}(Y)|\mathcal{M}(X))$ is of order at most d, with equality if and only if $\mathcal{M}(Y)|\mathcal{M}(X)$ is a Galois extension. The claim follows from these facts. □

Combining the theorem with Theorem 1.5.4 we obtain:

Corollary 3.3.10 *Let X be a connected compact Riemann surface. The category of compact Riemann surfaces equipped with a holomorphic map onto X is equivalent to that of finite continuous left* $\mathrm{Gal}\,(\overline{\mathcal{M}(X)}|\mathcal{M}(X))$*-sets.*

The case $X = \mathbf{P}^1(\mathbf{C})$ of the Theorem 3.3.7 is particularly interesting because of the following consequence of the Riemann Existence Theorem.

Proposition 3.3.11 *Let Y be a connected compact Riemann surface. There exists a non-constant holomorphic map $Y \to \mathbf{P}^1(\mathbf{C})$. Consequently $\mathcal{M}(Y)$ is a finite extension of $\mathbf{C}(t)$.*

Proof By the Riemann existence theorem $\mathcal{M}(Y)$ contains a non-constant function f. Define a map $\phi_f : Y \to \mathbf{P}^1(\mathbf{C})$ by

$$\phi_f(y) = \begin{cases} f(y) & y \text{ is not a pole of } f \\ \infty & y \text{ is a pole of } f. \end{cases}$$

For each $y \in Y$ choose a complex chart $g : U \to \mathbf{C}$ around y so that f is holomorphic on $U \setminus \{y\}$. Recall that the two standard complex charts on $\mathbf{P}^1(\mathbf{C})$ are given by z and $1/z$, respectively. If f is holomorphic at y, then $z \circ \phi_f \circ g^{-1}$ is holomorphic on $g(U)$. If not, then $(1/z) \circ \phi_f \circ g^{-1}$ maps $g(U \setminus \{y\})$ to a bounded open subset of $\mathbf{C}$, so $(1/z) \circ \phi_f \circ g^{-1}$ extends to a holomorphic function on $g(U)$ by Riemann's Removable Singularity Theorem (see e.g. [80], Theorem 10.20). This proves that ϕ_f is holomorphic. The second statement follows from Proposition 3.3.5 and the well-known fact that $\mathcal{M}(\mathbf{P}^1(\mathbf{C})) \cong \mathbf{C}(t)$. □

Corollary 3.3.12 *The contravariant functor $Y \mapsto \mathcal{M}(Y)$, $\phi \mapsto \phi^*$ induces an anti-equivalence between the category of connected compact Riemann surfaces with non-constant holomorphic maps and that of fields finitely generated over $\mathbf{C}$ of transcendence degree 1.*

Recall that a finitely generated field of transcendence degree 1 over $\mathbf{C}$ is just a finite extension of the rational function field $\mathbf{C}(t)$. We chose the above formulation of the corollary in order to emphasize that the morphisms in the category under consideration are $\mathbf{C}$-algebra homomorphisms and *not* $\mathbf{C}(t)$-algebra homomorphisms.

Proof Essential surjectivity follows from Theorem 3.3.7 applied with $X = \mathbf{P}^1(\mathbf{C})$, taking the proposition into account. To prove full faithfulness, i.e. the bijectivity of the map $\mathrm{Hom}(Y, Z) \to \mathrm{Hom}(\mathcal{M}(Z), \mathcal{M}(Y))$ we first choose a homomorphic map $\phi_f : Z \to \mathbf{P}^1(\mathbf{C})$ inducing a $\mathbf{C}$-algebra homomorphism $\mathbf{C}(t) \to \mathcal{M}(Z)$. This enables us to consider the elements of $\mathrm{Hom}(Y, Z)$ as morphisms of spaces over $\mathbf{P}^1(\mathbf{C})$, and those of $\mathrm{Hom}(\mathcal{M}(Z), \mathcal{M}(Y))$ as $\mathbf{C}(t)$-algebra homomorphisms. We can then apply the fully faithfulness part of Theorem 3.3.7. □

Remark 3.3.13 The corollary is often summarized in the concise statement that 'compact Riemann surfaces are algebraic'. In fact, it can be shown that compact Riemann surfaces are always holomorphically isomorphic to smooth projective complex algebraic curves, and that holomorphic maps between them all come from algebraic morphisms. Here, however, one has to allow smooth curves in projective 3-space as well, not just the plane curves mentioned at the beginning of this chapter. For a vast generalization in higher dimension, see Theorem 5.7.4 and the subsequent discussion.

3.4 The absolute Galois group of $\mathbf{C}(t)$

Corollary 3.3.10 bears a close resemblance to Corollary 2.3.9, except that we now allow branched covers as well. But as we have seen in Proposition 3.2.6, each finite branched cover of the compact Riemann surface X restricts to a cover over a cofinite open subset $X' \subset X$. Therefore it is natural to expect that $\widehat{\pi_1(X', x)}$ with some base point x is isomorphic to a quotient of $\mathrm{Gal}\,(\overline{\mathcal{M}(X)}|\mathcal{M}(X))$ in a natural way. The following theorem confirms this intuition.

Theorem 3.4.1 *Let X be a connected compact Riemann surface, and let X' be the complement of a finite set of points in X. Let $K_{X'}$ be the composite in a fixed algebraic closure $\overline{\mathcal{M}(X)}$ of $\mathcal{M}(X)$ of all finite subextensions arising from holomorphic maps of connected compact Riemann surfaces $Y \to X$ that restrict to a cover over X'. The field extension $K_{X'}|\mathcal{M}(X)$ is Galois, and its Galois group is isomorphic to the profinite completion of $\pi_1(X', x)$ for some base point $x \in X'$.*

The key point is the following property of the extension $K_{X'}$.

Lemma 3.4.2 *Every finite subextension of $K_{X'}|\mathcal{M}(X)$ comes from a connected compact Riemann surface that restricts to a cover over X'.*

Proof First we show that given two subextensions $L_i|\mathcal{M}(X)$ $(i = 1, 2)$ coming from compact connected Riemann surfaces $Y_i \to X$ that restrict to covers Y_i' over X', their compositum in $\overline{\mathcal{M}(X)}$ comes from a compact Riemann surface $Y_{12} \to X$ with the same property. For this consider the fibre product of covers $Y_1' \times_{X'} Y_2'$ introduced at the end of Chapter 2. It is a cover of X', whence a compact Riemann surface $Y_{12} \to X$ restricting to $Y_1' \times_{X'} Y_2'$ over X' by Theorem 3.2.7. We claim that its ring of meromorphic functions is isomorphic to the finite étale $\mathcal{M}(X)$-algebra $L_1 \otimes_{\mathcal{M}(X)} L_2$. Indeed, the latter algebra represents the functor on the category of $\mathcal{M}(X)$-algebras associating with an algebra R the set of $\mathcal{M}(X)$-algebra homomorphisms $L_1 \times L_2 \to R$. On the other hand, the fibre product represents the functor on the category of X'-spaces that maps a space Y the set of pairs of morphisms $(\phi : Y \to Y_1', \psi : Y \to Y_2')$ compatible with the projections to X'. The anti-equivalence of categories obtained by a successive application of Theorems 3.3.7 and 3.2.7 transforms these functors to each other, whence the claim. Now connected components of $Y_1 \times_X Y_2$ correspond exactly to direct factors of $L_1 \otimes_{\mathcal{M}(X)} L_2$, both corresponding to the factorization of a minimal polynomial of a generator of $L_1|\mathcal{M}(X)$ into irreducible factors over L_2. But when we look at the fixed embeddings of the L_i into $\overline{\mathcal{M}(X)}$, the component coming from one of these factors becomes exactly the composite $L_1 L_2$, and we are done.

The above argument shows that $K_{X'}$ can be written as a union of a tower of finite subextensions $L_1 \subset L_1 L_2 \subset L_1 L_2 L_3 \subset \cdots$ of $\mathcal{M}(X)$ coming from Riemann surfaces that restrict to a cover over X'. To conclude the proof of the lemma we show that if this property holds for a finite subextension $K_{X'} \supset L \supset \mathcal{M}(X)$, it holds for all subextensions $L \supset K \supset \mathcal{M}(X)$ as well. This is an easy counting argument: according to Proposition 3.3.5, if $L = \mathcal{M}(Y)$ and $K = \mathcal{M}(Z)$, each point of X' has $[L : \mathcal{M}(X)]$ preimages in Y and at most $[K : \mathcal{M}(X)]$ preimages in L, whereas each point of Z has at most $[L : K]$ preimages in Y (taking Theorem 3.3.7 into account). Therefore we must have equality everywhere, and Z must restrict to a cover over X'. □

Proof of Theorem 3.4.1 To see that $K_{X'}$ is Galois over $\mathcal{M}(X)$ it suffices to remark that if $L|\mathcal{M}(X)$ comes from a Riemann surface that restricts to a cover over X', then the same holds for all Galois conjugates of L by construction. We now realize $\widehat{\pi_1(X', x)}$ as a quotient of $\mathrm{Gal}\,(K_{X'}|\mathcal{M}(X))$ as follows. Each finite quotient of $\widehat{\pi_1(X', x)}$ corresponds to a finite Galois cover of X', which in turn corresponds to a finite Galois branched cover of X by Proposition 3.2.9, and finally to a finite Galois subextension of $K_{X'}|\mathcal{M}(X)$. By Theorem 3.3.7 and the lemma above we get in this way a bijection between isomorphic finite quotients of $\widehat{\pi_1(X', x)}$ and $\mathrm{Gal}\,(K_{X'}|\mathcal{M}(X))$, respectively, and moreover this bijection is seen to be compatible with taking towers of covers on the one side and field extensions on the other. The theorem follows by passing to the inverse limit. □

Remark 3.4.3 Since by Proposition 3.2.6 every holomorphic map $Y \to X$ of connected compact Riemann surfaces restricts to a cover over a suitable X', Theorem 3.3.7 shows that each finite subextension of $\overline{\mathcal{M}(X)}$ is contained in some extension of the form $K_{X'}$ considered above. Thus the Galois group $\mathrm{Gal}\,(\overline{\mathcal{M}(X)}|\mathcal{M}(X))$ is isomorphic to the inverse limit of the natural inverse system of groups formed by the $\mathrm{Gal}\,(K_{X'}|\mathcal{M}(X))$ with respect to the inclusions $K_{X'} \supset K_{X''}$ coming from $X' \subset X''$. Each of the groups $\mathrm{Gal}\,(K_{X'}|\mathcal{M}(X))$ is isomorphic to the profinite completion of the fundamental group of the Riemann surface X', and an explicit presentation for these fundamental groups is known from topology. Thus it is possible to determine the absolute Galois group of $\mathcal{M}(X)$. The case $X = \mathbf{P}^1(\mathbf{C})$ will be treated in detail below.

Consider the case $X = \mathbf{P}^1(\mathbf{C})$. As already mentioned in the previous chapter (and will be stated in greater generality in Remark 3.6.4 below), given a finite set $\{x_1, \ldots, x_n\}$ of points of $\mathbf{P}^1(\mathbf{C})$, it is known from topology that the fundamental group of the complement with respect to some base point x has a presentation

$$\pi_1(\mathbf{P}^1(\mathbf{C}) \setminus \{x_1, \ldots, x_n\}, x) = \langle \gamma_1, \ldots, \gamma_n \mid \gamma_1 \cdots \gamma_n = 1 \rangle, \tag{3.1}$$

where each generator γ_i can be represented by a closed path around the point x_i passing through x. For $n > 1$ this group is isomorphic to the free group F_{n-1} on $n - 1$ generators (send γ_i to a free generator f_i of F_{n-1} for $i < n$ and γ_n to $(f_1 \cdots f_{n-1})^{-1}$). In particular, every finite group arises as a finite quotient of $\pi_1(\mathbf{P}^1(\mathbf{C}) \setminus \{x_1, \ldots, x_n\}, x)$ for sufficiently large n. Since the field $\mathcal{M}(\mathbf{P}^1(\mathbf{C}))$ is isomorphic to $\mathbf{C}(t)$, Theorem 3.4.1 implies the corollary:

Corollary 3.4.4 *Every finite group occurs as the Galois group of some finite Galois extension $L|\mathbf{C}(t)$.*

We have proven more than what is stated in the corollary, namely that every finite group G that can be generated by $n - 1$ elements arises as the automorphism group of a Galois branched cover $p : Y \to \mathbf{P}^1(\mathbf{C})$ ramified above at most n points $x_1, \ldots, x_n$. We note for later use the following complement.

Proposition 3.4.5 *Under the surjection $\pi_1(\mathbf{P}^1(\mathbf{C}) \setminus \{x_1, \ldots, x_n\}, x) \twoheadrightarrow G$ the image of each topological generator γ_i generates the stabilizer of a point in the fibre $p^{-1}(x_i)$.*

Proof To ease notation, set $X := \mathbf{P}^1(\mathbf{C}) \setminus \{x_1, \ldots, x_n\}$. According to Theorem 2.3.4 the surjection $\pi_1(X, x) \twoheadrightarrow G$ is induced by a morphism of covers $p_x : \widetilde{X}_x \to p^{-1}(X)$ from the universal cover $\widetilde{X}_x$, which in turn corresponds to a point $y \in p^{-1}(x)$. If x' is another base point, the choice of a path between x and x' induces an isomorphism $\pi_1(X, x) \cong \pi_1(X, x')$, whence a surjection $\pi_1(X, x') \twoheadrightarrow G$ and a point $y' \in \pi^{-1}(x')$. Choose an open neighbourhood U_i of x_i so that $p^{-1}(U_i)$ is a finite disjoint union of open sets over which p is isomorphic to the branched cover $z \mapsto z^{e_j}$ of the unit disc. Modifying y if necessary we may assume $x \in U_i \setminus \{x_i\}$. Denote by V_i the component of $p^{-1}(U_i)$ containing y, and by y_i the unique preimage of x_i in V_i. By continuity the stabilizer of y_i in G must map V_i onto itself, so it is a subgroup of the cyclic group $\mathrm{Aut}(V_i|U_i)$. The latter group is generated by the image of γ_i in G by Example 2.4.12. The action of γ_i stabilizes y_i, so the stabilizer of y_i is the whole of $\mathrm{Aut}(V_i|U_i)$. □

Corollary 3.4.4 describes the finite quotients of Gal $(\overline{\mathbf{C}(t)}|\mathbf{C}(t))$, but does not determine the profinite group itself. As the corollary indicates, it should be free in an approprate sense. Here is a formal definition.

Definition 3.4.6 Let X be a set, and let $F(X)$ be the free group with basis X. The *free profinite group* $\widehat{F}(X)$ with basis X is defined as the inverse limit formed by the natural system of quotients $F(X)/U$, where $U \subset F(X)$ is a normal subgroup of finite index containing all but finitely many elements of X.

Remarks 3.4.7

1. If X is finite, then $\widehat{F}(X)$ is just the profinite completion of $F(X)$. In this case the cardinality r of X is called the *rank* of $\widehat{F}(X)$.
2. The inclusion map $i : X \to \widehat{F}(X)$ is characterized by the following universal property: given a profinite group G and a map $\lambda : X \to G$ such that every open normal subgroup of G contains all but finitely many elements of $\lambda(X)$, there is a unique morphism $\lambda_F : \widehat{F}(X) \to G$ of profinite groups making the diagram

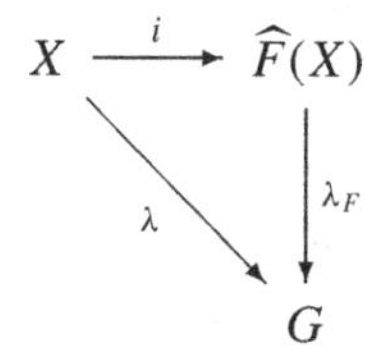

commute. This universal property may also be taken as a definition of $\widehat{F}(X)$. By writing G as an inverse limit of finite groups G/V and λ as an inverse limit of the composite maps $X \to G \to G/V$ one sees that it is enough to require the universal property for G finite.

Theorem 3.4.8 (Douady) *There is an isomorphism of profinite groups*

$$\mathrm{Gal}\,(\overline{\mathbf{C}(t)}|\mathbf{C}(t)) \cong \widehat{F}(\mathbf{C})$$

of the absolute Galois group of $\mathbf{C}(t)$ *with the free profinite group on the set* $\mathbf{C}$ *of complex numbers.*

Proof Let $S \subset \mathbf{C}$ be a finite set of r points. Applying Theorem 3.4.1 with $X_S := \mathbf{P}^1(\mathbf{C}) \setminus (S \cup \infty)$ in place of X' we obtain a quotient Gal $(K_{X_S}|\mathbf{C}(t))$ of Gal $(\overline{\mathbf{C}(t)}|\mathbf{C}(t))$ that is isomorphic to the free profinite group on r generators by (3.1), one generator γ_x for each $x \in S$. For a finite subset $S \subset T \subset \mathbf{C}$ giving rise to $X_T = \mathbf{P}^1(\mathbf{C}) \setminus (T \cup \infty)$ we have a natural inclusion $K_{X_S} \subset K_{X_T}$ of Galois extensions of $\mathbf{C}(t)$, whence a surjection $\lambda_{ST} :=$ Gal $(K_{X_T}|\mathbf{C}(t)) \twoheadrightarrow$ Gal $(K_{X_S}|\mathbf{C}(t))$ by infinite Galois theory. By theorem 3.4.1, this map comes from a natural map of fundamental groups $\pi_1(X_T, x_0) \to \pi_1(X_S, x_0)$ for some base point x_0 after taking profinite completion, and moreover $\lambda_{ST}(\gamma_x) = 1$ for $x \in T \setminus S$. The groups $\mathrm{Gal}(K_{X_S}|\mathbf{C}(t))$ together with the maps λ_{ST} form an inverse system indexed by the system of finite subsets of $\mathbf{C}$ partially ordered by inclusion. The inverse limit is $\mathrm{Gal}(\overline{\mathbf{C}(t)}|\mathbf{C}(t))$, because by Theorem 3.3.7 and Proposition 3.2.6 every finite subextension of $\overline{\mathbf{C}(t)}|\mathbf{C}(t)$ is contained in K_{X_S} for sufficiently large S, so the intersection of the open normal subgroups $\mathrm{Gal}(\overline{\mathbf{C}(t)}|K_{X_S})$ must be trivial. The theorem now follows from the purely group-theoretic proposition below. □

Proposition 3.4.9 *Let X be a set, and $\mathcal{S}$ the system of finite subsets $S \subset X$ partially ordered by inclusion. Let (G_S, λ_{ST}) be an inverse system of profinite groups indexed by $\mathcal{S}$ satisfying the following conditions:*

1. *The λ_{ST} are surjective for all $S \subset T$.*
2. *Each G_S has a system $\{g_x : x \in S\}$ of elements so that the map $\widehat{F}(S) \to G_S$ induced by the map $x \to g_x$ is an isomorphism, and moreover for every $S \subset T$ we have $\lambda_{ST}(g_x) = 1$ for $x \notin S$.*

Then $\varprojlim G_S$ is isomorphic to $\widehat{F}(X)$.

For the proof we need three lemmas.

Lemma 3.4.10 *The proposition is true in the case when $G_S = \widehat{F}(S)$ for all $S \in \mathcal{S}$, and $\lambda_{ST} : \widehat{F}(T) \to \widehat{F}(S)$ is the map characterized by*

$$\lambda_{ST}(x) = \begin{cases} x & x \in S \\ 1 & x \in T \setminus S \end{cases}$$

for all $S \subset T$.

Proof We check the property of Remark 3.4.7 (2). First observe that there is a natural injection $\hat{\imath} : X \to \varprojlim \widehat{F}(S)$ sending $x \in X$ to $(x_S)_{S \in \mathcal{S}}$, where $x_S = x$ for $x \in S$ and $x_S = 1$ otherwise. It generates a dense subgroup in $\varprojlim \widehat{F}(S)$, so given a map $\lambda : X \to G$ with G finite, an extension $\varprojlim \widehat{F}(S) \to G$ must be unique if it exists. But since G is finite, we must have $\overleftarrow{\lambda}(x) = 1$ for all but finitely many $x \in X$, so λ factors through the image of $\hat{\imath}(X)$ in some quotient $\widehat{F}(S)$, which is none but S. The existence then follows from the freeness of $\widehat{F}(S)$. □

Before the next lemma we introduce some terminology. A subset S of a free profinite group $\widehat{F}(X)$ is called a *basis* if each open normal subgroup of $\widehat{F}(X)$ contains all but finitely many elements of S, and moreover the map $i_F : \widehat{F}(S) \to \widehat{F}(X)$ extending the natural inclusion $i : S \to \widehat{F}(X)$ as in Remark 3.4.7 (2) is an isomorphism.

Lemma 3.4.11 *In a free profinite group $\widehat{F}(X)$ of finite rank r every system $S \subset \widehat{F}(X)$ of r elements that topologically generate $F(X)$ is a basis of $\widehat{F}(X)$.*

Proof By assumption the map $i_F : \widehat{F}(S) \to \widehat{F}(X)$ is surjective, so it is enough to show injectivity. For each $n > 0$ consider the sets $Q_n(S)$ (resp. $Q_n(X)$) of open normal subgroups of index n in $\widehat{F}(S)$ (resp. $\widehat{F}(X)$). As $\widehat{F}(S)$ and $\widehat{F}(X)$ are both profinite free of rank r, these sets have the same finite cardinality (bounded by $(n!)^r$). By surjectivity of i_F the map $Q_n(X) \to Q_n(S)$ sending $U \subset \widehat{F}(X)$ to $i_F^{-1}(U)$ is injective, hence bijective. It follows that ker (i_F) is contained in all subgroups in $Q_n(S)$, for all $n > 0$. As $\widehat{F}(S)$ is profinite, this means ker $(i_F) = \{1\}$. □

Lemma 3.4.12 *An inverse system $(X_\alpha, \phi_{\alpha\beta})$ of nonempty compact topological spaces is nonempty.*

Proof Consider the subsets $X_{\lambda\mu} \subset \prod X_\alpha$ consisting of the sequences (x_α) satisfying $\phi_{\lambda\mu}(x_\mu) = x_\lambda$ for a fixed pair $\lambda \leq \mu$. These are closed subsets of the product, and their intersection is precisely $\varprojlim X_\alpha$. Furthermore, the directedness of the index set implies that finite intersections of the $X_{\lambda\mu}$ are nonempty. Since $\prod X_\alpha$ is compact by Tikhonov's theorem, it ensues that $\varprojlim X_\alpha$ is nonempty. □

Proof of Proposition 3.4.9 For each $S \in \mathcal{S}$ denote by r the cardinality of S and by $B_S \subset G_S^r$ the set of all r-tuples that satisfy condition 2 of the proposition. If $g = (g_1, \ldots, g_r) \in G_S^r$ is such that each open neighbourhood of g meets B_S, then $g \in B_S$ by continuity. This means that $B_S \subset G_S^r$ is a closed subset, hence it is compact by Corollary 1.3.9 and Tikhonov's theorem. By conditions 1 and 2 together with Lemma 3.4.11 the λ_{ST} induce maps $B_S \to B_T$ for all pairs $S \subset T$. We thus obtain an inverse system of nonempty compact spaces indexed by $\mathcal{S}$; its inverse limit is nonempty by the lemma above. By construction, an element of $\varprojlim B_S$ induces an isomorphism of the inverse system of Lemma 3.4.10 with (G_S, λ_{ST}). The proposition now follows from that lemma. □

3.5 An alternate approach: patching Galois covers

We now present another approach to the proof of Corollary 3.4.4 based on a nowadays commonly used technique known as patching, pioneered by David Harbater. The specific argument we shall present is an adaptation of a rigid analytic method due to Florian Pop [75] in the complex setting; we are grateful to him for explaining it to us.

We begin by some purely topological constructions that could have figured in earlier chapters. The first is about patching together topological covers.

Lemma 3.5.1 *Let X be a locally connected topological space, and let $\{U_i : i \in I\}$ be an open covering of X. Assume given covers $p_i : Y_i \to U_i$ for $i \in I$, together with isomorphisms*

$$\theta_{ij} : p_i^{-1}(U_i \cap U_j) \xrightarrow{\sim} p_j^{-1}(U_i \cap U_j)$$

for each pair $i \neq j$ satisfying $\theta_{jk} \circ \theta_{ij} = \theta_{ik}$ over $U_i \cap U_j \cap U_k$. Then there exists a cover $p : Y \to X$ with $p^{-1}(U_i)$ isomorphic to Y_i as a space over U_i for each $i \in I$. It is unique up to unique isomorphism.

If moreover X and Y are connected and the Y_i are Galois covers of X with group G, then $p : Y \to X$ is also a Galois cover of X with group G.

Proof In Construction 2.7.8 we have already seen how to patch locally constant sheaves together. Therefore the dictionary between covers and locally constant sheaves (Theorem 2.5.9) yields the first part of the lemma. For the second part notice first that in the above construction automorphisms of the covers $p_i : Y_i \to X$ also patch together to automorphisms of X; in particular we obtain an injective map $G \to \mathrm{Aut}(Y|X)$. Since each $\mathrm{Aut}(Y_i|X)$ is transitive on the fibres of p_i, so is $\mathrm{Aut}(Y|X)$. By the connectedness assumption on Y we thus obtain that it is Galois over X. Finally, restriction to the Y_i shows that $Y/G \cong X$, whence it follows that $G \cong \mathrm{Aut}(Y|X)$. □

We also need another topological construction, that of *induced covers*. This concept is analogous to that of induced representations in algebra.

Construction 3.5.2 Let G be a group, and $H \subset G$ a subgroup. Assume moreover given a space $p : Y \to X$ over X such that H is isomorphic to a subgroup of $\mathrm{Aut}(Y|X)$. We construct a space $\mathrm{Ind}_H^G(Y)$ over X such that G is isomorphic to a subgroup of $\mathrm{Aut}(\mathrm{Ind}_H^G(Y)|X)$. Moreover, in the case $H \cong \mathrm{Aut}(Y|X)$ we shall actually have $G \cong \mathrm{Aut}(\mathrm{Ind}_H^G(Y)|X)$.

Consider the left coset space G/H as a discrete topological space, and define $\mathrm{Ind}_H^G(Y)$ to be the topological product $(G/H) \times Y$. The projections on each component equip it with the structure of a space over X. The G-action on $\mathrm{Ind}_H^G(Y)$ is defined as follows. Fix a system of left representatives $\{g_i : i \in G/H\}$ for G mod H. For $g \in G$ and $i \in G/H$ we find $j \in G/H$ and $h \in H$ such that $gg_i = g_j h$. The action of G on $\mathrm{Ind}_H^G(Y) = (G/H) \times Y$ is then defined by $g(i, y) = (j, hy)$. As $h \in \mathrm{Aut}(Y|X)$ by assumption, this is indeed an automorphism of $\mathrm{Ind}_H^G(Y)$ as a space over X. Note that this G-action depends on the choice of the g_i, but the space $\mathrm{Ind}_H^G(Y)$ itself does not.

It is immediate from the construction that if $Y \to X$ is a cover, then so is $\mathrm{Ind}_H^G(Y) \to X$, and if Y is Galois over X with group H, then $\mathrm{Aut}(\mathrm{Ind}_H^G(Y)|X) \cong G$. For this reason $\mathrm{Ind}_H^G(Y)$ is called the *induced G-cover* from Y. Note, however, that it is not connected when $H \neq G$, and therefore it is not a Galois cover.

Using the above patching and induction techniques, we now prove:

Proposition 3.5.3 *Let G be a finite group, and $g_1, \ldots, g_n$ a system of generators of G. Fix points $x_1, \ldots, x_n$ on the complex projective line $\mathbf{P}^1(\mathbf{C})$. There exists a finite Galois branched cover $Y \to \mathbf{P}^1(\mathbf{C})$ with group G such that each x_i is the image of a branch point whose stabilizer is generated by g_i.*

Notice that we do not claim that *all* branch points lie above the x_i. In fact, the construction will yield n other points $y_1, \ldots, y_n$ above which the cover is branched. Still, by the same arguments as in the previous section, the proposition implies Corollary 3.4.4, i.e. that every finite group arises as a Galois group over $\mathbf{C}(t)$.

Proof By an adequate choice of the complex coordinate z we may assume that none of the x_i lies at infinity. For each x_i choose a small open disc D_i of radius r_i centred around x_i in such a way that the D_i are all disjoint. Denote by G_i the cyclic subgroup generated by g_i in G, and by k_i the order of g_i. Fix for each i a point $y_i \in \mathbf{C}$ with $0 < |x_i - y_i| < r_i/2$, and consider the rational function

$$z \mapsto f_i(z) := \frac{y_i z^{k_i} + x_i}{z^{k_i} + 1}.$$

It induces a holomorphic map $\phi_i : \mathbf{P}^1(\mathbf{C}) \to \mathbf{P}^1(\mathbf{C})$ mapping 0 to x_i and ∞ to y_i. Since f_i is the composite of the map $z \mapsto z^{k_i}$ with a complex automorphism of $\mathbf{P}^1(\mathbf{C})$, the branch points of ϕ_i are 0 and ∞, lying above x_i and y_i, respectively. The map $\phi_i : \phi_i^{-1}(D_i) \to D_i$ is by construction a Galois branched cover with group G_i; it restricts to a Galois cover over $\ddot{D}_i := D_i \setminus \{x_i, y_i\}$. Let $Y_i \to \ddot{D}_i$ be the G-cover obtained by inducing this cover from G_i to G as in Construction 3.5.2.

For each i denote by B_i the *closed* disc of radius $r_i/2$ around x_i; it contains y_i in its interior. Since $\mathbf{P}^1(\mathbf{C}) \setminus B_i$ is simply connected, the restriction of the Galois cover $\phi_i : \mathbf{P}^1(\mathbf{C}) \setminus \{0, \infty\} \to \mathbf{P}^1(\mathbf{C}) \setminus \{x_i, y_i\}$ to $\mathbf{P}^1(\mathbf{C}) \setminus B_i$ is trivial. Therefore the restriction of the G-cover $Y_i \to \ddot{D}_i$ to the annulus $D_i \setminus B_i$ is trivial as well, being induced from a trivial cover. Now denote by U the open subset of $\mathbf{P}^1(\mathbf{C})$ obtained by removing all the B_i, and consider the trivial G-cover $G \times U \to U$. By what we have just said, the restrictions of $Y_i \to \ddot{D}_i$ and $G \times U \to U$ to $\ddot{D}_i \cap U = D_i \setminus B_i$ are both trivial G-covers, and are therefore isomorphic. We can now apply Lemma 3.5.1 to the covering of $X := \mathbf{P}^1(\mathbf{C}) \setminus \{x_1, \ldots, x_n, y_1, \ldots, y_n\}$ formed by the $\ddot{D}_i$ and U and patch the covers $Y_i \to \ddot{D}_i$ and $G \times U \to U$ together in a G-cover $Y' \to X$ (notice that the triple intersections are empty). It extends to a finite branched cover $Y \to X$ by Theorem 3.2.7.

As the statement about the g_i results from the construction, it remains to show that $Y \to X$ is a Galois branched cover, which boils down to the connectedness of Y'. Write $Y_i = (G/G_i) \times \phi_i^{-1}(\ddot{D}_i)$ using the definition of induced covers. If $\bar{e}$ denotes the class of the unit element $e \in G$ in G/G_i, we see from the construction that the action of $g_i \in G_i$ maps the component $\{\bar{e}\} \times \phi_i^{-1}(\ddot{D}_i)$ of Y_i onto itself. Since the component $\{\bar{e}\} \times \phi_i^{-1}(\ddot{D}_i)$ meets the component $\{e\} \times U$ of $G \times U$ in Y', we conclude applying g_i that it also meets $\{g_i\} \times U$. It follows that $\{e\} \times U$ and $\{g_i\} \times U$ lie in the same connected component of Y'. Iterating the argument we find that $\{e\} \times U$ and $\{g\} \times U$ lie in the same component of Y' for an arbitrary product $g = g_1^{s_1} \cdots g_n^{s^n}$ of the g_i. But the g_i generate G by assumption, hence all components of $G \times U$ lie in the same component of Y'. This shows that Y' has only one connected component. □

Remark 3.5.4 The advantage of the above method is that it works in a more general setting, that of complete valued fields. For instance, the same argument can be used to prove the following interesting theorem originally due to D. Harbater [31]: *Every finite group arises as the Galois group of a finite Galois extension of $L|\mathbf{Q}_p(t)$, with $\mathbf{Q}_p$ algebraically closed in L.* Here $\mathbf{Q}_p$ is the field of *p-adic numbers*, i.e. the fraction field of the ring $\mathbf{Z}_p$ encountered in Example 1.3.4 (4).

3.6 Topology of Riemann surfaces

In order to give a reasonably complete treatment of the theory of covers of Riemann surfaces we have to mention several topological results that are proven by methods different from those encountered above. Since this material is well documented in several introductory textbooks in topology, we shall mostly review the results without proofs, the book of Fulton [26] being our main reference. All Riemann surfaces in this section will be assumed to be connected.

We begin with the topological classification of compact Riemann surfaces. This is a very classical result stemming from the early days of topology and is proven by a method commonly called as 'cutting and pasting' (see [26], Theorem 17.4).

Theorem 3.6.1 *Every compact Riemann surface is homeomorphic to a torus with g holes.*

Recall that the simplest way to conceive a torus with g holes is to take a two-dimensional sphere and attach g 'handles' on it. This includes the case $g = 0$, where we just mean the 2-sphere. The number g of is called the *genus* of the Riemann surface.

Remarks 3.6.2

1. A more rigorous way of defining a torus with g holes (for $g > 0$) is by taking a *connected sum* of g usual tori. This is done as follows: one first takes two copies of the usual torus, cuts out a piece homeomorphic to a closed disc from each of them, and then patches the two pieces together by identifying the boundaries of the two discs just cut out. In this way one obtains a torus with two holes; the general case is done by iterating the procedure.

   Another construction generalizes the fact that we may obtain the usual torus by identifying opposite sides of a square (with the same orientation). In the general case one takes a regular $4g$-gon and labels its sides clockwise by $a_1, b_1, a_1^{-1}, b_1^{-1}, \ldots, a_g, b_g, a_g^{-1}, b_g^{-1}$. Here the notation means that we consider the a_i, b_i with clockwise orientation and the a_i^{-1}, b_i^{-1} with counterclockwise orientation. Now identify each a_i with a_i^{-1} and b_i with b_i^{-1} taking care of the chosen orientations (see the very suggestive drawings on pp. 240–241 of [26]). In this way one gets a sphere with g handles, and the sides $a_i, b_i, a_i^{-1}, b_i^{-1}$ of our initial polygon get mapped to closed paths all going through a common point x.
2. What the theorem really uses about the topology of compact Riemann surfaces is that they are *orientable topological manifolds of dimension* 2. The topological manifold structure is obtained by considering the complex charts as homeomorphisms of some neighbourhood of each point with an open subset of $\mathbf{R}^2$. Orientability can be expressed in this case by remarking that if $f_i : U_i \to \mathbf{C}$ $(i = 1, 2)$ are some complex charts on our Riemann surface, then the map $f_1^{-1} \circ f_2$ regarded as a real differentiable map from $\mathbf{R}^2$ to $\mathbf{R}^2$ has a positive Jacobian determinant at each point; this is a consequence of the Cauchy–Riemann equations.

The second representation of tori with g holes in Remark 3.6.2 (1) makes it possible to compute the fundamental group of a compact Riemann surface of genus g. Here it is convenient to take as a base point the point x where all the closed paths coming from the $a_i, b_i, a_i^{-1}, b_i^{-1}$ meet. The homotopy classes of these paths then generate the fundamental group. More precisely, one proves:

Theorem 3.6.3 *The fundamental group of a compact Riemann surface X of genus g has a presentation of the form*

$$\pi_1(X, x) = \langle a_1, b_1, \ldots, a_g, b_g \mid [a_1, b_1] \cdots [a_g, b_g] = 1 \rangle,$$

where the brackets $[a_i, b_i]$ denote the commutators $a_i b_i a_i^{-1} b_i^{-1}$.

For the proof see e.g. [26], Proposition 17.6. It uses the definition of the fundamental group in terms of closed paths and the van Kampen theorem.

Remark 3.6.4 By the same method that proves the theorem one can also determine the fundamental group of the complement of $n+1$ points $x_0, \ldots, x_n$ in a compact Riemann surface X of genus g. Here one has to add one generator γ_i for each x_i represented by a closed path going through x and turning around x_i. We get a presentation of $\pi_1(X \setminus \{x_1, \ldots, x_n\}, x)$ by

$$\langle a_1, b_1, \ldots, a_g, b_g, \gamma_0, \ldots, \gamma_n \mid [a_1, b_1] \cdots [a_g, b_g]\gamma_0 \cdots \gamma_n = 1 \rangle.$$

The special case $g = 0$ has already turned up in previous sections.

Remark 3.6.5 Realizing the groups described in the theorem as automorphism groups of the universal cover gives rise to a fascinating classical theory known as the theory of *uniformization*; see Chapter IX of [95] for a nice introduction.

The main result here, originating in work by Riemann and proven completely by Poincaré and Koebe, is that a simply connected Riemann surface is isomorphic as a complex manifold to the projective line $\mathbf{P}^1(\mathbf{C})$, the complex plane $\mathbf{C}$ or the open unit disc D (see e.g. [23], Theorem 27.6). Now one can produce compact Riemann surfaces as quotients of the above as follows. In the case of $\mathbf{P}^1(\mathbf{C})$ there is no quotient other than itself, for an automorphism of $\mathbf{P}^1(\mathbf{C})$ is known to have a fixed point (Exercise 2). For $\mathbf{C}$, one can prove that the only even action on it with compact quotient is the one by $\mathbf{Z}^2$ as in the second example of Chapter 2, Example 2.1.8, so the quotient is a torus $\mathbf{C}/\Lambda$; this is in accordance with the case $g = 1$ of the theorem. All other compact Riemann surfaces are thus quotients of the open unit disc D by some even group action. Poincaré studied such actions and showed that they come exactly from transformations mapping a_i to a_i^{-1} and b_i to b_i^{-1} in a $4g$-gon with sides labelled as above; the only difference is that in this case the sides of the polygon are not usual segments but circular arcs inscribed into the unit disc, for he worked in the model of the hyperbolic plane named after him.

We finally discuss triangulations of compact Riemann surfaces. Recall the definition:

Definition 3.6.6 Let X be a compact topological manifold of dimension 2. A *triangulation* consists of a finite system $\mathcal{T} = \{T_1, \ldots, T_n\}$ of closed subsets of X whose union is the whole of X, and homeomorphisms $\phi_i : \Delta \xrightarrow{\sim} T_i$, where Δ is the unit triangle in $\mathbf{R}^2$. The T_i are called the *faces* of the triangulation, and the images of the edges (resp. vertices) of Δ *edges* (resp. *vertices*) of the triangulation. These are subject to the following conditions: each vertex (resp. edge) of $\mathcal{T}$ contained in a face T_i should be the image of a vertex (resp. edge) of Δ by ϕ_i, and moreover any two different faces must be either disjoint, or intersect in a single vertex, or else intersect in a single edge.

Examples 3.6.7 We describe triangulations of compact Riemann surfaces of genus 0 and 1.

1. The two-dimensional sphere (which is the underlying topological space of $\mathbf{P}^1_{\mathbf{C}}$) has several natural triangulations; one of them is cut out by the equator and two meridians.
2. A triangulation of the complex torus $\mathbf{C}/\Lambda$ may be easily obtained from its description as a square with opposite edges identified. Divide first the square into nine subsquares by dividing each edge in three, and then divide each subsquare in two triangles by the diagonal from the upper left to the lower right corner. After identification of the edges of the original square these induce a triangulation of the torus.

We now prove:

Proposition 3.6.8 *Every compact Riemann surface has a triangulation.*

The proposition is an immediate consequence of Example 3.6.7 (1), of Proposition 3.3.11 and the following lemma.

Lemma 3.6.9 *Let $\phi : Y \to X$ be a branched cover of compact Riemann surfaces (e.g. a holomorphic map). Than every triangulation of X can be lifted canonically to a triangulation of Y.*

Before giving the proof, note the obvious fact that given a triangulation $\mathcal{T}$ of a compact topological surface X and a point $x \in X$ which is not a vertex of $\mathcal{T}$, the triangulation can be refined in a canonical way to a triangulation whose set of vertices is that of $\mathcal{T}$ with x added: take the face $\phi_i(\Delta)$ containing x (if x happens to lie on an edge, take both faces meeting at that edge), consider the natural subdivision of Δ given by joining $\phi_i^{-1}(x)$ to the vertices and replace ϕ_i by its restrictions to the smaller triangles arising from the subdivision (which are of course homeomorphic to Δ).

Proof By refining the triangulation as above if necessary, we may assume that in the given triangulation of X the set S_0 of vertices contains all images of branch points. Hence the restriction of ϕ to $X \setminus \phi^{-1}(S_0)$ is a cover. As the subset $\Delta' \subset \Delta$ obtained by omitting the vertices is simply connected, the restriction of the cover $\phi : Y \to X$ above each $\phi_i(\Delta')$ is trivial. Therefore the restriction of each ϕ_i to Δ' can be canonically lifted to each sheet of the cover; we may lift those vertices of $\phi_i(\Delta)$ as well which are not images of branch points. By comparing with the triangulation of X we see that away from the branch points we have defined a triangulation of Y. Finally, the local form of Lemma 3.2.1 shows that in the neighbourhood of branch points we obtain a triangulation by adding the branch point as a vertex. □

Given a triangulation $\mathcal{T}$ of a compact Riemann surface X, denote by S_0, S_1, S_2 the set of vertices, edges and faces of $\mathcal{T}$, respectively, and write s_i for the cardinality of S_i.

Definition 3.6.10 The integer $\chi_X := s_0 - s_1 + s_2$ is called the *Euler characteristic* of X.

To justify the definition, one has to verify that χ_X is an invariant of X itself, and does not depend on the triangulation. Indeed, one checks immediately that χ_X does not change if we refine a triangulation by the process described above. From this the invariance of χ_X follows by choosing common refinements of two triangulations.

Proposition 3.6.11 *Let $\phi : Y \to X$ be a holomorphic map of compact Riemann surfaces having degree d as a branched cover. The Euler characteristics χ_X and χ_Y of X and Y are related by the formula*

$$\chi_Y = d \cdot \chi_X - \sum_y (e_y - 1)$$

where the sum is over the branch points of ϕ and e_y is the ramification index at the branch point $y \in Y$.

Proof In the process of lifting a triangulation to a branched cover as in the lemma the number of edges of the lifted triangulation is ds_1 and the number of its faces is ds_2. Those points of S_0 which are not in the image of the branch locus have d preimages as well but with each branch point the number of preimages diminishes by $e_y - 1$. □

Now it is a known topological fact that the Euler characteristic of a torus with g holes is $2 - 2g$ (see [26], p. 244; the cases $g = 0, 1$ may be read off from the above example). Hence the proposition implies:

Corollary 3.6.12 (Riemann–Hurwitz Formula) *The formula of the proposition can be rewritten as*

$$2g_Y - 2 = d(2g_X - 2) + \sum_y (e_y - 1)$$

where g_X and g_Y are the genera of X and Y, respectively.

The formula is extremely useful in practice, as it puts constraints on branched covers of compact Riemann surfaces. As an example, note that if $X = \mathbf{P}^1(\mathbf{C})$, $d = 2$ and there are four branch points (necessarily with ramification index 2), then $g_Y = 1$ and Y is a torus. Another famous application is:

Corollary 3.6.13 *If X is a compact Riemann surface of genus $g > 0$, there are no non-constant holomorphic maps $\mathbf{P}^1(\mathbf{C}) \to X$.*

Proof Otherwise the right-hand side of the Riemann–Hurwitz formula would be non-negative and the left-hand side -2. □

Remark 3.6.14 Combining the last corollary with Theorem 3.3.7 one obtains the following purely algebraic fact: *Every subfield $\mathbf{C} \subset K \subset \mathbf{C}(t)$ with $[\mathbf{C}(t) : K] < \infty$ is isomorphic to* $\mathbf{C}(t)$. This is in fact true for an arbitrary field k in place of $\mathbf{C}$ and is known as Lüroth's Theorem (see [106], §73). Using the techniques of subsequent chapters it is possible to extend the above proof to the general case.

Exercises

1. Let $\phi : Y \to X$ be a non-constant holomorphic map of Riemann surfaces, with X connected. Show that for all $x \in X$ we have

$$\sum_{y \in \phi^{-1}(x)} e_y = n,$$

where n is the cardinality of the fibres not containing branch points, and e_y is the ramification index at $y \in Y$.

2. Consider a non-constant holomorphic map $\phi : \mathbf{P}^1(\mathbf{C}) \to \mathbf{P}^1(\mathbf{C})$.
   (a) Show that there exists a unique rational function $f \in \mathbf{C}(t)$ such that $\phi = \phi_f$ as in the proof of Proposition 3.3.11.
   (b) Relate the branch points of ϕ_f to zeros and poles of f, and determine the ramification indices.
   (c) Show that ϕ_f is a holomorphic automorphism of $\mathbf{P}^1(\mathbf{C})$ if and only if $f = (at + b)(ct + d)^{-1}$ with some $a, b, c, d \in \mathbf{C}$ satisfying $ad - bc \neq 0$. [*Hint:* Observe that if ϕ_f is an automorphism, f can only have a single zero and pole, and these must be of order 1.]
   (d) Deduce that a holomorphic automorphism of $\mathbf{P}^1(\mathbf{C})$ has one or two fixed points.
3. Let $Y \to X$ be a holomorphic map of compact Riemann surfaces, restricting to a cover $Y' \to X'$ outside the branch points.
   (a) Show that the étale $\mathcal{M}(X)$-algebra $\mathcal{M}(Y)$ is isomorphic to a finite direct sum of copies of $\mathcal{M}(X)$ if and only if the cover $Y' \to X'$ is trivial.
   (b) Using Chapter 2, Exercise 2 give another proof of the fact that in the anti-equivalence of Theorem 3.3.7 Galois branched covers correspond to Galois extensions.
4. Let X be a compact connected Riemann surface, and $\phi : X \to \mathbf{P}^1(\mathbf{C})$ a holomorphic map. Show that if ϕ is not an isomorphism, then its branch points cannot all lie above the same point of $\mathbf{P}^1(\mathbf{C})$. [*Hint:* Use the Riemann–Hurwitz formula.]

5. Let $n > 2$ be an even integer, and consider the dihedral group

$$D_n = \langle \sigma, \tau \mid \sigma^n = \tau^2 = 1,\ \sigma\tau = \tau\sigma^{n-1} \rangle.$$

   Show that every complex torus $X = \mathbf{C}/\Lambda$ has a Galois branched cover $Y \to X$ with group D_n with exactly n branch points, all lying above the same point of X.
6. Let $\phi : \mathbf{C}/\Lambda \to \mathbf{C}/\Lambda'$ be a holomorphic map of complex tori.
   (a) Show that ϕ has no branch points.
   (b) Denoting by $p : \mathbf{C} \to \mathbf{C}/\Lambda$, $p' : \mathbf{C} \to \mathbf{C}/\Lambda'$ the natural projections, show that there exists a holomorphic map $\widetilde{\phi} : \mathbf{C} \to \mathbf{C}$ with $p' \circ \widetilde{\phi} = \phi \circ p$.
   (c) Check that setting $\widetilde{\phi}(\infty) = \infty$ defines an extension of $\widetilde{\phi}$ to a holomorphic map $\mathbf{P}^1(\mathbf{C}) \to \mathbf{P}^1(C)$, and show that this map is in fact a holomorphic automorphism of $\mathbf{P}^1(\mathbf{C})$. [*Hint:* Use Exercise 4.]
   (d) Establish a correspondence between holomorphic maps $\phi : \mathbf{C}/\Lambda \to \mathbf{C}/\Lambda'$ and pairs $(a, b) \in \mathbf{C}^2$ with $a \neq 0$ and $a\Lambda + b \subset \Lambda'$. [*Hint:* Use Exercise 2.]

4

Fundamental groups of algebraic curves

In the previous chapter the Riemann Existence Theorem created a link between the category of compact connected Riemann surfaces and that of finite extensions of $\mathbf{C}(t)$. This hints at a possibility of developing a theory of the fundamental group in a purely algebraic way. We shall now present such a theory for curves over an arbitrary perfect base field, using a modest amount of algebraic geometry. Over the complex numbers the results will be equivalent to those of the previous chapter, but a new and extremely important feature over an arbitrary base field k will be the existence of a canonical quotient of the algebraic fundamental group isomorphic to the absolute Galois group of k. In fact, over a subfield of $\mathbf{C}$ we shall obtain an extension of the absolute Galois group of the base field by the profinite completion of the topological fundamental group of the corresponding Riemann surface over $\mathbf{C}$. This interplay between algebra and topology is a source for many powerful results in recent research. Among these we shall discuss applications to the inverse Galois problem, Belyi's theorem on covers of the projective line minus three points and some advanced results on 'anabelian geometry' of curves.

Reading this chapter requires no previous acquaintance with algebraic geometry. We shall, however, use some standard results from commutative algebra that we summarize in the first section. The next three sections contain foundational material, and the discussion of the fundamental group itself begins in Section 4.5.

4.1 Background in commutative algebra

We collect here some standard facts from algebra needed for subsequent developments. The reader is invited to use this section as a reference and consult it only in case of need. In what follows, the term 'ring' means a commutative ring with unit.

Recall that given an extension of rings $A \subset B$, an element $b \in B$ is said to be *integral* over A if it is a root of a *monic* polynomial $x^n + a_{n-1}x^{n-1} + \cdots + a_0$ in $A[x]$. The *integral closure* of A in B consists of the elements of B integral over A; if this is the whole of B, then one says that the extension $A \subset B$ is *integral* or that B is *integral over* A. Finally, A is *integrally closed* in B if it equals its integral closure in B. In the special case when A is an integral domain and B is the fraction field of A one says that A is *integrally closed.* The basic properties of integral extensions may be summarized as follows.

Facts 4.1.1 Let $A \subset B$ be an extension of rings.

1. An element $b \in B$ is integral over A if and only if the subring $A[b]$ of B is finitely generated as an A-module.
2. The integral closure of A in B is a subring of B, and moreover it is integrally closed in B.
3. Given a tower extensions $A \subset B \subset C$ with $A \subset B$ and $B \subset C$ integral, the extension $A \subset C$ is also integral.
4. If B is integral over A and $P \subset A$ is a prime ideal, there exists a prime ideal $Q \subset B$ with $Q \cap A = P$. Here P is a maximal ideal in A if and only if Q is a maximal ideal in B.

All these facts are proven in [48], Chapter VII, §1, or [2], Chapter 5, 5.1–5.10.

Example 4.1.2 A unique factorization domain A is integrally closed. Indeed, if an element a/b of the fraction field (with a, b coprime) satisfies a monic polynomial equation over A of degree n, then after multiplication by b^n we see that a^n should be divisible by b, which is only possible when b is a unit.

Assume now that $A \subset B$ is an integral extension of integrally closed domains such that moreover the induced extension $K \subset L$ of fraction fields is Galois with finite Galois group G. Then B is stable by the action of G on L, being the integral closure of A in L. Given a maximal ideal $P \subset A$, denote by S_P the set of maximal ideals $Q \subset B$ with $Q \cap A = P$. The group G acts on the finite set S_P: each $\sigma \in G$ sends $Q \in S_P$ to the prime ideal $\sigma(Q) \in S_P$.

Fix $Q \in S_P$, and denote by D_Q its stabilizer in G for the above action. The action of each $\sigma \in D_Q$ on B induces an automorphism $\bar{\sigma}$ of $\kappa(Q) := B/Q$ fixing $\kappa(P) := A/P$ elementwise. Moreover, the map $\sigma \mapsto \bar{\sigma}$ is a homomorphism $D_Q \to \mathrm{Aut}(\kappa(Q)|\kappa(P))$. Its kernel I_Q is a normal subgroup of D_Q called the *inertia subgroup* at Q.

Facts 4.1.3 In the situation above the following statements hold.

1. The group G acts transitively on the set S_P; in particular, S_P is finite.
2. The subgroups D_Q and I_Q for $Q \in S_P$ are all conjugate in G.
3. It the extension $\kappa(Q)|\kappa(P)$ is separable, then it is a Galois extension and the homomorphism $D_Q/I_Q \to \mathrm{Gal}\,(\kappa(Q)|\kappa(P))$ defined above is an isomorphism.
4. If the order of I_Q is prime to the characteristic of $\kappa(P)$, then I_Q is cyclic.

Statement (1) is [48], Chapter VII, Proposition 2.1, and (2) results from (1). Statement (3) is proven in [48], Chapter VII, Proposition 2.5. Finally, statement (4) results from [69], Theorem 9.12 (and the discussion preceding it).

We now state an important result concerning the finiteness of integral closure.

Facts 4.1.4 Let A be an integral domain with fraction field K, and let $L|K$ be a finite extension. Assume that either

(a) A is integrally closed and $L|K$ is a separable extension, or
(b) A is a finitely generated algebra over a field.

Then the integral closure of A in L is a finitely generated A-module.

For the proof of part (a), see e.g. [2], Corollary 5.17; for (b), see [18], Corollary 13.13.

An integral domain A is called a *Dedekind ring* if A is Noetherian (i.e. all of its ideals are finitely generated), integrally closed, and all nonzero prime ideals in A are maximal. Basic examples of Dedekind rings are polynomial rings in one variable over a field, the ring **Z** of integers and, more generally, the integral closure of **Z** in a finite extension of **Q**.

Facts 4.1.5 Let A be a Dedekind ring.

1. Every nonzero ideal $I \subset A$ decomposes uniquely as a product $I = P_1^{e_1} \cdots P_r^{e_r}$ of powers of prime ideals P_i.
2. For a prime ideal $P \subset A$ the localization A_P is a principal ideal domain.

Recall that the localization A_P means the fraction ring of A with respect to the multiplicatively closed subset $A \setminus P$, which in our case is the subring of the fraction field of A constisting of fractions with denominator in $A \setminus P$. For a proof, see e.g. [2], Theorem 9.3 and Corollary 9.4.

Note that in view of Facts 4.1.1 (2), (4) and 4.1.4 (a) the integral closure of a Dedekind ring in a finite separable extension of its fraction field is again a Dedekind ring. We then have the following consequence of the above facts.

Proposition 4.1.6 *Let A be a Dedekind ring with fraction field K, and let B be the integral closure of A in a finite separable extension $L|K$. For a nonzero prime ideal $P \subset A$ consider the decomposition $PB = Q_1^{e_1} \cdots Q_r^{e_r}$ in B. Then*

$$\sum_{i=1}^{r} e_i[\kappa(Q_i) : \kappa(P)] = [L : K].$$

Proof By the Chinese Remainder Theorem and Fact 4.1.5 (1) we have an isomorphism

$$B/PB \cong B/Q_1^{e_1} \oplus \cdots \oplus B/Q_r^{e_r}. \tag{4.1}$$

Since each Q_i generates a principal ideal (q_i) in the localization B_{Q_i} by Fact 4.1.5 (2), the map $b \mapsto q_i^j b$ induces isomorphisms $\kappa(Q_i) = B/Q_i \xrightarrow{\sim} Q_i^j/Q_i^{j+1}$ for all $0 \leq j \leq e_i - 1$. It follows that the left-hand side of the formula of the proposition equals the dimension of the $\kappa(P)$-vector space B/PB. Choose elements $t_1 \ldots, t_n \in B$ whose images B/PB form a basis of this vector space. By a form of Nakayama's lemma ([48], Chapter X, Lemma 4.3) the t_i generate the finitely generated A_P-module $B \otimes_A A_P$, hence they generate the K-vector space L. It remains to see that there is no nontrivial relation $\sum a_i t_i = 0$ with $a_i \in K$. Assume there is one. By multiplying with a suitable generator of the principal ideal PA_P we may assume the a_i lie in A_P and not all of them are in PA_P. Reducing modulo PA_P we then obtain a nontrivial relation with coefficients in $\kappa(P)$, which is impossible. □

The integers e_i in formula (4.1) are called the *ramification indices* at the prime ideals Q_i lying above P. As it will turn out later, it is by no means accidental that we are using the same terminology as for branched topological covers.

Corollary 4.1.7 *Let $A \subset B$ be an integral extension of Dedekind rings such that the induced extension $K \subset L$ of fraction fields is a finite Galois extension with group G, and let P be a maximal ideal of A. Assume that the extensions $\kappa(Q_i)|\kappa(P)$ are separable for all $Q_i \in S_P$. Then the ramification indices e_i in the above formula are the same for all i, and they equal the order of the inertia subgroups at the Q_i.*

Proof It is enough to verify the second statement for the inertia subgroup at Q_1, the rest then follows from Proposition 4.1.3 (2). Let K_1 be the subfield of L fixed by D_{Q_1}, A_1 the integral closure of A in K_1 and $P_1 := Q_1 \cap A_1$. Since Q_1 is the only maximal ideal of B above P_1 by construction, the formula of the proposition reads $|D_{Q_1}| = e_1 \cdot [\kappa(Q_1) : \kappa(P_1)]$. On the other hand, Proposition 4.1.3 (3) implies $|D_{Q_1}| = |I_{Q_1}| \cdot [\kappa(Q_1) : \kappa(P_1)]$ (note that the extension $\kappa(Q_1)|\kappa(P_1)$ is separable, being a subextension of $\kappa(Q_1)|\kappa(P)$). The statement follows by comparing the two equalities. □

Before leaving this topic, we collect some facts about local Dedekind rings.

Fact 4.1.8 The following are equivalent for an integral domain A.

1. A is a local Dedekind ring.
2. A is a local principal ideal domain that is not a field.
3. A is a Noetherian local domain with nonzero principal maximal ideal.

For a proof, see e.g. [57], Theorem 11.2. Such rings are called *discrete valuation rings* in the literature.

Proposition 4.1.9 *Let A be a discrete valuation ring, and t a generator of its maximal ideal.*

1. *Every nonzero $a \in A$ can be written as $a = ut^n$ with some unit $u \in A$ and $n \geq 0$. Here n does not depend on the choice of t.*
2. *If x is an element of the fraction field K of A, then either x or x^{-1} is contained in A.*
3. *If $B \supset A$ is a discrete valuation ring with the same fraction field, then $A = B$.*

Proof The intersection of the ideals (t^n) is 0 (this follows, for instance, from the fact that A is a principal ideal domain). Thus for $a \neq 0$ there is a unique $n \geq 0$ with $a \in (t^n) \setminus (t^{n+1})$, whence (1). By (1), if x is an element of the fraction field, we may write $x = ut^n$ with a unit u and $t \in \mathbf{Z}$, whence (2). For statement (3), assume $b \in B$ is not a unit. Then $b \in A$, for otherwise we would have $b^{-1} \in A \subset B$ by (2), which is impossible. It moreover follows that $b \in (t)$ (otherwise it would be a unit), so using (1) we see that t cannot be a unit in B. It follows that non-units in A are non-units in B, from which we conclude by (2) that the units of B lie in A. □

The first two statements imply:

Corollary 4.1.10 *Every nonzero $a \in K$ can be written as $a = ut^n$ with some unit $u \in A$ and $n \in \mathbf{Z}$. Here n does not depend on the choice of the generator t.*

In the notation of the corollary, the rule $a \mapsto n$ defines a well-defined map $v : K^\times \to \mathbf{Z}$ called the *discrete valuation associated with A*. It is actually a homomorphism of abelian groups and satisfies $v(x + y) \geq \min(v(x), v(y))$. One often extends v to a map $K \to \mathbf{Z} \cup \{\infty\}$ by setting $v(0) = \infty$.

Finally, let A be a finitely generated algebra over a field k. Recall that A is Noetherian by the Hilbert basis theorem. If A is an integral domain, we define its *transcendence degree* over k to be that of its fraction field K. Recall that this is the largest integer d for which there exist elements $a_1, \ldots, a_d$ in K that are algebraically independent over k, i.e. they satisfy no nontrivial polynomial relations with coefficients in k. As K is finitely generated over k, such a d exists.

Fact 4.1.11 (Noether's Normalization Lemma) Let A be an integral domain finitely generated over a field k, of transcendence degree d. There exist algebraically independent elements $x_1, \ldots, x_d \in A$ such that A is finitely generated as a $k[x_1, \ldots, x_d]$-module.

See [48], Chapter VIII, Theorem 2.1 for a proof. Notice that in the situation above $k[x_1, \dots, x_d]$ is isomorphic to a polynomial ring.

Corollary 4.1.12 *If A is as above and $d = 1$, then every nonzero prime ideal in A is maximal.*

Hence if moreover A is integrally closed, it is a Dedekind ring.

Proof Let $P \subset A$ be a nonzero prime ideal, and use the normalization lemma to write A as an integral extension of the polynomial ring $k[x]$. The prime ideal $P \cap k[x]$ of $k[x]$ is nonzero, because if t is a nonzero element of P, it satisfies a monic polynomial equation $t^n + a_{n-1}t^{n-1} + \cdots + a_0 = 0$ with $a_i \in k[x]$. Here we may assume a_0 is nonzero, but it is an element of $P \cap k[x]$ by the equation. Now all nonzero prime ideals in $k[x]$ are generated by irreducible polynomials and hence they are maximal. Thus P is maximal by Fact 4.1.1 (4). □

Corollary 4.1.13 *Let A be an integral domain finitely generated over a field k, and let $M \subset A$ be a maximal ideal. Then A/M is a finite algebraic extension of k.*

Proof Apply Noether's Normalization Lemma to A/M. If A/M had positive transcendence degree d, it would be integral over the polynomial ring $k[x_1, \dots, x_d]$. This contradicts Fact 4.1.1 (4) (with $P = Q = 0$), because A/M is a field but the polynomial ring is not. □

Corollary 4.1.14 *Let k be algebraically closed. Every maximal ideal M in the polynomial ring $k[x_1, \dots, x_n]$ is of the form $(x_1 - a_1, \dots, x_n - a_n)$ with appropriate $a_i \in k$.*

Proof As k is algebraically closed, we have an isomorphism $k[x_1, \dots, x_n]/M \cong k$ by the previous corollary. Let a_i be the image of x_i in k via this isomorphism. Then M contains the maximal ideal $(x_1 - a_1, \dots, x_n - a_n)$, hence they must be equal. □

Corollary 4.1.15 *Let k be a field, and $\phi : A \to B$ a k-homomorphism of finitely generated k-algebras. If M is a maximal ideal in B, then $\phi^{-1}(M)$ is a maximal ideal in A.*

Proof By replacing A with $\phi(A)$ we may assume A is a subring of B and $\phi^{-1}(M) = M \cap A$. In the tower of ring extensions $k \subset A/(M \cap A) \subset B/M$ the field B/M is a finite extension of k by Corollary 4.1.13, so $A/(M \cap A)$ is an integral domain algebraic over k. By Fact 4.1.1 (4) it must be a field, i.e. $M \cap A$ is maximal in A. □

Corollaries 4.1.13 and 4.1.14 are weak forms of Hilbert's Nullstellensatz. Here is a statement that may be considered as a strong form. Recall that the *radical* $\sqrt{I}$ of an ideal I in a ring A consists of the elements $f \in A$ satisfying $f^m \in I$ with an appropriate $m > 0$.

Fact 4.1.16 Let A be an integral domain finitely generated over a field k, and let $I \subset A$ be an ideal. The radical $\sqrt{I}$ is the intersection of all maximal ideals containing I.

This follows from [48], Chapter IX, Theorem 1.5 in view of the previous corollary. See also [18], Corollary 13.12 combined with Corollary 2.12.

4.2 Curves over an algebraically closed field

We now introduce the main objects of study in this chapter in the simplest context, that of affine curves over an algebraically closed field. When the base field is $\mathbf{C}$, we shall establish a connection with the theory of Riemann surfaces.

We begin by defining affine varieties over an *algebraically closed field* k. To this end, let us identify points of *affine n-space* $\mathbf{A}^n$ over k with

$$\mathbf{A}^n(k) := \{(a_1, \dots, a_n) : a_i \in k\}.$$

Given an ideal $I \subset k[x_1, \dots, x_n]$, we define

$$V(I) := \{P \in \mathbf{A}^n(k) : f(P) = 0 \quad \text{for all } f \in I\}.$$

Definition 4.2.1 The subset $X := V(I) \subset \mathbf{A}^n(k)$ is called the *affine closed set* defined by I.

Remark 4.2.2 According to the Hilbert Basis Theorem there exist finitely many polynomials $f_1, \dots, f_m \in k[x_1, \dots, x_n]$ with $I = (f_1, \dots, f_m)$. Therefore

$$V(I) = \{P \in \mathbf{A}^n(k) : f_i(P) = 0 \quad i = 1, \dots, m\}.$$

Example 4.2.3 Let us look at the simplest examples. For $n = 1$ each ideal in $k[x]$ is generated by a single polynomial f; since k algebraically closed, f factors in linear polynomials $x - a_i$ with some $a_i \in k$. The affine closed set we obtain is a finite set of points corresponding to the a_i.

For $n = 2, m = 1$ we obtain the locus of zeros of a single two-variable polynomial f in $\mathbf{A}^2$: it is a *plane curve*. In general it may be shown that an affine closed set in $\mathbf{A}^2$ is always the union of a plane curve and a finite set of points.

The following lemma is an easy consequence of the definition; its proof is left to the readers.

Lemma 4.2.4 *Let I_1, I_2, I_λ $(\lambda \in \Lambda)$ be ideals in $k[x_1, \dots, x_n]$. Then*

(a) $I_1 \subseteq I_2 \Rightarrow V(I_1) \supseteq V(I_2)$;
(b) $V(I_1) \cup V(I_2) = V(I_1 \cap I_2) = V(I_1 I_2)$;
(c) $V(\langle I_\lambda : \lambda \in \Lambda \rangle) = \bigcap_{\lambda \in \Lambda} V(I_\lambda)$.

The last two properties imply that the affine closed sets may be used to define the closed subsets in a topology on $\mathbf{A}^n$ (note that $\mathbf{A}^n = V(0), \emptyset = V(1)$). This topology is called the *Zariski topology* on $\mathbf{A}^n$, and affine closed sets are equipped with the induced topology. A basis for the Zariski topology is given by the open subsets of the shape $D(f) := \{P \in \mathbf{A}^n : f(P) \neq 0\}$, where $f \in k[x_1, \dots, x_n]$ is a fixed polynomial. Indeed, each closed subset $V(I)$ is the intersection of subsets of the form $V(f)$.

If $I = \sqrt{I}$, then by Fact 4.1.16 it is the intersection of the maximal ideals containing it. These are of the form $(x_1 - a_1, \dots, x_n - a_n)$ with some $a_i \in k$ according to Corollary 4.1.14, therefore I consists precisely of those $f \in k[x_1, \dots, x_n]$ that vanish at all $P \in V(I)$. Thus in this case the ideal I and the set $X = V(I)$ determine each other, and we call X an *affine variety.*

Definition 4.2.5 If $X = V(I)$ is an affine variety, we define the *coordinate ring* of X to be the quotient $\mathcal{O}(X) := k[x_1, \dots, x_n]/I$. Its elements are called *regular functions on* X; the images $\bar{x}_i$ of the x_i are called the *coordinate functions* on X.

We may evaluate a regular function $f \in \mathcal{O}(X)$ at a point $P = (a_1, \dots, a_n) \in X$ by putting $f(P) := \tilde{f}(a_1, \dots, a_n)$ with a preimage $\tilde{f}$ of f in $k[x_1, \dots, x_n]$; the value does not depend on the choice of $\tilde{f}$.

Note that by definition the finitely generated k-algebra $\mathcal{O}(X)$ is reduced, i.e. it has no nilpotent elements. It may have zero-divisors, however; it is an integral domain if and only if I is a prime ideal. In that case we say that X is an *integral* affine variety over k.

Example 4.2.6 If we look back at the examples in 4.2.3, we see that for $n = 1$ the affine closed set we obtain is a variety if and only if the a_i are distinct, and it is integral if and only if it is a single point, i.e. $f = x - a_i$.

The affine plane curve $X = V(f) \subset \mathbf{A}^2$ is an integral variety if and only if f is irreducible.

We may use the notion of regular functions to define morphisms of affine varieties, and hence obtain a category.

Definition 4.2.7 Given an affine variety $Y = V(J)$, by a *morphism* or *regular map* $Y \to \mathbf{A}^m$ we mean an m-tuple $\phi = (f_1, \dots, f_m) \in \mathcal{O}(Y)^m$. Given an affine variety $X \subset \mathbf{A}^m$, by a morphism $\phi : Y \to X$ we mean a morphism $\phi : Y \to \mathbf{A}^m$ such that $\phi(P) := (f_1(P), \dots, f_m(P)) \in \mathbf{A}^m$ lies in X for all points $P \in Y$.

Example 4.2.8 If $X = V(f)$ is a plane curve, a morphism $X \to \mathbf{A}^1$ is defined by a polynomial $f_1 \in k[x_1, x_2]$. The polynomial $f_1 + fg$ defines the same morphism $X \to \mathbf{A}^1$ for all $g \in k[x_1, x_2]$. If $Y = V(h)$ is another plane curve, a morphism $Y \to X$ is defined by a pair of polynomials (h_1, h_2) such that $f \circ (h_1, h_2)$ is a multiple of h. Again, the h_i are determined up to adding multiples of h.

If $\phi : Y \to X$ is a morphism of affine varieties, there is an induced k-algebra homomorphism $\phi^* : \mathcal{O}(X) \to \mathcal{O}(Y)$ given by $\phi^*(f) = f \circ \phi$. Note that ϕ^* takes functions vanishing at a point $P \in X$ to functions vanishing at the points $\phi^{-1}(P)$, so the preimage in $\mathcal{O}(X)$ of the ideal of $\mathcal{O}(Y)$ corresponding to a point $Q \in \phi^{-1}(P)$ is precisely the ideal defined by P.

Remark 4.2.9 A morphism $\phi : Y \to X$ is continuous with respect to the Zariski topology. Indeed, it is enough to see that the preimage of each basic open set $D(f) \subset X$ is open in Y, which holds because $\phi^{-1}(D(f)) = D(\phi^*(f))$.

Proposition 4.2.10 *The maps $X \to \mathcal{O}(X)$, $\phi \to \phi^*$ induce an anti-equivalence between the category of affine varieties over k and that of finitely generated reduced k-algebras.*

Proof For fully faithfulness, let $\bar{x}_1, \dots, \bar{x}_m$ be the coordinate functions on X. Then $\phi^* \mapsto (\phi^*(\bar{x}_1), \dots, \phi^*(\bar{x}_m))$ defines an inverse for the map $\phi \mapsto \phi^*$. For essential surjectivity, simply write a finitely generated reduced k-algebra as a quotient $A \cong k[x_1, \dots, x_n]/I$. Then $X = V(I)$ is a good choice. □

To proceed further, we define regular functions and morphisms on open subsets of an integral affine variety X. First, the *function field* $K(X)$ of X is the fraction field of the integral domain $\mathcal{O}(X)$. By definition, an element of $K(X)$ may be represented by a quotient of polynomials f/g with $g \notin I$, with two quotients f_1/g_1 and f_2/g_2 identified if $f_1 g_2 - f_2 g_1 \notin I$.

Next let P be a point of X. We define the *local ring* $\mathcal{O}_{X,P}$ at P as the subring of $K(X)$ consisting of functions that have a representative with $g(P) \neq 0$. It is the same as the localization of $\mathcal{O}(X)$ by the maximal ideal corresponding to P. One thinks of it as the ring of functions 'regular at P'. For an open subset $U \subset X$ we define the *ring of regular functions on* U by

$$\mathcal{O}_X(U) := \bigcap_{P \in U} \mathcal{O}_{X,P},$$

the intersection being taken inside $K(X)$. The following lemma shows that for $U = X$ this definition agrees with the previous one.

Lemma 4.2.11 *For an integral affine variety X one has $\mathcal{O}(X) = \bigcap_{P \in X} \mathcal{O}_{X,P}$.*

Proof To show the nontrivial inclusion, pick $f \in \bigcap_P \mathcal{O}_{X,P}$, and choose for each P a representation $f = f_P/g_P$ with $g_P \notin P$. By our assumption on the g_P none of the maximal ideals of $\mathcal{O}(X)$ contains the ideal $I := \langle g_P : P \in X \rangle$ of $\mathcal{O}(X)$, so Fact 4.1.16 implies $I = \mathcal{O}(X)$. In particular, there exist $P_1, \dots, P_r \in X$ with $1 = g_{P_1} h_{P_1} + \cdots g_{P_r} h_{P_r}$ with some $h_{P_i} \in \mathcal{O}(X)$. Thus

$$f = \sum_{i=1}^{r} f g_{P_i} h_{P_i} = \sum_{i=1}^{r} (f_{P_i}/g_{P_i}) g_{P_i} h_{P_i} = \sum_{i=1}^{r} f_{P_i} h_{P_i} \in \mathcal{O}(X),$$

as required. □

Now given integral affine varieties X and Y and open subsets $U \subset X$, $V \subset Y$, we define a morphism $\phi : V \to U$ similarly as above: we consider X together with its embedding in $\mathbf{A}^m$, and we define ϕ as an m-tuple $\phi = (f_1, \dots, f_m) \in \mathcal{O}_Y(V)^m$ such that $\phi(P) := (f_1(P), \dots, f_m(P))$ lies in U for all points $P \in Y$. We say that ϕ is an isomorphism if it has a two-sided inverse.

We now restrict the category under consideration. First a definition:

Definition 4.2.12 The *dimension* of an integral affine k-variety X is the transcendence degree of its function field $K(X)$ over k.

Remark 4.2.13 We can give a geometric meaning to this algebraic notion as follows. First note that since $\mathcal{O}(\mathbf{A}_k^n) = k[x_1, \dots, x_n]$, affine n-space has dimension n, as expected. Next, let X be an integral affine variety of dimension n. The Noether Normalization Lemma (Fact 4.1.11) together with Proposition 4.2.10 shows that there is a surjective morphism $\phi : X \to \mathbf{A}^n$ so that moreover $\mathcal{O}(X)$ is a finitely generated $k[x_1, \dots, x_n]$-module. The latter property implies that ϕ has finite fibres. Indeed, if $P = (a_1, \dots, a_n)$ is a point in $\mathbf{A}^n$ and $M_P = (x_1 - a_1, \dots, x_n - a_n)$ the corresponding maximal ideal in $k[x_1, \dots, x_n]$, then $\mathcal{O}(X)/M_P\mathcal{O}(X)$ is a finite dimensional k-algebra, and as such has finitely many maximal ideals. Their preimages in $\mathcal{O}(X)$ correspond to the finitely many points in $\phi^{-1}(P)$. Thus n-dimensional affine varieties are 'finite over $\mathbf{A}^n$'.

Integral affine varieties of dimension 1 are called integral affine *curves*. The following lemma shows that their Zariski topology is particularly simple.

Lemma 4.2.14 *All proper Zariski closed subsets of an integral affine curve are finite.*

Proof Quite generally, in a Noetherian ring every proper ideal I satisfying $I = \sqrt{I}$ is an intersection of finitely many prime ideals (a consequence of primary decomposition; see e.g. [2], Theorem 7.13). Therefore the lemma follows from Corollary 4.1.12. □

We now impose a further restriction, this time of local nature.

Definition 4.2.15 A point P of an integral affine variety X is *normal* if the local ring $\mathcal{O}_{X,P}$ is integrally closed. We say that X is normal if all of its points are normal.

Remark 4.2.16 In fact, X is normal if and only if $\mathcal{O}(X)$ is integrally closed. Indeed, if $\mathcal{O}(X)$ is integrally closed, then so is each localization $\mathcal{O}_{X,P}$; the converse follows from Lemma 4.2.11.

Normality is again an algebraic condition, but in dimension 1 the geometric meaning is easy to describe. In this case normality means by definition that the $\mathcal{O}_{X,P}$ are discrete valuation rings. We first look at the key example of plane curves.

Example 4.2.17 Let $X = V(f) \subset \mathbf{A}^2$ be an integral affine plane curve. Write x and y for the coordinate functions on $\mathbf{A}^2$ and assume that P is a point such that one of the partial derivatives $\partial_x f(P)$, $\partial_y f(P)$ is nonzero; such a point is called a *smooth* point. Then $\mathcal{O}_{X,P}$ is a discrete valuation ring, i.e. P is a normal point.

To see this, we may assume after a coordinate transformation that $P = (0, 0)$ and $\partial_y f(P) \neq 0$. The maximal ideal M_P of $\mathcal{O}_{X,P}$ is generated by x and y. Regrouping terms in the equation f we may write $f = \phi(x)x + \psi(x, y)y$, where $\phi \in k[x]$ and $\psi \in k[x, y]$. The constant term of ψ is $\partial_y(P)$, which is nonzero by assumption. Thus in $\mathcal{O}_{X,P}$ we may write $y = gx$, where g is the image of $-\phi\psi^{-1}$ in $\mathcal{O}_{X,P}$, and hence $M_P = (x)$. We conclude by Fact 4.1.8.

We now show that in characteristic 0 every normal affine curve is locally isomorphic to one as in the above example.

Proposition 4.2.18 *Assume k is of characteristic 0, and let X be an integral affine curve. Every normal point P of X has a Zariski open neighbourhood isomorphic to an open neighbourhood of a smooth point on an affine plane curve.*

Proof The local ring $\mathcal{O}_{X,P}$ is a discrete valuation ring, so its maximal ideal is principal, generated by an element t. Since we are in characteristic 0, by the theorem of the primitive element we find $s \in K(X)$ such that $K(X) = k(t, s)$. Replacing s by st^m for m sufficiently large if necessary, we may assume $s \in \mathcal{O}_{X,P}$. Taking the minimal polynomial of s over $k(t)$ and multiplying by a common denominator of the coefficients we find an irreducible polynomial $f \in k[x, y]$ such that $f(t, s) = 0$ and moreover the fraction field of the ring

$k[x, y]/(f)$ is isomorphic to $K(X)$. It follows that the map $(t, s) \mapsto (x, y)$ defines an isomorphism of $K(X)$ onto the function field of the plane curve $V(f) \subset \mathbf{A}^2_k$. If we choose $U \subset X$ so that $t, s \in \mathcal{O}_X(U)$, then the above map defines a morphism $\rho : U \to V(f)$. Conversely, the map $x \mapsto t$, $y \mapsto s$ defines a morphism $V(f) \to X$ that is an inverse to ρ on $\rho(U)$; in particular, $\rho(U)$ is open in $V(f)$. We conclude that X and $V(f)$ contain the isomorphic open subsets U and $\rho(U)$, with U containing P.

We finally show that $(\partial_y f)(\rho(P)) \neq 0$. The image of P by ρ is a point of the form $(0, \alpha)$; by composing ρ with the map $(x, y) \mapsto (x, y - \alpha)$ we may assume $\rho(P) = (0, 0)$. Since t generates the maximal ideal of $\mathcal{O}_{X,P} \cong \mathcal{O}_{V(f),(0,0)}$, we find $\bar{a}, \bar{b} \in \mathcal{O}(V(f))$ with $\bar{b}((0, 0)) \neq 0$ and $s = (\bar{a}/\bar{b})t$. Lifting them to polynomials $a, b \in k[x, y]$, we get the equality $by = ax + cf$ in $k[x, y]$. Taking partial derivative with respect to y gives $(\partial_y b)y + b = (\partial_y a)x + (\partial_y c)f + c\partial_y f$. Evaluating at $(0, 0)$ we obtain $b(0, 0) = c(0, 0) \cdot \partial_y f(0, 0)$. Here the left-hand side is nonzero since $\bar{b}((0, 0)) \neq 0$, hence so is $\partial_y f(0, 0)$. □

Remarks 4.2.19

1. The only place in the above proof where we used the characteristic 0 assumption is where we applied the theorem of the primitive element. But if t is a generator of the maximal ideal of a normal point as in the above proof, the extension $K(X)|k(t)$ is always separable (see e.g. [100], Proposition II.1.4), and hence the theorem applies. Thus the proposition extends to arbitrary characteristic.
2. Readers should be warned that in dimension greater than 1 the normality condition is weaker than smoothness (which is in general a condition on the rank of the Jacobian matrix of the equations of the variety; see Definition 5.1.30 and the subsequent discussion).

In the case $k = \mathbf{C}$ the above considerations enable us to equip a normal affine curve X with the structure of a Riemann surface.

Construction 4.2.20 Let X be an integral normal affine curve over $\mathbf{C}$ and P a point of X. Choose a generator t of the maximal ideal of $\mathcal{O}_{X,P}$. By the discussion above we find an open neighbourhood U of P and a function $u \in \mathcal{O}_X(U)$ such that the map $(t, u) \mapsto (x, y)$ yields an isomorphism ρ of U onto a Zariski open subset of some $V(f) \subset \mathbf{A}^2_{\mathbf{C}}$ satisfying $(\partial_y f)(\rho(P)) \neq 0$. Equip $V(f)$ with the restriction of the complex topology of $\mathbf{C}^2$. As in Example 3.1.3 (4) we find a complex open neighbourhood V of $\rho(P)$ (which we may choose so small that it is contained in $\rho(U)$) where x defines a complex chart on $V(f)$. Now define a 'complex' topology on $\rho^{-1}(V)$ by pulling back the complex topology of V and declare $x \circ \rho$ to be a complex chart in the neighbourhood $\rho^{-1}(V)$ of P.

We contend that performing this construction for all $P \in X$ yields a well-defined topology and a complex atlas on X. Indeed, if $P' \in \rho^{-1}(V)$ is a point for which the complex chart is constructed via a morphism $\rho' : (t', s') \mapsto (x, y)$, the map $\tau : (t, s) \mapsto (t', s')$ defines an algebraic isomorphism between some Zariski open neighbourhoods of P and P'. The composite $\rho' \circ \tau \circ \rho^{-1}$ induces a holomorphic isomorphism between suitable small complex neighbourhoods of $\rho(P)$ and $\rho'(P)$ (an algebraic function regular at a point is always holomorphic in some neighbourhood). It follows that the topologies and the complex charts around P and P' are compatible.

Remarks 4.2.21

1. In fact, one sees that the complex chart $x \circ \rho$ in the neighbourhood of P viewed as a $\mathbf{C}$-valued function is nothing but t. For this reason generators of the maximal ideal of $\mathcal{O}_{X,P}$ are called *local parameters at P*.
2. Given a morphism $\phi : Y \to X$ of normal affine curves over $\mathbf{C}$, an examination of the above construction shows that ϕ is holomorphic with respect to the complex structures on Y and X.

4.3 Affine curves over a general base field

We now extend the theory of the previous section to an arbitrary base field. The main difficulty over a non-closed field is that there is no reasonable way to identify a variety with a point set. For instance, though the polynomial $f = x^2 + y^2 + 1$ defines a curve $V(f)_{\mathbf{C}}$ in $\mathbf{A}^2_{\mathbf{C}}$, it has no points with coordinates in $\mathbf{R}$. Still, it would make no sense to define the real curve defined by f to be the empty set. Furthermore, the 'coordinate ring' $\mathbf{R}[x, y]/(x^2 + y^2 + 1)$ still makes sense. If we tensor it with $\mathbf{C}$, we obtain the ring $\mathcal{O}(V(f)_{\mathbf{C}})$ whose maximal ideals are in bijection with the points of $V(f)_{\mathbf{C}}$ as defined in the previous section. These points come in conjugate pairs: each pair corresponds to a maximal ideal in $\mathbf{R}[x, y]/(x^2 + y^2 + 1)$.

If we examine the situation of the last section further, we see from Proposition 4.2.10 that the coordinate ring $\mathcal{O}(X)$ completely determines an affine variety X over the algebraically closed field. In particular, we may recover the Zariski topology: the open sets correspond to sets of maximal ideals not containing some ideal $I \subset \mathcal{O}(X)$. When X is integral, the function field, the local rings and the ring of regular functions on an open subset $U \subset X$ are all constructed out of $\mathcal{O}(X)$. Moreover, we see that for each pair $V \subset U$ of open subsets there are natural restriction homomorphisms $\mathcal{O}_X(U) \to \mathcal{O}_X(V)$, and thus the rule $U \mapsto \mathcal{O}_X(U)$ defines a presheaf of rings on X. It is immediate to check that the sheaf axioms are satisfied, so we obtain a sheaf of rings $\mathcal{O}_X$ on X, the *sheaf of regular functions*. To proceed further it is convenient to formalize the situation.

Definition 4.3.1 A *ringed space* is a pair $(X, \mathcal{F})$ consisting of a topological space X and a sheaf of rings $\mathcal{F}$ on X.

We now give the general definition of integral affine curves. This will be a special case of the definition of affine schemes to be discussed in the next chapter, but there are some simplifying features.

Construction 4.3.2 We define an integral affine curve over an arbitrary field k as follows. Start with an integral domain $A \supset k$ finitely generated and of transcendence degree 1 over k. By Corollary 4.1.12 every nonzero prime ideal in A is maximal. We associate a topological space X with A whose underlying set is the set of prime ideals of A, and we equip it with the topology in which the open subsets are X and those that do not contain a given ideal $I \subset A$. Note that all nonempty open subsets contain the point (0); it is called the *generic point* of X. The other points come from maximal ideals and hence are closed as one-point subsets; we call them *closed points*. By the same argument as in Corollary 4.2.14 the open subsets in X are exactly the subsets whose complement is a finite (possibly empty) set of closed points.

Given a point P in X, we define the local ring $\mathcal{O}_{X,P}$ as the localization A_P; note that for $P = (0)$ we obtain $\mathcal{O}_{X,(0)} = K(X)$, the fraction field of A. Finally, we put

$$\mathcal{O}_X(U) := \bigcap_{P \in U} \mathcal{O}_{X,P}$$

for an open subset $U \subset X$. As above, it defines a sheaf of rings on X. We define an integral affine curve over k to be a ringed space $(X, \mathcal{O}_X)$ constructed in the above way. We usually drop the sheaf $\mathcal{O}_X$ from the notation. When we would like to emphasize the relationship with A, we shall use the scheme-theoretic notation $X = \operatorname{Spec}(A)$.

Next we introduce morphisms for the curves just defined. They are to be morphisms of ringed spaces, whose general definition is as follows.

Definition 4.3.3 A morphism $(Y, \mathcal{G}) \to (X, \mathcal{F})$ of ringed spaces is a pair $(\phi, \phi^\sharp)$, where $\phi : Y \to X$ is a continuous map, and $\phi^\sharp : \mathcal{F} \to \phi_*\mathcal{G}$ a morphism of sheaves on X. Here $\phi_*\mathcal{G}$ denotes the sheaf on X defined by $\phi_*\mathcal{G}(U) = \mathcal{G}(\phi^{-1}(U))$ for all $U \subset X$; it is called the *pushforward* of $\mathcal{G}$ by ϕ.

In more down-to-earth terms, a morphism $Y \to X$ of integral affine curves is a continuous map $\phi : Y \to X$ of underlying spaces and a rule that to each regular function $f \in \mathcal{O}_X(U)$ defined over an open subset $U \subset X$ associates a function $\phi_U^\sharp(f)$ in $\mathcal{O}_Y(\phi^{-1}(U))$. One should think of $\phi_U^\sharp(f)$ as the composite $f \circ \phi$.

Remark 4.3.4 In the case when k is algebraically closed, this definition is in accordance with that of the previous section. Indeed, the morphisms defined there are continuous maps (Remark 4.2.9) and induce maps $\phi^\sharp$ of sheaves via the rule $f \mapsto f \circ \phi$. Conversely, to see that a morphism $\phi : X \to \mathbf{A}_k^n$ in the new sense induces a morphism as in the previous section it is enough to consider the n-tuple $(\phi^\sharp(x_1), \ldots, \phi^\sharp(x_n))$.

We now establish an analogue of Proposition 4.2.10 for affine curves. To begin with, a finitely generated integral domain A of transcendence degree 1 over a field determines an integral affine curve $X = \mathrm{Spec}\,(A)$; conversely, an integral affine curve X gives rise to an A as above by setting $A = \mathcal{O}_X(X)$. By construction, these two maps are inverse to each other. We shall also use the notation $\mathcal{O}(X)$ instead of $\mathcal{O}_X(X)$. This is in accordance with the notation of the previous chapter, and we may also call $\mathcal{O}(X)$ the coordinate ring if X.

For affine curves $X = \mathrm{Spec}\,(A)$ and $Y = \mathrm{Spec}\,(B)$ a morphism $\phi : X \to Y$ induces a ring homomorphism $\phi_X^\sharp : A \to B$ given by $\mathcal{O}(X) \to (\phi_* \mathcal{O}_Y)(X) = \mathcal{O}(Y)$. Associating a morphism of curves with a ring homomorphism is a bit more complicated.

Lemma 4.3.5 *Given a homomorphism $\rho : A \to B$ with A and B as above, there is a unique morphism* $\mathrm{Spec}\,(\rho) : Y \to X$ *such that the homomorphism* $(\mathrm{Spec}\,(\rho))_X^\sharp : \mathcal{O}(X) \to \mathcal{O}(Y)$ *equals ρ.*

Proof For a prime ideal $P \subset B$ the ideal $\rho^{-1}(P) \subset A$ is a prime ideal (indeed, the map $A/(\rho^{-1}(P)) \to B/P$ is injective, hence $A/(\rho^{-1}(P))$ is an integral domain). This defines a map of sets $\mathrm{Spec}\,(\rho) : Y \to X$ that is easily seen to be a continuous map of topological spaces. But in our situation we can say more. There are two cases.

Case 1: ρ is injective. In this case A is a subring of B via ρ, and moreover by Corollary 4.1.15 if P is a maximal ideal in B, then $\rho^{-1}(P) = P \cap A$ is a maximal ideal in A. Of course, we have $\rho^{-1}((0)) = (0)$.

Case 2: ρ is not injective. As we are dealing with curves, the ideal $M := \ker(\rho)$ is maximal in A, and so $\rho^{-1}(P) = M$ for all prime ideals $P \subset B$. This corresponds to a 'constant morphism' $Y \to \{M\}$.

We define morphisms of sheaves $\mathrm{Spec}\,(\rho)^\sharp : \mathcal{O}_X \to \mathrm{Spec}\,(\rho)_* \mathcal{O}_Y$ in each case. In Case 1 we have an inclusion of function fields $K(X) \subset K(Y)$ and also of localizations $A_{(P \cap A)} \subset B_P$ for each maximal ideal $P \subset B$. By taking intersections this defines maps $\mathcal{O}_X(U) \to \mathcal{O}_Y(\mathrm{Spec}\,(\rho)^{-1}(U))$ for each open set $U \subset X$; for $U = X$ we get $\rho : A \to B$ by the same argument as in Lemma 4.2.11. In Case 2 we define $\mathcal{O}_X(U) \to \mathcal{O}_Y(\mathrm{Spec}\,(\rho)^{-1}(U))$ to be the composite

$$\mathcal{O}_X(U) \to A_M \to A_M/MA_M \xrightarrow{\sim} A/M \to B$$

if $M \in U$, and to be 0 otherwise. The reader will check that this indeed yields a morphism of sheaves. □

The lemma and the arguments preceding it now imply:

Proposition 4.3.6 *The assignments* $A \mapsto \mathrm{Spec}\,(A)$, $\rho \mapsto \mathrm{Spec}\,(\rho)$ *and* $X \mapsto \mathcal{O}(X)$, $\phi \mapsto \phi_X^\sharp$ *yield mutually inverse contravariant functors between the category of integral domains finitely generated and of transcendence degree 1 over a field, and that of integral affine curves.*

Note that the conclusion here is stronger than in Proposition 4.2.10, because here we say that the two categories are actually *anti-isomorphic*: there is an arrow-reversing bijection between objects and morphisms. In Proposition 4.2.10 this was only true up to isomorphism, because an affine variety as defined there may have several embeddings in affine spaces.

We now discuss an important construction related to extensions of the base field.

Construction 4.3.7 Let $X = \mathrm{Spec}\,(A)$ be an integral affine curve over a field k, and $L|k$ a field extension for which the tensor product $A \otimes_k L$ is an integral domain. Then the integral affine curve $X_L = \mathrm{Spec}\,(A \otimes_k L)$ is defined. We call the resulting curve over L the *base change* of X to L. There is a natural morphism $X_L \to X$ corresponding by the previous proposition to the map $A \to A \otimes_k L$ sending a to $a \otimes 1$.

Assume that $A \otimes_k \bar{k}$ is an integral domain for an algebraic closure $\bar{k}|k$; in this case X is called *geometrically integral*. Then $A \otimes_k L$ is an integral domain for all algebraic extensions $L|k$, so that the above assumption on L is satisfied. Thus for a fixed algebraic extension $L|k$ the rule $X \mapsto X_L$ defines a functor from the category of geometrically integral affine k-curves to that of integral affine L-curves.

Example 4.3.8 We can now discuss the $\mathbf{R}$-curve with equation $x^2 + y^2 + 1 = 0$ rigorously. It is defined as $X := \mathrm{Spec}\,(\mathbf{R}[x, y]/(x^2 + y^2 + 1))$. The closed points of X correspond to the maximal ideals in $\mathbf{R}[x, y]$ containing $(x^2 + y^2 + 1)$; for each such ideal M we must have $\mathbf{R}[x, y]/M \cong \mathbf{C}$, as $\mathbf{C}$ is the only nontrivial finite extension of $\mathbf{R}$ and X has no points over $\mathbf{R}$. Under the base change morphism $X_\mathbf{C} \to X$ there are two closed points lying above each closed point of X, because $\mathbf{C} \otimes_\mathbf{R} \mathbf{C} \cong \mathbf{C} \oplus \mathbf{C}$. If we make $\mathrm{Gal}\,(\mathbf{C}|\mathbf{R})$ act on the tensor product via its action on the second term, then on the right-hand side the resulting action interchanges the components.

To give a concrete example, the ideal $M = (x^2, y^2 + 1) \subset \mathbf{R}[x, y]$ contains $(x^2 + y^2 + 1)$, hence defines a point of X. The maximal ideals of $\mathbf{C}[x, y]$ lying above M are $(x, y + i)$ and $(x, y - i)$, corresponding to the points $(0, -i)$ and $(0, i)$. They are indeed conjugate under the Galois action.

We say that an integral affine curve is *normal* if its local rings are integrally closed. As in the previous section, this is equivalent to requiring that the coordinate ring $\mathcal{O}(X)$ is integrally closed.

We now prove an analogue of Theorem 3.3.7 for integral affine curves. For this we have to restrict the morphisms under consideration. We say that a morphism $\phi : Y \to X$ of integral affine curves is *finite* if $\mathcal{O}(Y)$ becomes a finitely generated $\mathcal{O}(X)$-module via the map $\phi_X^\sharp : \mathcal{O}(X) \to \mathcal{O}(Y)$. A finite morphism has finite fibres, by the same argument as in Remark 4.2.13. This property is shared by proper holomorphic maps of Riemann surfaces.

Remark 4.3.9 A finite morphism of integral affine curves is always surjective. Indeed, Case 2 of the proof of Lemma 4.3.5 cannot occur for a finite morphism, and in Case 1 we may apply Fact 4.1.1 (4). An example of a non-finite morphism is given by the inclusion $\mathbf{A}_k^1 \setminus \{0\} \to \mathbf{A}_k^1$, corresponding to the natural ring homomorphism $k[x] \to k[x, x^{-1}]$ (over any field k).

Now assume given a finite morphism $Y \to X$ integral affine curves. We have just remarked that the corresponding homomorphism $\mathcal{O}(Y) \to \mathcal{O}(X)$ of coordinate rings is injective, whence an inclusion of function fields $\phi^* : K(X) \subset K(Y)$. As the morphism is finite, this must be a finite extension.

Theorem 4.3.10 *Let X be an integral normal affine curve. The rule $Y \mapsto K(Y)$, $\phi \mapsto \phi^*$ induces an anti-equivalence between the category of normal integral affine curves equipped with a finite morphism $\phi : Y \to X$ and that of finite field extensions of the function field $K(X)$.*

Proof For essential surjectivity take a finite extension $L|K(X)$ and apply Fact 4.1.4 (*b*) with $A = \mathcal{O}(X)$. It implies that the integral closure B of $\mathcal{O}(X)$ in L is a finitely generated k-algebra, which is also integrally closed by Fact 4.1.1 (2). As L is finite over $K(X)$, it is still of transcendence degree 1. Applying Proposition 4.3.6 to the ring extension $\mathcal{O}(X) \subset B$ we obtain an integral affine curve $Y = \mathrm{Spec}\,(B)$ and a morphism $\phi : Y \to X$ inducing the ring inclusion $\mathcal{O}(X) \subset B$ above. Again by Fact 4.1.4 (*b*) the morphism ϕ is finite. Fully faithfulness is proven by a similar argument as in Theorem 3.3.7. □

The affine curve Y constructed in the first part of the proof is called the *normalization* of X in L.

Examples 4.3.11

1. The theorem is already interesting over an algebraically closed field k. For instance if we take $X = \mathbf{A}^1_k$ and $L = k(x)[y]/(y^2 - f)$, where $f \in k[x]$ is of degree at least 3 having no multiple roots, then the normalization of $\mathbf{A}^1_k$ in L is the normal affine plane curve $V(y^2 - f) \subset \mathbf{A}^2_k$.
2. Over a non-closed field we get other kinds of examples as well. If we assume that X is geometrically integral, then for every finite extension $L|k$ we may look at the normalization of X in $L \otimes_k K(X)$. It will be none other than the base change X_L, because tensorizing with L does not affect integral closedness.

Remark 4.3.12 The concept of normalization is also interesting for a non-normal integral affine curve X. Taking the integral closure B of $\mathcal{O}(X)$ in $K(X)$ yields via Proposition 4.3.6 a normal integral affine curve $\widetilde{X}$ with function field $K(X)$ that comes equipped with a finite surjective morphism $\widetilde{X} \to X$. This implies a characterization of normality: an integral affine curve X is normal if and only if every finite morphism $\phi : Y \to X$ inducing an isomorphism $\phi^* : K(X) \xrightarrow{\sim} K(Y)$ is an isomorphism. As in the proof of Proposition 4.2.18 one sees that the condition $\phi^* : K(X) \xrightarrow{\sim} K(Y)$ can be rephrased by saying that ϕ is an isomorphism over an open subset. So the criterion becomes: X is normal if and only if every finite surjective morphism $Y \to X$ inducing an isomorphism over an open subset is in fact an isomorphism.

4.4 Proper normal curves

When one compares the theory developed so far with the theory of finite covers of Riemann surfaces, it is manifest that our presentation is incomplete at one point: the preceding discussion does not include the case of *compact* Riemann surfaces, only those with some points deleted. For instance, we have an algebraic definition of the affine line, but not that of the projective line. We now fill in this gap by considering proper normal curves.

We shall give the scheme-theoretic definition, which is in fact quite close to what Zariski and his followers called an 'abstract Riemann surface'. Its starting point is the study of the local rings $\mathcal{O}_{X,P}$ of an integral normal affine curve X over a field k. They are all discrete valuation rings having the same fraction field, namely the function field $K(X)$ of X, and they all contain the ground field k.

Lemma 4.4.1 *The local rings of an integral normal affine curve X are exactly the discrete valuation rings R with fraction field $K(X)$ containing $\mathcal{O}(X)$.*

Proof If R is such a ring, the intersection of its maximal ideal M with $\mathcal{O}(X)$ is nonzero, for otherwise the restriction of the projection $R \to R/M$ to $\mathcal{O}(X)$ would be injective, and the field R/M would contain $K(X)$, which is absurd. Thus $M \cap \mathcal{O}(X)$ is a maximal ideal in $\mathcal{O}(X)$, and R contains the local ring $\mathcal{O}_{X,P}$. But then by Proposition 4.1.9 (3) we have $R = \mathcal{O}_{X,P}$. □

We now consider the simplest example.

Example 4.4.2 The rational function field $k(x)$ is the function field of the affine line $\mathbf{A}^1_k$ over k; we have $\mathcal{O}(\mathbf{A}^1_k) = k[x]$. But $k(x)$ is also the fraction field of the ring $k[x^{-1}]$, which we may view as the coordinate ring of another copy of $\mathbf{A}^1_k$ with coordinate function x^{-1}. By Proposition 4.1.9 (3) every discrete valuation ring $R \supset k$ with fraction field $k(x)$ contains either x or x^{-1}, and hence by the preceding discussion R is a local ring of one of the two copies of $\mathbf{A}^1_k$. In fact, there is only one localization of $k[x^{-1}]$ that does not contain x: the localization at the ideal (x^{-1}). Thus there is only one discrete valuation ring R as above that is not a local ring on the first copy of $\mathbf{A}^1_k$; it corresponds to the 'point at infinity'. The whole discussion is parallel to the construction of the complex structure on the Riemann surface $\mathbf{P}^1(\mathbf{C})$ in Example 3.1.3 (2): there we took a copy of $\mathbf{C}$ around 0, another copy around ∞, and outside these two points we identified the two charts via the isomorphism $z \mapsto z^{-1}$. Thus we may regard the discrete valuation rings $R \supset k$ with fraction field $k(x)$ as the *local rings of the projective line over k*.

We can generalize the example as follows. Given a normal integral affine curve X, we may use the Noether Normalization Lemma (Fact 4.1.11) to find a regular function $f \in \mathcal{O}(X)$ such that $\mathcal{O}(X)$ is a finitely generated module over $k[f]$. Let X^- be the normal affine curve corresponding to the integral closure of $k[f^{-1}]$ in $K(X)$. By Lemma 4.4.1 and the above example every discrete valuation ring $R \supset k$ with fraction field $K(X)$ is a local ring of either X or X^-. Moreover, there are only finitely many R that are not local rings of X, namely the localizations of $\mathcal{O}(X^-)$ at the finitely many maximal ideals lying above $(f^{-1}) \subset k[f^{-1}]$. Informally speaking, we may view the set of the above R as the local rings on a 'curve' obtained by 'gluing X and X^- together'.

We now give a formal definition that is independent of the choice of the function f above.

Construction 4.4.3 Let k be a field, and $K|k$ a finitely generated field extension of transcendence degree 1. Let X^K be the set of discrete valuation rings with fraction field K containing k. Endow X^K with the topology in which the proper closed subsets are the finite subsets. Define a sheaf of rings on X^K by the formula $\mathcal{O}^K(U) = \bigcap_{R \in U} R$ for an open subset $U \subset X^K$. We call the ringed space $(X^K, \mathcal{O}^K)$ an *integral proper normal curve* over k with function field K.

A morphism $Y^L \to X^K$ of proper normal curves is again defined as a morphism of ringed spaces. The preceding discussion shows that every integral proper normal curve has an open covering (as a ringed space) by two integral affine normal curves. The learned reader will recognize that this is the extra ingredient needed to define a scheme.

Remark 4.4.4 It can be shown that a proper normal curve comes from a *projective curve* in the same way as its affine open subsets come from affine curves. We explain the necessary notions very briefly over an algebraically closed field k. One identifies points of *projective n-space* $\mathbf{P}^n_k$ over k with $(n+1)$-tuples $(a_0, \dots, a_n) \in k^{n+1} \setminus \{(0, \dots, 0)\}$, modulo the equivalence relation identifying two $(n+1)$-tuples $(a_0, \dots, a_n)$ and $(b_0, \dots, b_n)$ if there exists $\lambda \in k^\times$ with $a_i = \lambda b_i$ for all i. A *projective variety* X over k is then a subset of some $\mathbf{P}^n(k)$ given by the locus of common zeros of a finite system of *homogeneous* polynomials. If these polynomials generate a prime ideal $I(X)$ in $k[x_0, \dots, x_n]$ we say that the variety is *integral*. The subring of the fraction field of $k[x_0, \dots, x_n]/I(X)$ that can be represented by quotients of homogeneous polynomials of the *same degree* is the function field $K(X)$ of X. The local ring $\mathcal{O}_{X,P}$ of a point $P \in X$ is the subring of $K(X)$ consisting of elements that can be represented by a function with nonvanishing denominator at P; the sheaf $\mathcal{O}_X$ is defined as in the affine case. The integral projective variety X is a curve if $K(X)$ is of transcendence degree 1; it is a normal curve if moreover all the $\mathcal{O}_{X,P}$ are discrete valuation rings.

There are two basic facts about normal integral projective curves. Firstly, if X is such a curve, then every discrete valuation ring $R \supset k$ with fraction field $K(X)$ is a local ring of some point $P \in X$, and consequently the pair $(X, \mathcal{O}_X)$ is isomorphic to the proper normal curve $(X^{K(X)}, \mathcal{O}^{K(X)})$ as a ringed space. Secondly, every integral proper normal curve $(X^K, \mathcal{O}^K)$ arises from a normal projective curve in this way; it is then necessarily unique up to isomorphism. These statements are proven e.g. in [34], Chapter I, §6.

Given a surjective morphism $Y^L \to X^K$ of integral proper normal curves, the field L is a finite extension of K, since both are finitely generated of transcendence degree 1 over K. Fixing K, we obtain in this way a contravariant functor.

Proposition 4.4.5 *The above functor induces an anti-equivalence between the category of integral proper normal curves equipped with a morphism onto X^K and that of finite field extensions of K.*

Proof Given a finite extension $L|K$, we may use the discussion after Example 4.4.2 to cover X^K (resp. X^L) by two open subsets X_+^K and X_-^K (resp. X_+^L and X_-^L) arising as normalizations of two overlapping copies of $\mathbf{A}_k^1$ in K (resp. L). The morphisms of affine curves $X_+^L \to X_+^K$ and $X_-^L \to X_-^K$ arising from this construction glue together to a morphism $\phi : X^L \to X^K$. To see that it does not depend on the choice of the open coverings, it suffices to remark that a point $S \in X^L$ viewed as a discrete valuation ring gets mapped to $S \cap K$ by the above construction, which determines ϕ uniquely. □

We call an open subset U^K of an integral proper normal curve X^K *affine* if $\mathcal{O}^K(U^K)$ is a finitely generated k-algebra. The ringed space $(U_K, \mathcal{O}^K|_{U_K})$ is by construction the same as the integral affine curve corresponding to $\mathcal{O}^K(U^K)$ via Proposition 4.3.6. Conversely, we have seen above that the set of local rings of an integral normal affine curve with function field K is a nonempty open subset of X^K. From these facts and the proposition above we deduce:

Proposition 4.4.6 *The category of integral affine normal curves is equivalent to that of affine open subsets of integral proper normal curves.*

In particular, every integral affine normal curve X can be embedded as an affine open subset in an integral proper normal curve X^K, and every morphism $Y \to X$ of integral affine normal curves extends uniquely to a morphism $Y^L \to X^K$ of proper normal curves.

It is a nonobvious fact that every open subset of X^K other than X^K is affine, but we shall not need this.

From now on we drop the annoying superscript K from the notation when discussing proper normal curves.

A morphism $\phi : Y \to X$ of proper normal curves is *finite* if for all affine open subsets $U \subset X$ the preimage $\phi^{-1}(U) \subset Y$ is affine, and moreover $\phi_*\mathcal{O}(U)$ is a finitely generated $\mathcal{O}(U)$-module. The restriction of ϕ to each $\phi^{-1}(U)$ may be identified with the finite morphism of affine curves corresponding to the k-algebra homomorphism $\mathcal{O}(U) \to \phi_*\mathcal{O}(U)$. It thus follows from Remark 4.3.9 that a finite morphism is always surjective. Conversely, we have:

Proposition 4.4.7 *A surjective morphism $\phi : Y \to X$ of proper normal curves is always finite.*

Proof Let $U \subset X$ be an affine open subset. The points of $\phi^{-1}(U)$ are the discrete valuation rings R with fraction field L containing $\mathcal{O}(U)$. Since each R is integrally closed, it contains the integral closure B of $\mathcal{O}(U)$ which is none but the coordinate ring of the normalization V of U in L by the proof of Theorem 4.3.10. Lemma 4.4.1 then allows us to identify V with $\phi^{-1}(U)$, so the latter is indeed an affine open subset finite over U. □

According to Propositions 4.4.5 and 4.4.7, given an integral proper normal curve X with function field K and an element $f \in K$ transcendental over k, the inclusion $k(f) \subset K$ corresponds to a finite surjective morphism $\phi_f : X \to \mathbf{P}^1_k$, where $\mathbf{P}^1_k$ is considered as a proper normal curve with function field $k(f)$. This is to be compared with Proposition 3.3.11: in fact, when k is algebraically closed and X is realized as a projective curve, one may check that $\phi_f(P) = f(P)$ for all points $P \in X$ where $f \in \mathcal{O}_{X,P}$ and $\phi_f(P) = \infty$ otherwise (see Exercise 6). As in Corollary 3.3.12, one then obtains:

Corollary 4.4.8 *Mapping an integral proper normal curve over k to its function field induces an anti-equivalence between the category of integral proper normal curves with finite surjective morphisms and that of finitely generated field extensions of k having transcendence degree 1.*

4.5 Finite branched covers of normal curves

We can now finally discuss our central topic in this chapter, the analogue of topological covers for normal algebraic curves. We first treat the case of integral affine curves. Let us begin with some terminology: a finite morphism of integral affine curves is called *separable* if the field extension $K(Y)|K(X)$ induced by ϕ is separable.

Definition 4.5.1 Let $\phi : Y \to X$ be a finite separable morphism of integral affine curves, corresponding to an inclusion of rings $A \subset B$ via Proposition 4.3.6. We say that ϕ is *étale* over a closed point $P \in X$ if B/PB is a finite étale algebra over the field $\kappa(P) = A/P$. It is étale over an open subset $U \subset X$ if it is étale over all $P \in U$.

Using some commutative algebra, we can give an equivalent definition under the additional assumption that X and Y are *normal*. Recall that in this case the rings A and B above are Dedekind rings, so we have a decomposition $PB = P_1^{e_1} \cdots P_r^{e_r}$ by Fact 4.1.5 (1). The maximal ideals $P_i \subset B$ correspond to the direct summands of the $\kappa(P)$-algebra B/PB, and geometrically to the points of the fibre $\phi^{-1}(P)$. Étaleness above P means that $e_i = 1$ for all i, and that the field extensions $\kappa(P_i)|\kappa(P)$ are separable, where $\kappa(P_i) := B/P_i$. In other words:

Lemma 4.5.2 *A morphism $\phi : Y \to X$ of integral normal affine curves is étale above P if and only if a generator of the maximal ideal of $\mathcal{O}_{X,P}$ generates the maximal ideal of $\mathcal{O}_{Y,P_i}$ for each i, and the field extensions $\kappa(P_i)|\kappa(P)$ are separable.*

Proof This follows from the above discussion, together with the facts that ϕ induces inclusions of discrete valuation rings $\mathcal{O}_{X,P} \subset \mathcal{O}_{Y,P_i}$ for each i, and $\mathcal{O}_{Y,P_i}$ is the localization of B by P_i. □

Remark 4.5.3 The $\kappa(P)$-algebra B/PB should be interpreted as the ring of regular functions on the fibre of ϕ over the point P. The fact that some e_i above is greater than 1 means that there are nilpotent functions on the fibre.

In sheaf-theoretic language, one can check that $B/PB \cong (\phi_*\mathcal{O}_Y)_P \otimes_{\mathcal{O}_{X,P}} \kappa(P)$, where $(\phi_*\mathcal{O}_Y)_P$ is the stalk of the direct image sheaf $\phi_*\mathcal{O}_Y$ at P. In this interpretation, the separability of ϕ means that $(\phi_*\mathcal{O}_Y)_{(0)}$ is a separable field extension of $\mathcal{O}_{X,(0)} = K(X)$, and therefore an étale $K(X)$-algebra.

The above abstract notions are enlightened by the following key example.

Example 4.5.4 Consider the map $\rho_n : \mathbf{A}^1_{\mathbf{C}} \to \mathbf{A}^1_{\mathbf{C}}$ given by $x \to x^n$ for some $n > 0$. The coordinate ring of $\mathbf{A}^1$ over $\mathbf{C}$ is $\mathbf{C}[x]$, and the morphism corresponding to ρ_n by Proposition 4.3.6 is given by the inclusion $\mathbf{C}[x^n] \to \mathbf{C}[x]$. A closed point $a \in \mathbf{A}^1(\mathbf{C})$ corresponds to the maximal ideal $M_a = (x - a)$. To check whether ρ_n is étale over the point a, we take a primitive n-th root of unity ω and an n-th root $\sqrt[n]{a}$ of a, and compute

$$\mathbf{C}[x]/(x^n - a)\mathbf{C}[x] \cong \begin{cases} \prod_{i=0}^{n-1} \mathbf{C}[x]/(x - \omega^i \sqrt[n]{a}) \cong \mathbf{C}^n & a \neq 0 \\ \mathbf{C}[x]/(x^n) & a = 0. \end{cases}$$

In the first case we indeed obtain a finite étale algebra of dimension n over $\mathbf{C}$. On the other hand, for $a = 0$ we obtain a $\mathbf{C}$-algebra containing nilpotents, which therefore cannot be étale. The nilpotent functions on the fibre over 0 reflect the property that the fibre is degenerate.

Remark 4.5.5 The above example is the algebraic analogue of the local branching behaviour of the morphism $z \to z^n$ of Riemann surfaces. In fact, the same argument shows more generally that if k is algebraically closed and $\rho_f : \mathbf{A}^1_k \to \mathbf{A}^1_k$ is the morphism coming from the k-homomorphism $k[x] \to k[x]$ mapping x to $f \in k[x]$, then ρ_f is étale above a point P if and only if every preimage Q of P satisfies $f'(Q) \neq 0$.

We can generalize this comparison with the theory over $\mathbf{C}$ as follows. Assume given a finite morphism $\phi : Y \to X$ of integral normal affine curves over $\mathbf{C}$ coming from a ring homomorphism $A \to B$, and equip X and Y with the complex structures defined in Construction 4.2.20. Let $P \in X$ be a closed point, and consider the decomposition $PB = P_1^{e_1} \cdots P_r^{e_r}$ discussed above.

Proposition 4.5.6 *With notations as above, the integer e_i is the same as the ramification index at P_i of ϕ considered as a holomorphic map. In particular, ϕ as an algebraic map is étale above P if and only if as a holomorphic map it restricts to a cover over a complex neighbourhood of P.*

Proof It suffices to prove the first statement. If t is a local parameter at P, then in $\mathcal{O}_{X,P_i}$ we have $t = g_i t_i^{e_i}$ with some local parameter t_i and element g_i with $g_i(P_i) \neq 0$. So in the complex charts on Y and X defined by t and t_i, respectively, the map ϕ looks like $t_i \mapsto g_i t_i^{e_i}$. As in the proof of Proposition 3.2.1 we may replace t_i by a complex chart such that the local form of ϕ becomes $z_i \mapsto z_i^{e_i}$, so e_i is indeed the ramification index at P_i. □

Remark 4.5.7 Comparing the above proposition with Theorem 3.2.7 we see that a finite morphism of normal affine curves must be *proper* as a holomorphic map.

Example 4.5.8 A key example occurring over a non-closed field is the following. Assume that X is a geometrically integral affine curve over a field k (see Construction 4.3.7). Given a finite separable extension $L|k$, the base change morphism $X_L \to X$ is finite and étale over the whole of X. Indeed, it is finite as $L|k$ is finite, and moreover it is étale over each $P \in X$ because $\mathcal{O}(X_L)/P\mathcal{O}(X_L) \cong \kappa(P) \otimes_k L$, which is indeed a product of finite separable field extensions of $\kappa(P)$.

The next proposition is an algebraic reformulation of the property that a proper holomorphic map of Riemann surfaces restricts to a cover outside a discrete closed set of points.

Proposition 4.5.9 *Let $\phi : Y \to X$ be a finite separable morphism of integral affine curves. There is a nonempty open subset $U \subset X$ such that ϕ is étale over all $P \in U$.*

The proof uses a lemma that will also serve later.

Lemma 4.5.10 *Let $\phi : Y \to X$ and $\psi : Z \to Y$ be finite separable morphisms of integral affine curves, and let P be a point of X. If ϕ is étale over P and ψ is étale over all points of Y lying above P, then $\psi \circ \phi$ is étale over P. If moreover X, Y and Z are normal, then the converse also holds.*

Proof Write $A = \mathcal{O}(X)$, $B = \mathcal{O}(Y)$ and $C = \mathcal{O}(Z)$. Then ϕ (resp. ψ) corresponds to the inclusion of rings $A \subset B$ (resp. $B \subset C$) via Proposition 4.3.6. We have

$$C/PC \cong C \otimes_A \kappa(P) \cong C \otimes_B (B \otimes_A \kappa(P)). \tag{4.2}$$

By assumption (and the Chinese Remainder Theorem) here $B \otimes_A \kappa(P)$ is isomorphic to the direct product of the residue fields $\kappa(Q)$, where Q runs over the points of Y lying above P, and these fields are separable over $\kappa(P)$. Thus $C/PC \cong \prod C \otimes_B \kappa(Q)$, and again by assumption each of the components is a direct product of finite separable extensions of $\kappa(Q)$. The first statement follows.

For the converse, note that by the normality assumption Proposition 4.1.6 applies and shows that ϕ is étale over P if and only if the $\kappa(Q)$ are separable extensions of $\kappa(P)$ for all Q lying above P, and moreover the sum of the degrees $[\kappa(Q) : \kappa(P)]$ equals $[K(Y) : K(X)]$. It then follows from formula (4.2) and a simple degree count that if one of these properties fails for ϕ, it also fails for $\psi \circ \phi$. Since $\psi \circ \phi$ was assumed to be étale over P, this cannot happen, so ϕ is étale over P. Once we know this, a similar reasoning shows that ψ must be étale above each Q as well. □

In the proof of the proposition we shall only use the first statement of the lemma, which holds without the assumption of normality.

Proof of Proposition 4.5.9 We keep the notation $A = \mathcal{O}(X)$, $B = \mathcal{O}(Y)$ from the previous proof. As ϕ is finite, viewing A as a subring of B we find finitely many elements $f_1, \dots, f_r \in B$ integral over A such that $B = A[f_1, \dots, f_r]$. Consider the chain

$$A \subset A[f_1] \subset A[f_1, f_2] \subset \cdots \subset A[f_1, \dots, f_{r-1}] \subset B$$

and the chain of morphisms of curves corresponding to it via Proposition 4.3.6. By induction on r using Lemmas 4.2.14 and 4.5.10 we reduce to the case $r = 1$, i.e. $B = A[t]/(F)$ with a monic polynomial $F \in A[t]$ satisfying $F(f_1) = 0$. Here F is also the minimal polynomial of f_1 over $K(X)$, so its derivative F' must be prime to F in the ring $K(X)[t]$ since f_1 is contained in the separable extension $K(Y)$ of $K(X)$. We then find polynomials $G_1, G_2 \in K(X)[t]$ satisfying $G_1F + G_2F' = 1$. Multiplying with a common denominator $g \in A$ of the coefficients of G_1 and G_2 we obtain polynomials $H_1 = gG_1$, $H_2 = gG_2 \in A[t]$ with $H_1F + H_2F' = g$. We claim that $U = D(g) \subset X$ is a good choice. Indeed, assume P is a maximal ideal in A with $g \notin P$. The image $\bar{F}$ of F in $\kappa(P)[t]$ cannot have multiple roots in an algebraic closure of $\kappa(P)$, because by reducing $H_1F + H_2F' = g$ mod P we obtain $\bar{H}_1\bar{F} + \bar{H}_2\bar{F}' \neq 0$, so $\bar{F}(\alpha) = 0$ implies $\bar{F}'(\alpha) \neq 0$. By the Chinese Remainder Theorem the $\kappa(P)$-algebra $B/PB \cong \kappa(P)[t]/(F)$ is isomorphic to a finite direct product of field extensions $\kappa(P)[t]/(\bar{F}_i)$, where $\bar{F} = \bar{F}_1 \cdots \bar{F}_s$ is the decomposition of $\bar{F}$ in irreducible components. By the above none of the $\bar{F}_i$ has multiple roots, so B/PB is indeed an étale $\kappa(P)$-algebra. □

We may call a morphism $Y \to X$ as in the proposition a *finite branched cover*. It is a *Galois branched cover* if the field extension $K(Y)|K(X)$ induced by ϕ^* is Galois. According to Fact 4.1.3 (1), if moreover the curves are normal, then the Galois group acts transitively on the fibres of ϕ. Combining Lemma 4.5.2 with Corollary 4.1.7 we then obtain the following group-theoretic criterion for étaleness.

Proposition 4.5.11 *Let $\phi : Y \to X$ be a finite Galois branched cover of normal integral affine curves defined over a perfect field k. The map ϕ is étale over a point P of X if and only if the inertia subgroups I_{Q_i} are trivial for all points Q_i of Y lying above P.*

One obtains an infinite version of the above proposition as follows. Fix a separable closure K_s of $K(X)$, and choose an infinite tower $K_1 \subset K_2 \subset \cdots$ of finite Galois extensions of $K(X)$ whose union is K_s. Let A_j be the integral closure of $A = \mathcal{O}(X)$ in K_j, and P_j a maximal ideal of A_j with $P_j \cap \mathcal{O}(X) = P$ such that $P_{j+1} \cap A_j = P_j$ for all j. The corresponding inertia subgroups I_j form an inverse system whose inverse limit is a closed subgroup I_P in $\text{Gal}(K_s|K(X))$. The subgroup I_P depends on the choice of the system (P_j), but it follows from Fact 4.1.3 (2) that by varying the (P_j) we obtain conjugate subgroups in $\text{Gal}(K_s|K(X))$.

Corollary 4.5.12 *Let $\phi : Y \to X$ be as in the previous corollary, and assume $K(Y) \subset K_s$. Then Y is étale over P if and only if the image of the subgroup I_P defined above is trivial in* $\text{Gal}(K(Y)|K(X))$.

Proof Choose j so large that $K(Y) \subset K_j$, and set $Q := P_j \cap \mathcal{O}(Y)$. By construction the inertia subgroup at Q is the image of I_P in $\text{Gal}(k(Y)|k(X))$. The corollary now follows from the previous one together with Fact 4.1.3 (2). □

We now extend the above notions to integral proper normal curves. Note first that the notion of a separable morphism carries over immediately to the proper case, as it only depends on the function fields. Then define a finite separable morphism $\phi : Y \to X$ of proper normal curves to be *étale* above a point P of X if there is an affine open subset $U \subset X$ containing P such that the finite morphism of affine curves $\phi^{-1}(U) \to U$ induced by ϕ is étale above P. Lemma 4.5.2 implies that this definition does not depend on the choice of U. If ϕ is étale above all P, we say that ϕ is a *finite étale morphism.*

Using this definition of étaleness we can extend the above theory from the affine to the proper case by choosing affine open coverings of proper normal curves. In particular, Proposition 4.5.9 implies that a finite separable morphism of integral proper normal curves is étale above an open subset. Lemma 4.5.2

and Corollary 4.5.12 immediately generalize to proper normal curves, since they are of local nature.

Finally, assume $k = \mathbf{C}$. We may then equip an integral proper normal curve X with a complex structure by taking affine open coverings and using Construction 4.2.20. The resulting Riemann surface $X(\mathbf{C})$ is *compact* because X has a finite morphism ϕ_f onto $\mathbf{P}^1_{\mathbf{C}}$, the Riemann surface $\mathbf{P}^1(\mathbf{C})$ is compact, and the holomorphic map coming from ϕ_f is proper by Remark 4.5.7. Combining Proposition 4.4.5 with Theorem 3.3.7 and Corollary 3.3.12, as well as applying Proposition 4.5.6 yields:

Proposition 4.5.13 *Let X be an integral proper normal curve over $\mathbf{C}$ with function field K. Both of the first two categories below are anti-equivalent to the third one:*

1. *Integral proper normal curves equipped with a finite morphism onto X.*
2. *Compact connected Riemann surfaces equipped with a proper holomorphic map onto $X(\mathbf{C})$.*
3. *Finite extensions of K.*

Moreover, a finite morphism $Y \to X$ is étale above a point $P \in X$ if and only if the induced holomorphic map $Y(\mathbf{C}) \to X(\mathbf{C})$ restricts to a cover in a neighbourhood of P.

4.6 The algebraic fundamental group

We now use the results of the previous section to define the algebraic fundamental group of an open subset in an integral proper normal curve. By Proposition 4.4.6 this will also yield a definition of the fundamental group of an integral affine normal curve. The procedure is completely parallel to the one proving Theorem 3.4.1 on Riemann surfaces. We first prove an analogue of Lemma 3.4.2.

Proposition 4.6.1 *Let k be a perfect field, X an integral proper normal k-curve with function field K, and $U \subset X$ a nonempty open subset. Denote by K_s a fixed separable closure of K. The composite K_U of all finite subextensions $L|K$ of K_s so that the corresponding finite morphism of proper normal curves is étale above all $P \in U$ is a Galois extension of K, and each finite subextension of $K_U|K$ comes from a curve étale over U.*

Proof To prove that K_U is Galois over K we have to check that it is stable by the action of Gal $(K_s|K)$, which is an easy consequence of its definition. As in the proof of Lemma 3.4.2, for the statement concerning finite subextensions it is enough to check the following two properties:

(*a*) If a finite subextension $L|K$ comes from a curve étale over U, then so does every subfield $L \supset L' \supset K$.
(*b*) If $M|K$ is another finite subextension coming from a curve étale over U, then so does the compositum LM in K_s.

Property (*a*) follows from Lemma 4.5.10. It is then enough to check property (*b*) when L and M are Galois over K, in which case the criterion of Corollary 4.5.12 applies, and we conclude by the equality $\mathrm{Gal}\,(K_s|LM) = \mathrm{Gal}\,(K_s|L) \cap \mathrm{Gal}\,(K_s|M)$. □

Remark 4.6.2 It is possible to prove the proposition in a way completely analogous to the proof of Lemma 3.4.2. There are some technical difficulties, though: for instance, one needs a definition of the fibre product of two finite branched covers of X. This we shall define in the next chapter, but an elementary presentation of the argument would be cumbersome. We have resorted to the above Galois-theoretic approach instead.

Definition 4.6.3 In the situation of the proposition we define the *algebraic fundamental group* $\pi_1(U)$ of U to be the Galois group $\mathrm{Gal}\,(K_U|K)$.

By definition $\pi_1(U)$ is a profinite group. It depends on the choice of the separable closure K_s which plays the role of a base point, just as in the Galois theory of fields. We shall discuss the role of the base point in more detail in the next chapter.

We now come to the main result of this section. Before stating it, we define a (not necessarily integral) proper normal curve to be a finite disjoint union of integral proper normal curves. The notion of morphism extends to these in an obvious way. Also, we define their ring of rational functions as the direct product of the function fields of their components. A finite morphism $\phi : Y \to X$ of a proper normal curve onto an integral proper normal curve equips the ring of rational functions on Y with the structure of a finite dimensional K-algebra; we say that the morphism is *separable* if this algebra is étale over K.

Theorem 4.6.4 *Let X be an integral proper normal curve over a perfect field k, and let $U \subset X$ be a nonempty open subset. The category of proper normal curves Y equipped with a finite separable morphism $\phi : Y \to X$ étale over U is equivalent to the category of finite continuous left $\pi_1(U)$-sets.*

Proof Apply Theorem 4.4.5, Proposition 4.6.1 and Theorem 1.5.4. □

Let now U be an integral normal *affine* curve over a perfect field k. By Proposition 4.4.6 we may realize U as an affine open subset of a proper normal curve X, and therefore by Lemma 4.6.1 the fundamental group $\pi_1(U)$ is defined;

it does not depend on the embedding of U in X. The theorem together with the last statement of Proposition 4.4.6 and Proposition 4.4.7 then implies:

Corollary 4.6.5 *The category of normal affine curves V equipped with a finite étale morphism $V \to U$ is equivalent to the category of finite continuous left $\pi_1(U)$-sets.*

Here, of course, a normal affine curve is defined to be a finite disjoint union of integral normal affine curves; morphisms extend as in the proper case. We could have proven the corollary directly without embedding U in the proper curve X, thereby circumventing the theory of proper curves. This embedding will be, however, crucial in proving Theorem 4.6.7 below.

Remark 4.6.6 In the situation of the above corollary let $\widetilde{A}$ be the integral closure of $A := \mathcal{O}(U)$ in K_U. For a finite subextension $K(V)$ of $K_U|K(X)$ coming from a finite étale cover $V \to U$ we have $\mathcal{O}(V) = \widetilde{A} \cap K(V)$, and for each maximal ideal $M \subset \widetilde{A}$ the intersection $M \cap \mathcal{O}(V)$ defines a closed point of $\mathcal{O}(V)$. Although the k-algebra $\widetilde{A}$ is not finitely generated, it is the union of the finitely generated k-algebras $\mathcal{O}(V)$, and the set $\widetilde{U}$ of its maximal ideals may be identified with the inverse limit of the natural inverse system formed by the closed points of each V. We may even equip $\widetilde{U}$ with the inverse limit topology, and view it as an affine 'pro-algebraic curve' by defining a sheaf of rings $\mathcal{O}_{\widetilde{U}}$ on it in the usual way: localizing $\widetilde{A}$ by a maximal ideal $\widetilde{Q} \subset \widetilde{A}$ we get the local ring $\mathcal{O}_{\widetilde{U},\widetilde{Q}}$ of the pro-point $\widetilde{Q}$, and for an open subset $\widetilde{V} \subset \widetilde{U}$ we let $\mathcal{O}_{\widetilde{U}}(\widetilde{V})$ be the intersection of the rings $\mathcal{O}_{\widetilde{U},\widetilde{Q}}$ for $\widetilde{Q} \in \widetilde{V}$. We obtain a 'pro-étale cover' of U which is the algebraic analogue of the universal cover in topology. Note that it carries a natural action by $\pi_1(U)$.

If we embed U in a proper normal curve X, we may perform a similar construction for X by taking an affine open cover and considering the normalizations of its elements in finite subextensions of $K_U|K$. We obtain a ringed space $(\widetilde{X}, \mathcal{O}_{\widetilde{X}})$ which is a 'profinite branched cover' of X 'pro-étale over U'. It carries a natural action of $\pi_1(U)$, but it also has 'pro-points at infinity' lying above points in $X \setminus U$. The action of $\pi_1(U)$ on the latter captures a lot of information about U.

In the case $k = \mathbf{C}$ our discussion was completely parallel with the theory for Riemann surfaces discussed in the previous chapter. This yields the following important structure theorem for the algebraic fundamental group.

Theorem 4.6.7 *Let X be an integral proper normal curve over $\mathbf{C}$, and let $U \subset X$ be an open subset. The algebraic fundamental group $\pi_1(U)$ is isomorphic to the profinite completion of the topological fundamental group of the Riemann surface associated with U. Hence as a profinite group it has a* *presentation*

$$\langle a_1, b_1, \ldots, a_g, b_g, \gamma_1, \ldots, \gamma_n \mid [a_1, b_1] \cdots [a_g, b_g]\gamma_1 \cdots \gamma_n = 1 \rangle,$$

where n is the number of points of X lying outside U, and g is the genus of the compact Riemann surface associated with X.

Proof In view of Proposition 4.5.13, the extension $K_U|K$ of Proposition 4.6.1 is isomorphic to the extension $K_{X'}|\mathcal{M}(X)$ of Theorem 3.4.1, when X' is taken to be the Riemann surface associated with U. The Galois group of the former extension is $\pi_1(U)$ by definition, and that of the latter is the profinite completion of the topological fundamental group of X' by Theorem 3.4.1. The last statement then follows from Remark 3.6.4. □

Remark 4.6.8 Let γ_i be one of the above generators, and let G be a finite quotient of $\pi_1(U)$ corresponding to a finite Galois branched cover $Y \to X$. By Proposition 3.4.5 the image of γ_i generates the cyclic stabilizer of a point Q_i of Y lying above a point $P_i \in X \setminus U$. If we make G vary among the finite quotients of $\pi_1(U)$, we obtain a coherent system of points Q_i which define a point $\widetilde{Q}$ of the profinite branched cover $\widetilde{X}$ of Remark 4.6.6 that lies above P_i. By construction, its stabilizer $I_{\widetilde{Q}}$ under the action of $\pi_1(U)$ is the procyclic subgroup generated by γ_i. In particular, it is isomorphic to $\widehat{\mathbf{Z}}$.

A surprising fact is that the above presentation for $\pi_1(U)$ holds over an arbitrary algebraically closed field of characteristic 0. To derive it we first have to discuss base change for proper normal curves.

Construction 4.6.9 Let X be an integral proper normal curve over a field k. Denote by K its function field, and let $L|k$ be a field extension. Recall that K is a finite extension of the rational function field $k(t)$, so $K \otimes_k L$ is a finite dimensional $L(t)$-algebra. Assume it is in fact a finite direct product of fields L_i. Each L_i is then finitely generated and of transcendence degree 1 over L, hence corresponds to an integral proper normal curve X_i over L. We define the base change X_L to be the disjoint union of the X_i. There is a natural morphism $X_L \to X$ of proper normal curves.

The assumption on $K \otimes_k L$ is satisfied when $L|k$ is a separable algebraic extension, or when k is algebraically closed. In the latter case $K \otimes_k L$ is in fact a field for all $L \supset k$.

If $U \subset X$ is an open subset, we define U_L to be the inverse image of U in X_L. In the case when U is affine this is in accordance with Construction 4.3.7. Indeed, there we worked under the assumption that $\mathcal{O}(U) \otimes_k L$ is an integral domain, which holds if and only if $K \otimes_k L$ is a field. Then by construction $U_L = \operatorname{Spec}(\mathcal{O}(U) \otimes_k L)$.

We can now state the following nontrivial theorem whose proof we leave to the next chapter (see Remark 5.7.8).

Theorem 4.6.10 *Let $k \subset L$ be an extension of algebraically closed fields of characteristic 0, X an integral proper normal curve over k, and $U \subset X$ an open subset. The base change functor $Y \mapsto Y_L$ induces an equivalence between the finite covers of X étale over U and those of X_L étale over U_L.*

Consequently there is an isomorphism $\pi_1(U_L) \xrightarrow{\sim} \pi_1(U)$.

Corollary 4.6.11 *Let k be an algebraically closed field of characteristic 0, X an integral proper normal curve over k, and $U \subset X$ an open subset. Then $\pi_1(U)$ has a presentation as in Theorem 4.6.7.*

Examples 4.6.12 Let k be an algebraically closed field of characteristic 0.

1. By Theorems 4.6.7 and 4.6.10 we have $\pi_1(\mathbf{P}^1_k) = \pi_1(\mathbf{A}^1_k) = \{1\}$.
2. The same theorems show that $\pi_1(\mathbf{P}^1_k \setminus \{0, \infty\}) \cong \widehat{\mathbf{Z}}$. Thus for each $n > 0$ there is a unique isomorphism class of finite Galois covers of $\mathbf{P}^1_k$ with group $\mathbf{Z}/n\mathbf{Z}$ that are étale outside 0 and ∞. Such a cover is given by the normalization of $\mathbf{A}^1_k$ in the cyclic Galois extension of $k(t)$ defined by the equation $x^n = t$.
3. The group $\pi_1(\mathbf{P}^1_k \setminus \{0, 1, \infty\})$ is the free profinite group on two generators. Thus every finite group that may be generated by two elements is the Galois group of a finite Galois cover of $\mathbf{P}^1_k$ étale outside 0, 1 and ∞. It is known from the classification of finite simple groups that all of them can be generated by two elements – hence all of them arise as quotients of $\pi_1(\mathbf{P}^1_k \setminus \{0, 1, \infty\})$.

4.7 The outer Galois action

Our examples in the last section concerned curves over algebraically closed fields. We now turn to non-closed base fields where a crucial new feature appears: the absolute Galois group of the base field arises as a canonical quotient of the algebraic fundamental group.

To explain this, let X be an integral proper normal curve over a perfect field k. Fix an algebraic closure of k, and denote as usual by K the function field of X. We assume that X is *geometrically integral*, which means that $K \otimes_k \bar{k}$ is a field. This is then the function field of the base change $X_{\bar{k}}$ of X to $\bar{k}$, and also that of $U_{\bar{k}}$ for an affine open subset $U \subset X$. The affine curve $U_{\bar{k}}$ is then integral, i.e. this definition is coherent with the earlier notion of geometric integrality encountered in Construction 4.3.7.

Let K_s be a separable closure of K containing $\bar{k}$. The field $K \otimes_k \bar{k}$ identifies with the composite $K\bar{k}$ of K and $\bar{k}$ in K_s. It may also be described as the composite in K_s of the function fields of all base changes X_L, with $L|k$ finite. The morphisms $X_L \to X$ are finite étale for all $L|k$ by the same argument as in Example 4.5.8. It follows that $K\bar{k}$ is contained in the subfield $K_U \subset K_s$ of Proposition 4.6.1 for all open subsets $U \subset X$. By construction there is a canonical isomorphism $\mathrm{Gal}\,(K(X_L)|k(X)) \cong \mathrm{Gal}\,(L|k)$ for each L, where $K(X_L)$ is the function field of X_L, whence an isomorphism $\mathrm{Gal}\,(K\bar{k}|K) \cong \mathrm{Gal}\,(\bar{k}|k)$. In conclusion, $\mathrm{Gal}\,(\bar{k}|k)$ arises as a quotient of $\pi_1(U)$ for all open $U \subset X$.

Proposition 4.7.1 *Let X be a geometrically integral proper normal curve over a perfect field k, and $U \subset X$ an open subset (possibly equal to X). There is an exact sequence of profinite groups*

$$1 \to \pi_1(U_{\bar{k}}) \to \pi_1(U) \to \mathrm{Gal}\,(\bar{k}|k) \to 1.$$

Proof By the above discussion it remains to identify the kernel of the map $\pi_1(U) \to \mathrm{Gal}\,(K\bar{k}|K)$ with $\pi_1(U_{\bar{k}})$. A finite quotient G of the latter group corresponds to a finite Galois extension K_0 of $K\bar{k}$ that is the function field of a finite Galois branched cover $Y_0 \to X_{\bar{k}}$ étale over $U_{\bar{k}}$. Let $f \in (K\bar{k})[t]$ be a minimal polynomial for this field extension. We may find a finite extension $L|k$ contained in $\bar{k}$ so that the coefficients of f lie in KL and the finite extension $L_0|KL$ defined by f is Galois with group G. By construction we have $L_0\bar{k} = K_0$, and moreover L_0 is the function field of a finite branched cover $Y \to X_L$ so that $Y_{\bar{k}} \cong Y_0$. It then follows from the definition of étaleness that Y must be étale over U_L. As X_L is finite étale above X, the composite map $Y \to X_L \to X$ realizes Y as a finite branched cover of X étale above U. Therefore $L_0 \subset K_U$, where K_U is as in Proposition 4.6.1. Moreover, we have $L_0 \cap K\bar{k} = KL$, which allows us to identify $G = \mathrm{Gal}\,(L_0|KL)$ with a finite quotient of $\ker(\pi_1(U) \to \mathrm{Gal}\,(K\bar{k}|K))$. On the other hand, as L_0 was shown to be a subfield of K_U, so is the composite $L_0\bar{k} = K_0$, and therefore by making G vary among the finite quotients of $\pi_1(U_{\bar{k}})$ we see that $K_{U_{\bar{k}}} \subset K_U$. It follows that there is a surjection from $\mathrm{Gal}\,(K_U|K\bar{k}) = \ker(\pi_1(U) \to \mathrm{Gal}\,(K\bar{k}|K))$ onto $\mathrm{Gal}\,(K_{U_{\bar{k}}}|K\bar{k}) = \pi_1(U_{\bar{k}})$. We have just seen that for each finite quotient G of the latter we may find a finite quotient of the former mapping isomorphically onto G via the above surjection, which shows that the map is an isomorphism. □

Now recall that quite generally given an exact sequence

$$1 \to N \to G \to \Gamma \to 1$$

of profinite groups, the action of G on the normal subgroup N via conjugation yields a continuous homomorphism $G \to \mathrm{Aut}(N)$. Its restriction to N takes values in the normal subgroup $\mathrm{Inn}(N) \subset \mathrm{Aut}(N)$ of inner automorphisms, i.e. those that come from conjugation by an element of N. Denote the quotient $\mathrm{Aut}(N)/\mathrm{Inn}(N)$ by $\mathrm{Out}(N)$; it is the group of *outer automorphisms* of N. By passing to the quotient we thus obtain a continuous homomorphism $\Gamma \to \mathrm{Out}(N)$.

Applying the above to the exact sequence of the proposition we obtain a continuous homomorphism

$$\rho_U : \mathrm{Gal}\,(\bar{k}|k) \to \mathrm{Out}(\pi_1(U_{\bar{k}})).$$

The group $\pi_1(U_{\bar{k}})$ is called the *geometric fundamental group* of U, and ρ_U the *outer Galois action* on the geometric fundamental group.

We now investigate the action of $\pi_1(U)$ on the space $\widetilde{X}$ introduced in Remark 4.6.6 above. Let $\widetilde{Q}$ be a pro-point of $\widetilde{X}$ lying above a point P of X, and $D_{\widetilde{Q}}$ its stabilizer in $\pi_1(U)$. The residue field $\kappa(\widetilde{Q}) = \mathcal{O}_{\widetilde{X},\widetilde{Q}}/\widetilde{Q}\mathcal{O}_{\widetilde{X},\widetilde{Q}}$ is an algebraic closure of $\kappa(P) = \mathcal{O}_{X,P}/P\mathcal{O}_{X,P}$. By the same argument as before the statement of Fact 4.1.3, we have a homomorphism $D_{\widetilde{Q}} \to \mathrm{Gal}\,(\overline{\kappa(P)}|\kappa(P))$, and one sees using Fact 4.1.3 (3) that it is surjective. Its kernel $I_{\widetilde{Q}}$ is called the *inertia group at* $\widetilde{Q}$.

By the previous proposition (and its proof) we may view $\widetilde{X}$ as a profinite Galois branched cover of $X_{\bar{k}}$ with Galois group $\pi_1(U_{\bar{k}})$, a normal subgroup in $\pi_1(U)$.

Lemma 4.7.2 *Let P be a closed point of X with $\kappa(P) \cong k$. The stabilizer of a point $\widetilde{Q}$ of $\widetilde{X}$ lying above P in $\pi_1(U_{\bar{k}})$ equals its inertia group $I_{\widetilde{Q}}$ in $\pi_1(U)$.*

Proof The subgroups in question consist of those elements of the stabilizer $D_{\widetilde{Q}}$ that are in the kernel of the natural projection $D_{\widetilde{Q}} \to \mathrm{Gal}\,(\bar{k}|k)$. □

A closed point P as in the statement of the lemma is called a *k-rational point* of X. It has the following more transparent interpretation. Take an affine open $U \subset X$ containing P, and view the coordinate ring $\mathcal{O}(U)$ as a quotient of the polynomial ring $k[x_1, \ldots, x_n]$ for suitable n. The point P identifies with a maximal ideal of $\mathcal{O}(U)$, and there is a k-isomorphism $\kappa(P) \cong k$ if and only if the preimage of P in $k[x_1, \ldots, x_n]$ is of the form $(x_1 - a_1, \ldots, x_n - a_n)$ with some $a_i \in k$. This means that if we choose an embedding of U in affine space $\mathbf{A}^n_k$, then after base change to $\bar{k}$ we obtain a single point of $\mathbf{A}^n_{\bar{k}}$ lying above our closed point P, and moreover it has coordinates in k.

Corollary 4.7.3 *If U contains a k-rational point, then the exact sequence of Proposition 4.7.1 splits, and $\pi_1(U)$ is a semidirect product of $\pi_1(U_{\bar{k}})$ with* $\mathrm{Gal}\,(\bar{k}|k)$.

Proof Let $\widetilde{Q}$ be a pro-point of $\widetilde{U}$ lying above a k-rational point of U, and $\bar{Q}$ its image in $U_{\bar{k}}$. By definition of $\pi_1(U_{\bar{k}})$ it must have trivial stabilizer in $\pi_1(U_{\bar{k}})$. The lemma then implies that $I_{\widetilde{Q}}$ is trivial, so $D_{\widetilde{Q}} \subset \pi_1(U)$ maps isomorphically onto $\mathrm{Gal}\,(\bar{k}|k)$ by the projection $\pi_1(U) \to \mathrm{Gal}\,(\bar{k}|k)$, whence the required splitting. □

Example 4.7.4 Let us consider the case $U = \mathbf{P}^1_k \setminus \{0, \infty\}$, k of characteristic 0. We have seen that $\pi_1(U_{\bar{k}}) \cong \widehat{\mathbf{Z}}$ is commutative, hence we have a true action of $\mathrm{Gal}\,(\bar{k}|k)$ on $\pi_1(U_{\bar{k}})$, not just an outer action. For each $n > 0$ the quotient $\pi_1(U_{\bar{k}})/n\pi_1(U_{\bar{k}}) \cong \mathbf{Z}/n\mathbf{Z}$ can be identified with the Galois group of the extension $K_n|\bar{k}(t)$ defined by the equation $x^n - t$ by Example 4.6.12 (2). Moreover, the action of $\mathrm{Gal}\,(\bar{k}|k)$ on $\pi_1(U_{\bar{k}})/n\pi_1(U_{\bar{k}})$ is the one coming from the extension of profinite groups

$$1 \to \mathrm{Gal}\,(K_n|\bar{k}(t)) \to \mathrm{Gal}\,(K_n|k(t)) \to \mathrm{Gal}\,(\bar{k}|k) \to 1.$$

A generator of $\mathrm{Gal}\,(K_n|\bar{k}(t))$ is given by sending 1 to the automorphism mapping a fixed n-th root $\sqrt[n]{t}$ of t to $\omega_n\sqrt[n]{t}$, where $\omega_n \in \bar{k}$ is a primitive n-th root of unity. The action of $\sigma \in \mathrm{Gal}\,(\bar{k}|k)$ on $\mathrm{Gal}\,(K_n|\bar{k}(t))$ sends this automorphism to $\sqrt[n]{t} \mapsto \sigma(\omega_n)\sqrt[n]{t}$.

The actions of $\mathrm{Gal}\,(\bar{k}|k)$ on the quotients $\pi_1(U_{\bar{k}})/n\pi_1(U_{\bar{k}})$ are compatible for different n. This translates to the following. By the above, defining an isomorphism $\pi_1(U_{\bar{k}}) \xrightarrow{\sim} \mathrm{Gal}\,(\bigcup K_n|\bar{k}(t))$ corresponds to fixing a choice of a primitive n-th root of unity ω_n for each n, with the property that for all pairs (n, m) with $n|m$ we have $\omega_m^{m/n} = \omega_n$. The action of $\sigma \in \mathrm{Gal}\,(\bar{k}|k)$ then corresponds to sending a system (ω_n) of roots of unity as above to the system $\sigma(\omega_n)$.

Via the fixed isomorphisms $\widehat{\mathbf{Z}} \xrightarrow{\sim} \mathrm{Gal}\,(\bigcup K_n|\bar{k}(t)) \xrightarrow{\sim} \pi_1(U_{\bar{k}})$ we get a continuous homomorphism $\mathrm{Gal}\,(\bar{k}|k) \to \mathrm{Aut}(\widehat{\mathbf{Z}}) \cong \widehat{\mathbf{Z}}^\times$, inducing maps $\mathrm{Gal}\,(\bar{k}|k) \to (\mathbf{Z}/n\mathbf{Z})^\times$ for each n. It is called the *cyclotomic character* of $\mathrm{Gal}\,(\bar{k}|k)$.

Remark 4.7.5 In the situation of Lemma 4.7.2 the action of $\mathrm{Gal}\,(\bar{k}|k)$ on $I_{\widetilde{Q}} \cong \widehat{\mathbf{Z}}$ coming from the extension

$$1 \to I_{\widetilde{Q}} \to D_{\widetilde{Q}} \to \mathrm{Gal}\,(\bar{k}|k) \to 1$$

is also given by the cyclotomic character in characteristic 0. Indeed, each degree n finite extension $K_n|K\bar{k}$ contained in the fixed field of $I_{\widetilde{Q}}$ is given by an n-th root of a local parameter at the unique point $\overline{Q}$ of $X_{\bar{k}}$ above P, by Kummer theory and the fact that the ramification index at the unique point of the curve corresponding to K_n lying above $\overline{Q}$ must be n. This being said, the same argument as above applies.

We conclude this section by a famous theorem of Belyi stating that for $k = \mathbf{Q}$ and $U = \mathbf{P}^1 \setminus \{0, 1, \infty\}$ the outer representation ρ_U is faithful. He derived this fact from the following result that is very interesting in its own right.

Theorem 4.7.6 (Belyi) *Let X be an integral proper normal curve defined over an algebraically closed field k of characteristic 0. There exists a morphism $X \to \mathbf{P}^1_k$ étale over $\mathbf{P}^1_k \setminus \{0, 1, \infty\}$ if and only if X can be defined over $\overline{\mathbf{Q}}$.*

Here the condition that X may be defined over $\overline{\mathbf{Q}}$ means that there exists a curve X_0 defined over $\overline{\mathbf{Q}}$ such that X comes from X_0 by base change from $\overline{\mathbf{Q}}$ to k.

Proof The 'only if' part follows from Theorem 4.6.10 (to be proven in the next chapter). Belyi proved the 'if' part as follows. In any case there is a morphism $p : X \to \mathbf{P}^1$ defined over $\overline{\mathbf{Q}}$ and étale above the complement of a finite set S of closed points. The idea is to compose p with suitable morphisms $\mathbf{P}^1 \to \mathbf{P}^1$ that reduce the size of S.

In a first step, we reduce to the case where S consists of $\mathbf{Q}$-rational points. For this, let P be a point for which the degree $n = [\kappa(P) : \mathbf{Q}]$ is maximal among the points in S. Choose a minimal polynomial $f \in \mathbf{Q}(t)$ for a generator of the extension $\kappa(P)|\mathbf{Q}$ and consider the map $\phi_f : \mathbf{P}^1 \to \mathbf{P}^1$ attached to f. Since f has coefficients in $\mathbf{Q}$, this map is defined over $\mathbf{Q}$, and as it is given by a polynomial, it restricts to a map $\mathbf{A}^1 \to \mathbf{A}^1$ that we denote in the same way. By Remark 4.5.5 the map ϕ_f is étale outside $\phi_f(S_f)$, where S_f is the set of points $Q \in \mathbf{A}^1_{\overline{\mathbf{Q}}}$ with $f'(Q) = 0$. Therefore the composite $\phi_f \circ p$ is étale outside the set $S' = \phi_f(S) \cup \{\infty\} \cup \phi_f(S_f)$. Now ∞ has degree 1 over $\mathbf{Q}$; the points of S_f, and hence of $\phi_f(S_f)$ have degree at most $n - 1$; finally, those in $\phi_f(S)$ have degree at most n. But since $\phi_f(P) = 0$, there are *strictly less* points of degree exactly n in S' than in S. Replacing p by $\phi_f \circ p$ and S by S' we may continue this procedure until we arrive at $n = 1$.

So assume all points in S are $\mathbf{Q}$-rational. If S consists of at most three points, we are done by composing with an automorphism of $\mathbf{P}^1$; otherwise we may assume S contains $0, 1, \infty$ and at least one more $\mathbf{Q}$-rational point α. The idea again is to compose p with a map $\phi_f : \mathbf{P}^1 \to \mathbf{P}^1$ associated with a well-chosen rational function f. This time we seek f in the form $x^A(x-1)^B$ with some nonzero integers A, B. Outside 0, 1 and ∞ it restricts to a morphism of affine curves $\mathbf{A}^1_{\mathbf{Q}} \setminus \{0, 1\} \to \mathbf{A}^1_{\mathbf{Q}}$ corresponding to the homomorphism $\mathbf{Q}[x] \to \mathbf{Q}[x][(x(x-1))^{-1}]$ sending x to f. As above, $\phi_f \circ p$ will be étale outside $\phi_f(S)$ together with the images of those points in $\mathbf{A}^1 \setminus \{0, 1\}$ where the derivative f'

vanishes. These are given by the equation $Ax^{A-1}(x-1)^B + Bx^A(x-1)^{B-1} = 0$, or else $A(x-1) + Bx = 0$. We therefore have only one such point, namely $x = A/(A+B)$. So if we choose A and B so that $\alpha = A/(A+B)$, then $\phi_f \circ p$ will be étale everywhere outside $\phi_f(S)$. But $\phi_f(S)$ contains strictly less points than S, because $\phi_f(\{0, 1, \infty\}) \subset \{0, \infty\}$. We may then continue the procedure until $\phi_f(S)$ has at most three elements. □

A function $f = x^A(x-1)^B$ as in the above proof is called a *Belyi function.* It is often multiplied with the constant $(A+B)^{A+B}A^{-A}B^{-B}$ which ensures that $\alpha = A/(A+B)$ gets mapped to 1.

We can now give the promised application concerning the outer Galois action on the fundamental group of $\mathbf{P}^1_{\overline{\mathbf{Q}}} \setminus \{0, 1, \infty\}$.

Theorem 4.7.7 *The outer Galois representation*

$$\rho_{\mathbf{P}^1\setminus\{0,1,\infty\}} : \mathrm{Gal}\,(\overline{\mathbf{Q}}|\mathbf{Q}) \to \mathrm{Out}(\pi_1(\mathbf{P}^1_{\overline{\mathbf{Q}}} \setminus \{0, 1, \infty\})$$

has trivial kernel.

Proof We use the shorthands U for $\mathbf{P}^1_{\mathbf{Q}} \setminus \{0, 1, \infty\}$ and $\overline{U}$ for $U_{\overline{\mathbf{Q}}}$. Assume that ρ_U has a nontrivial kernel, fixing a (possibly infinite) extension $L|\mathbf{Q}$. The representation $\rho_{U_L} : \mathrm{Gal}\,(L) \to \mathrm{Out}\,(\pi_1(\overline{U}))$ is trivial. Recall that ρ_{U_L} comes from the short exact sequence

$$1 \to \pi_1(\overline{U}) \to \pi_1(U_L) \to \mathrm{Gal}\,(\overline{\mathbf{Q}}|L) \to 1$$

via the conjugation action of $\pi_1(U_L)$ on $\pi_1(\overline{U})$. The triviality of ρ_{U_L} means that every automorphism of $\pi_1(\overline{U})$ induced by conjugating with an element $x \in \pi_1(U_L)$ equals the conjugation automorphism by an element $y \in \pi_1(\overline{U})$. This implies that $y^{-1}x$ is in the centralizer C of $\pi_1(\overline{U})$ in $\pi_1(U_L)$, and therefore the latter group is generated by C and $\pi_1(\overline{U})$. But by Example 4.6.12 (3) the group $\pi_1(\overline{U})$ is a free profinite group on two generators, and as such has trivial centre. Hence C and $\pi_1(\overline{U})$ have trivial intersection, which implies that $\pi_1(U_L)$ is actually their direct product. Whence a quotient map $\pi_1(U_L) \to \pi_1(\overline{U})$ which is a retraction of the natural inclusion. Considering finite continuous $\pi_1(\overline{U})$-sets and applying Theorem 4.6.4 we conclude that every finite étale cover of $\overline{U}$ comes by base change from a cover of U_L. By Belyi's theorem this then means that every integral proper normal curve defined over $\overline{\mathbf{Q}}$ can in fact be defined over L. But there are counter-examples to this latter assertion; see the facts below. □

Facts 4.7.8 Let L be a subfield of $\overline{\mathbf{Q}}$. A geometrically integral proper normal curve E defined over L is an *elliptic curve* if it has an L-rational point P and is of genus 1 (meaning that the compact Riemann surface coming from $E_{\mathbf{C}}$ has genus 1 in the sense of Section 3.6). Then it is known that there is an embedding $E \to \mathbf{P}^2_L$ whose image is defined by an equation of the form $y^2z = x^3 + Axz^2 + Bz^3$ with $A, B \in L$, the point $(0, 1, 0)$ being the image of P. The *j-invariant*

$$j(E) := 1728\frac{4A^3}{4A^3 + 27B^2} \in L$$

is preserved by all $\overline{\mathbf{Q}}$-isomorphisms $E_{\overline{\mathbf{Q}}} \cong X'$, where X' is a projective plane curve over $\overline{\mathbf{Q}}$ given by an equation of the above shape. Furthermore, for every $j \in \overline{\mathbf{Q}}$ there exists an elliptic curve E' over $\overline{\mathbf{Q}}$ as above with $j(E') = j$. For these facts see e.g. [100], Proposition III.3.1 and §3.1.

Fix now an elliptic curve E' over $\overline{\mathbf{Q}}$ such that $j(E') \notin L$. We contend that there is no proper normal curve X over L with $X_{\overline{\mathbf{Q}}} \cong E'$. Indeed, were there such a curve, it would be of genus 1, and then it is known ([100], Ex. 10.3) that there is an elliptic curve E over L canonically attached to X, its *Jacobian*, satisfying $X_{\overline{\mathbf{Q}}} \cong E_{\overline{\mathbf{Q}}}$. We would then have an isomorphism $E_{\overline{\mathbf{Q}}} \cong E'$, which would contradict the assumption about the j-invariant of E'.

Remarks 4.7.9

1. The significance of Theorem 4.7.7 lies in the fact that it embeds the group $\mathrm{Gal}(\overline{\mathbf{Q}}|\mathbf{Q})$, which is of arithmetic nature, in the outer automorphism group of a group coming from topology. It is the starting point of Grothendieck's fascinating theory of *dessins d'enfants*; see [85] for a comprehensive introduction.
2. Theorem 4.7.6 also has important applications in Diophantine geometry. For instance, Elkies [19] used it to deduce Mordell's Conjecture (now Faltings' Theorem) from the *abc* conjecture. See also [6], Chapter 12 for further discussion.

4.8 Application to the inverse Galois problem

We now discuss a spectacular application of the methods developed so far. It concerns the *regular inverse Galois problem* over $\mathbf{Q}$ that may be stated as follows.

Problem 4.8.1 *Let G be a finite group. Construct a finite Galois extension $K|\mathbf{Q}(t)$ regular over $\mathbf{Q}$ with* $\mathrm{Gal}(K|\mathbf{Q}(t)) \cong G$.

The regularity condition means that there is no subextension in $K|\mathbf{Q}(t)$ of the form $L(t)$, where L is a nontrivial extension of $\mathbf{Q}$.

A positive solution to the regular inverse Galois problem implies a positive solution to the inverse Galois problem over $\mathbf{Q}$ because of the following well-known result.

Fact 4.8.2 Consider a finite regular Galois extension $K|\mathbf{Q}(t)$ with Galois group G. Let $x^m + a_{m-1}x^{m-1} + \cdots + a_0$ be a minimal polynomial for this extension, with $a_i \in \mathbf{Q}(t)$. There exist infinitely many $\alpha \in \mathbf{Q}$ such that none of the a_i has denominator vanishing at α, and $x^m + a_{m-1}(\alpha)x^{m-1} + \cdots + a_0(\alpha) \in \mathbf{Q}[x]$ defines a Galois extension of $\mathbf{Q}$ with group G.

This is a somewhat sharpened form of Hilbert's Irreducibility Theorem. (The original form only states that there are infinitely many α for which the polynomial $x^m + a_{m-1}(\alpha)x^{m-1} + \cdots + a_0(\alpha)$ remains irreducible.) For proofs, see e.g. Serre's books [91] or [93]. They show that the α can in fact be chosen to be integers, or even prime numbers.

Problem 4.8.1 is largely open at the present day and is the subject of intense research. Towards the end of the 1970s Belyi, Fried, Matzat and later Thompson independently developed a method based on the theory of the algebraic fundamental group that yields a positive solution for many of the finite simple groups. We now explain the basic idea of the construction, relying largely on the exposition of Serre in [93].

The starting point is that, as we have seen in Corollary 3.4.4, every finite group G occurs as a Galois group over $\mathbf{C}(t)$. More precisely, we have shown that if G can be generated by $n-1$ elements, then G is isomorphic to a finite quotient of the topological fundamental group

$$\pi_1^{\mathrm{top}}(\mathbf{P}^1(\mathbf{C}) \setminus \{P_1, \ldots, P_n\}) = \langle \gamma_1, \ldots, \gamma_n \mid \gamma_1 \cdots \gamma_n = 1 \rangle.$$

Thus giving a surjective homomorphism $\phi : \pi_1^{\mathrm{top}}(\mathbf{P}^1(\mathbf{C}) \setminus \{P_1, \ldots, P_n\}) \twoheadrightarrow G$ is equivalent to specifying an n-tuple $(g_1, \ldots, g_n) \in G^n$ with $g_1 \ldots g_n = 1$ such that the g_i generate G. We shall call n-tuples satisfying these two properties *generating.*

Since the above holds with an arbitrary choice of the points P_i, we may assume that the P_i are $\mathbf{Q}$-rational. By Theorems 4.6.7 and 4.6.10 the profinite completion $\pi(n)$ of $\pi_1^{\mathrm{top}}(\mathbf{P}^1(\mathbf{C}) \setminus \{P_1, \ldots, P_n\})$ is isomorphic to the algebraic fundamental group $\pi_1(\mathbf{P}^1_{\overline{\mathbf{Q}}} \setminus \{P_1, \ldots, P_n\})$, and the surjection ϕ induces a continuous surjection $\pi(n) \twoheadrightarrow G$. Moreover, with the notation $\Pi(n) := \pi_1(\mathbf{P}^1_{\mathbf{Q}} \setminus \{P_1, \ldots, P_n\})$ we have the basic exact sequence

$$1 \to \pi(n) \to \Pi(n) \to \mathrm{Gal}\,(\overline{\mathbf{Q}}|\mathbf{Q}) \to 1.$$

Our task is then to extend the surjection $\phi : \pi(n) \twoheadrightarrow G$ to a continuous homomorphism $\tilde{\phi} : \Pi(n) \to G$. It will be automatically surjective (since ϕ is), and hence will define a finite Galois extension $K|\mathbf{Q}(t)$, because by construction $\Pi(n)$ is a quotient of $\mathrm{Gal}\,(\overline{\mathbf{Q}(t)}|\mathbf{Q}(t))$. The regularity of the extension $K|\mathbf{Q}(t)$ over $\mathbf{Q}$ will follow from the assumption that the restriction of $\tilde{\phi}$ to $\pi(n)$ is already surjective.

To construct $\tilde{\phi}$ we first formulate an abstract group-theoretic lemma. Assume given a profinite group Γ, a closed normal subgroup $N \subset \Gamma$, and a finite group G. The set $\mathrm{Hom}(N, G)$ of *continuous* homomorphisms $N \to G$ is equipped with a left action of G given by $(g, \phi) \mapsto {}_g\phi$, where ${}_g\phi(n) = g\phi(n)g^{-1}$ for all $n \in N$. There is also a right action of Γ on $\mathrm{Hom}(N, G)$ given by $(\phi, \sigma) \mapsto \phi_\sigma$, where $\phi_\sigma(n) = \phi(\sigma n \sigma^{-1})$ for all $n \in N$, and the two actions are compatible, i.e. ${}_g(\phi_\sigma) = ({}_g\phi)_\sigma$.

Lemma 4.8.3 *In the above situation let $S \subset \mathrm{Hom}(N, G)$ be a subset stable by the actions of G and Γ, such that moreover G acts freely and transitively on S. Then every $\phi \in S$ extends to a continuous homomorphism $\tilde{\phi} : \Gamma \to G$.*

Proof Let ϕ be an element of S. For each $\sigma \in \Gamma$ there exists $g_\sigma \in G$ such that $\phi(\sigma n \sigma^{-1}) = g_\sigma \phi(n) g_\sigma^{-1}$ for all $n \in N$, because S is stable by Γ and G acts transitively on S. Moreover, such a g_σ is unique, because the action of G on S is free. We contend that the formula $\tilde{\phi}(\sigma) := g_\sigma$ for $\sigma \in \Gamma$ defines the required extension. Indeed, the compatibility of the actions of G and Γ implies that $\tilde{\phi}$ is a homomorphism, and moreover for $\sigma \in N$ we have $g_\sigma = \phi(\sigma)$, since ϕ is a homomorphism. For continuity it is enough to show by finiteness of G that $\tilde{\phi}$ has open kernel. But $\ker(\tilde{\phi})$ consists of those $\sigma \in \Gamma$ that leave ϕ invariant, so it is open by continuity of ϕ. □

We would like to apply the lemma in the situation where $N = \pi(n)$ and $\Gamma = \Pi(n)$, so we have to specify a subset $S \subset \mathrm{Hom}(\pi(n), G)$ with the required properties. Recall that a surjection $\phi : \pi(n) \twoheadrightarrow G$ is determined by the elements $\phi(\gamma_i)$, which are to form a generating n-tuple. If S contains ϕ and is stable by the action of $\Pi(n)$, then it should contain all the homomorphisms ϕ_σ for $\sigma \in \Pi(n)$. In particular this should hold for $\sigma \in \pi(n)$, in which case $\phi_\sigma = {}_{\phi(\sigma)}\phi$. Conversely, for each $g \in G$ the map ${}_g\phi$ defines a continuous surjection $\pi(n) \twoheadrightarrow G$. Thus it is natural to fix n conjugacy classes $C_1, \ldots, C_n$ in G, and look for S in the form

$$S = \{\phi \in \mathrm{Hom}(\pi(n), G) : \phi(\gamma_i) \in C_i \text{ and}$$
$$(\phi(\gamma_1), \ldots, \phi(\gamma_n)) \text{ is a generating } n\text{-tuple}\}.$$

By definition S is stable by the actions of G and $\pi(n)$. We now impose conditions on the C_i that force the other conditions of the lemma.

Definition 4.8.4 Let G be a finite group. An n-tuple $C_1, \ldots, C_n$ of conjugacy classes in G is called *rigid* if there exists a generating n-tuple $(g_1, \ldots, g_n)$ in G^n with $g_i \in C_i$, and moreover G acts transitively on the set of all such generating n-tuples.

By the above discussion, G acts transitively on S if and only if the C_i form a rigid system. Moreover, if G has trivial centre, then its action on S is also free. Indeed, in general if $\phi \in S$, $g \in G$ and each $g_i = \phi(\gamma_i)$ is invariant for conjugation by g, then g must lie in the centre of G since the g_i generate G.

We still have to present a criterion ensuring that S is stable by the action of $\Pi(n)$.

Definition 4.8.5 A conjugacy class C in a finite group G is called *rational* if $g \in C$ implies $g^m \in C$ for all $m \in \mathbf{Z}$ prime to the order of G.

For rational conjugacy classes the stability of the set S by the action of $\Pi(n)$ boils down to:

Lemma 4.8.6 *Assume that $C_1, \dots, C_n$ are rational conjugacy classes in a finite group G, and $\phi : \pi(n) \to G$ is a continuous homomorphism with $\phi(\gamma_i) \in C_i$ for all i. Then the same holds for ϕ_σ for all $\sigma \in \Pi(n)$.*

Proof By Remark 4.6.8 and Lemma 4.7.2 each γ_i topologically generates the inertia subgroup $I_{\widetilde{Q}_i} \subset \pi(n)$ of a point $\widetilde{Q}_i$ of the profinite branched cover $\widetilde{\mathbf{P}^1_{\mathbf{C}}}$ of Remark 4.6.6 above the point P_i. By definition of inertia subgroups, for every $\sigma \in \Pi(n)$ the conjugate $\sigma\gamma_i\sigma^{-1}$ lies in the inertia subgroup $I_{\sigma(\widetilde{Q}_i)}$ of another point $\sigma(\widetilde{Q}_i)$ above P_i. Since P_i is rational over $\mathbf{Q}$, there is a unique point $\overline{Q}_i$ of $X_{\bar{k}}$ lying above P_i. Hence both $I_{\widetilde{Q}_i}$ and $I_{\sigma(\widetilde{Q}_i)}$ are stabilizers of points above $\overline{Q}_i$ in $\pi(n)$, and as such they are conjugate in $\pi(n)$ as well. Thus $\phi(I_{\widetilde{Q}_i})$ and $\phi(I_{\sigma(\widetilde{Q}_i)})$ are conjugate cyclic subgroups in G, with respective generators $\phi(\gamma_i)$ and $\phi(\sigma\gamma_i\sigma^{-1})$. Therefore there is $m \in \mathbf{Z}$ prime to the order of G and $g \in G$ with $\phi(\sigma\gamma_i\sigma^{-1}) = g\phi(\gamma_i)^m g^{-1}$. As C_i is a rational conjugacy class, this shows that $\phi(\sigma\gamma_i\sigma^{-1}) \in C_i$, as required. □

We can summarize the result of the above discussion in the following theorem.

Theorem 4.8.7 *Let G be a finite group with trivial centre, and assume there exists a rigid system $C_1, \dots, C_n$ of rational conjugacy classes in G. For each n-tuple $P_1, \dots, P_n$ of $\mathbf{Q}$-rational points on $\mathbf{P}^1_{\mathbf{Q}}$ there is a surjection*

$$\pi_1(\mathbf{P}^1_{\mathbf{Q}} \setminus \{P_1, \dots, P_n\}) \twoheadrightarrow G$$

such that moreover the image of each canonical generator γ_i lies in C_i. In particular, G is the Galois group of a Galois extension of $\mathbf{Q}(t)$ regular over $\mathbf{Q}$.

Remark 4.8.8 The role played by rationality in the above proof can be made more precise as follows. Denoting by N the order of G, there is a natural action of $(\mathbf{Z}/N\mathbf{Z})^\times$ on G (as a set) via $(m, \gamma) \mapsto \gamma^m$; this also induces an action of $(\mathbf{Z}/N\mathbf{Z})^\times$ on the set of conjugacy classes of G. Rationality of a class C_i

then means that C_i is preserved by the action of $(\mathbf{Z}/N\mathbf{Z})^\times$. Now observe that $(\mathbf{Z}/N\mathbf{Z})^\times \cong \mathrm{Gal}\,(\mathbf{Q}(\mu_N)|\mathbf{Q})$, where as usual μ_N denotes the group of N-th roots of unity. Remark 4.7.5 then implies that the integer m at the end of the proof of Lemma 4.8.6 can be recovered as the image of σ by the composite map

$$\Pi(n) \to \mathrm{Gal}\,(\overline{\mathbf{Q}}|\mathbf{Q}) \xrightarrow{\chi} \widehat{\mathbf{Z}}^\times \twoheadrightarrow (\mathbf{Z}/N\mathbf{Z})^\times,$$

where χ denotes the cyclotomic character.

This interpretation opens the way for various generalizations of Theorem 4.8.7. For instance, a variant we shall use in Example 4.8.10 below is the following. Assume that G has a rigid system of conjugacy classes that are either rational or come in pairs (C_i, C_i') such that both C_i and C_i' are preserved by a subgroup $H_i \subset (\mathbf{Z}/N\mathbf{Z})^\times$ of index 2, and the induced action of $(\mathbf{Z}/N\mathbf{Z})^\times/H_i \cong \mathbf{Z}/2\mathbf{Z}$ interchanges C_i and C_i'. Then the conclusion of the theorem holds with the modification that the points P_i corresponding to the rational C_i are $\mathbf{Q}$-rational, and otherwise come in pairs (P_i, P_i') where $\kappa(P_i) \cong \kappa(P_i')$ is the subfield of $\mathbf{Q}(\mu_N)$ fixed by H_i and the action of $\mathrm{Gal}\,(\overline{\mathbf{Q}}|\mathbf{Q})$ on the points of $\mathbf{P}^1_{\overline{\mathbf{Q}}}$ interchanges P_i and P_i'. The proof is a straightforward modification of that of Lemma 4.8.6.

Example 4.8.9 The criterion of Theorem 4.8.7 is satisfied for many of the sporadic finite simple groups. For instance, Thompson has verified that the Monster has a rigid system of three rational conjugacy classes of orders 2, 3 and 29, respectively. Such a verification is of course far from trivial and often relies on the classification of finite simple groups. Concerning Thompson's theorem and further results about sporadic groups, see Section II.9 of [56].

We now present an example where the criterion of Theorem 4.8.7 does not apply directly but the variant of Remark 4.8.8 does.

Example 4.8.10 Let p be an odd prime such that 2 is not a square modulo p. We construct a rigid triple of conjugacy classes in the finite simple group $\mathrm{PSL}_2(\mathbf{F}_p)$. Each element of order p in $\mathrm{PSL}_2(\mathbf{F}_p)$ comes from a matrix in $\mathrm{SL}_2(\mathbf{F}_p)$ having an eigenvalue 1, so since it has determinant 1, its upper triangular form is

$$E_a := \begin{bmatrix} 1 & a \\ 0 & 1 \end{bmatrix}$$

for some $a \in \mathbf{F}_p^\times$. It follows that order p elements in $\mathrm{PSL}_2(\mathbf{F}_p)$ fall into two conjugacy classes pA and pB, depending on whether a is a square in $\mathbf{F}_p$ or not. Another calculation shows that the order 2 elements form a single conjugacy class $2A$. We contend that $(2A, pA, pB)$ is a rigid triple of conjugacy classes in $\mathrm{PSL}_2(\mathbf{F}_p)$. The matrices

$$M_1 = \begin{bmatrix} 1 & -1 \\ 2 & -1 \end{bmatrix}, \quad M_2 = \begin{bmatrix} 1 & 1 \\ 0 & 1 \end{bmatrix}, \quad M_3 = \begin{bmatrix} 1 & 0 \\ -2 & 1 \end{bmatrix}$$

represent elements of $2A$, pA, pB, respectively, the last one because it is conjugate to E_2 and 2 is not a square in $\mathbf{F}_p$ by assumption. Their product is the identity matrix, and the pair (M_2, M_3) generates $\mathrm{SL}_2(\mathbf{F}_p)$ because their powers are exactly the 2×2 elementary matrices (see [48], Chapter XIII, Lemma 8.1). To conclude it suffices to show that given a triple (N_1, N_2, N_3) of matrices representing a generating triple from $2A \times pA \times pB$, the pair (N_2, N_3) is conjugate in $\mathrm{SL}_2(\mathbf{F}_p)$ to (M_2, M_3). For this notice that both N_2 and N_3 have a one-dimensional eigenspace corresponding to the eigenvalue 1, and they are not the same since otherwise the N_i would not generate. Hence we may choose a basis of $\mathbf{F}_p^2$ consisting of an eigenvector of N_2 and an eigenvector of N_3. Passing to this basis the matrices become

$$N_2' = \begin{bmatrix} 1 & a \\ 0 & 1 \end{bmatrix}, \quad N_3' = \begin{bmatrix} 1 & 0 \\ b & 1 \end{bmatrix}.$$

By assumption we find $c \in \mathbf{F}_p$ with $c^2 = a$, therefore we have $DN_2'D^{-1} = M_2$, where D is the diagonal matrix with entries (c, c^{-1}). Conjugation by D preserves the shape of N_3', so we may assume $N_2' = M_2$. Finally the fact that $N_2N_3 = N_1^{-1}$ has order 4 implies $\mathrm{Tr}(N_2N_3) = \mathrm{Tr}(N_2'N_3') = 0$, whence $b = -2$ as required.

The conjugacy class $2A$ is rational. However, the classes pA and pB are not, because by the assumption that 2 is nonsquare modulo p the odd powers of each matrix E_a are in the same conjugacy class as E_a and the even powers are in the other one. But this shows that the pair (pA, pB) satisfies a property as in the criterion of Remark 4.8.8. Hence for p as above $\mathrm{PSL}_2(\mathbf{F}_p)$ occurs as a Galois group over $\mathbf{Q}(t)$.

Many of the finite simple groups have been realized as Galois groups over $\mathbf{Q}(t)$ by a variant of the rigidity method like in the above example. For the state of the art in 1998, see the book of Malle and Matzat [56].

4.9 A survey of advanced results

In this final section we give an overview of some of the major results currently known about the structure of fundamental groups of integral normal curves. In the first part we assume that the base field k is algebraically closed.

A fundamental invariant of an integral proper normal curve X over k is its *genus* g. Over $k = \mathbf{C}$ this is the same as the genus of the associated Riemann surface introduced in Section 3.6. For definitions that work in general, see [34] or [95]. An open subcurve $U \subset X$ different from X is isomorphic to an affine curve (see e.g. [34], Exercise IV.1.3). The fundamental invariants of U are g and the number n of closed points in $X \setminus U$.

The first main theorem says that these invariants completely determine the maximal quotient of $\pi_1(U)$ which is 'prime to the characteristic of k'. In order to state it precisely, we need to introduce some terminology. Given a prime p and a group G, the inverse limit of the natural inverse system of the finite quotients of G having order prime to p (resp. a power of p) is called the *profinite p'-completion* (resp. *profinite p-completion*) of G. If G is moreover profinite, these are the *maximal prime-to-p quotient* $G^{(p')}$ and *maximal pro-p quotient* $G^{(p)}$ of G, respectively. We extend the notion of profinite p'-completion to $p = 0$ by defining it as the usual completion.

Theorem 4.9.1 (Grothendieck) *Let k be an algebraically closed field of characteristic $p \geq 0$, and let X be an integral proper normal curve of genus g over k. Let $U \subset X$ be an open subcurve (possibly equal to X), and $n \geq 0$ the number of closed points in $X \setminus U$. Then $\pi_1(U)^{(p')}$ is isomorphic to the profinite p'-completion of the group*

$$\Pi_{g,n} := \langle a_1, b_1, \dots, a_g, b_g, \gamma_1, \dots, \gamma_n \mid [a_1, b_1] \cdots [a_g, b_g]\gamma_1 \cdots \gamma_n = 1 \rangle.$$

For $k = \mathbf{C}$ we have seen this in Theorem 4.6.7, and the case of a general k of characteristic 0 follows using Theorem 4.6.10. The main contribution of Grothendieck was in positive characteristic; we shall discuss it in more detail in Theorem 5.7.13 and Remark 5.7.16 (4) of the next chapter.

Remark 4.9.2 It is an interesting question whether one can prove Theorem 4.9.1 without recourse to transcendental techniques. It would be enough to show that the finite quotients are exactly the finite groups that can be generated by $2g + n$ elements satisfying a relation as above. Indeed, it can be proven in a purely algebraic way that for fixed $N > 0$ the group $\pi_1(X)^{(p')}$ has only finitely many quotients of order N (Lang–Serre [51], Abhyankar [1]). But a profinite group having this property is determined up to isomorphism by its finite quotients (see Lemma 5.7.7 in the next chapter).

Results in this direction are quite scarce. Borne and Emsalem have observed in a recent preprint [7] that the methods of Serre [92] can be used to determine the finite solvable quotients of $\pi_1(\mathbf{P}^1 \setminus \{0, 1, \infty\})^{(p')}$, and hence to describe the maximal pro-solvable quotient of $\pi_1(\mathbf{P}^1 \setminus \{0, 1, \infty\})^{(p')}$ in a purely algebraic way; see also [54] for further results in this direction. Also, Wingberg used methods of Galois cohomology and class field theory to show that for a normal curve U defined over $\overline{\mathbf{F}}_p$ the maximal pro-ℓ quotient of $\pi_1(U)$ has a presentation as in Theorem 4.9.1 for all primes $\ell \neq p$ (see [70], Theorem 10.1.2).

A special case of Theorem 4.9.1 says that in characteristic 0 the fundamental group of a proper normal curve is completely determined by the genus of the associated compact Riemann surface. In particular, there are many curves having the same fundamental group. However, over the algebraic closure of a finite field the situation is completely different, as a striking result of Tamagawa [103] (building upon earlier work by Raynaud, Pop and Saïdi) shows.

Theorem 4.9.3 (Tamagawa) *Let p be a prime number, and G a profinite group. There are only finitely many proper normal curves of genus $g \geq 2$ over $\overline{\mathbf{F}}_p$ whose fundamental group is isomorphic to G.*

Of course, the theorem is only interesting for those G that actually arise as the fundamental group of some curve as above.

Now that we know the maximal prime-to-p quotient of the fundamental group, we may ask for the structure of its maximal pro-p quotient in characteristic $p > 0$. The answer is radically different in the proper and the affine cases.

In order to be able to state the result for proper curves, we need to recall some facts from algebraic geometry. With a proper normal curve over a field k one associates its Jacobian variety which is an abelian variety, i.e. a projective group variety over k (these are known to be commutative). The *p-rank* of an abelian variety A over an algebraically closed field k of characteristic $p > 0$ is the dimension of the $\mathbf{F}_p$-vector space given by the kernel of the multiplication-by-p map on the k-points of A. It is a non-negative integer bounded by $\dim A$.

Theorem 4.9.4 *For an integral normal curve X over an algebraically closed field k of characteristic $p > 0$ the group $\pi_1(X)^{(p)}$ is a free pro-p group. It is of finite rank equal to the p-rank of the Jacobian variety of X if X is proper, and of infinite rank equal to the cardinality of k if X is affine.*

Here one may define a free pro-p group of rank r as the maximal pro-p quotient of the free profinite group of rank r. The proper case was first proven by Shafarevich in [94] using the classical theory of Hasse–Witt matrices. Nowadays the theorem can be quickly derived using methods of étale cohomology; see the chapters by Gille and Bouw in [8] or [10], Theorem 1.9 (which only deals with the proper case).

Observe that Theorems 4.9.1 and 4.9.4 do not elucidate completely the structure of the fundamental group of an integral normal curve over an algebraically closed field of positive characteristic; this is still unknown at the present day. The theorems give, however, a good description of its maximal *abelian* quotient: this group is the direct sum of its maximal prime-to-p and pro-p quotients, and hence the previous two theorems together suffice to describe it.

By the previous theorem every finite p-group arises as a quotient of the fundamental group of a normal *affine* curve over an algebraically closed field of characteristic $p > 0$. More generally, we have the following important theorem, previously known as Abhyankar's Conjecture.

Theorem 4.9.5 (Raynaud, Harbater) *Let k be an algebraically closed field of characteristic $p > 0$, and let U be an affine curve over k arising from an integral proper normal curve $X \supset U$ of genus g by deleting n points.*

Every finite group G whose maximal prime-to-p quotient can be generated by $2g + n - 1$ elements arises as a quotient of $\pi_1(U)$.

Raynaud proved the crucial case $X = \mathbf{A}^1_k$ in his paper [79]. He constructed Galois covers of the affine line with prescribed group G satisfying the condition of the theorem by combining three different methods. The first came from an earlier paper [92] by Serre that handled the case of solvable G, and contained an inductive statement for extensions of groups satisfying the conclusion of the theorem by solvable groups. The second method used a patching technique like the one we encountered in Chapter 3, but in the setting of rigid analytic geometry. Finally, the third method exploited the theory of semi-stable curves. Soon after Raynaud's work Harbater reduced the general case to the case of the affine line in [32]; another proof for this reduction was given by Pop [76]. See also the chapters by Chambert-Loir and Saïdi in [8].

In the remainder of this section k will denote a perfect field, $\bar{k}$ a fixed algebraic closure, and X an integral proper normal curve over k. For an open subcurve $U \subset X$ we introduced after Proposition 4.7.1 an outer Galois representation

$$\rho_U : \mathrm{Gal}\,(\bar{k}|k) \to \mathrm{Out}(\pi_1(U_{\bar{k}})).$$

In a 1983 letter to Faltings (reprinted in [85]) Grothendieck proposed a conjectural theory named 'anabelian geometry' according to which the above representation should determine U when $\pi_1(U_{\bar{k}})$ is 'far from being abelian'. A satisfactory formulation of the conjectures in higher dimension is not known at present, but in the case of curves precise statements are known, and mostly proven. They concern *hyperbolic curves U*, i.e. integral normal curves such that $2g - 2 + n > 0$, where g is the genus of the compactification $X_{\bar{k}}$, and n is the number of closed points in $X_{\bar{k}} \setminus U_{\bar{k}}$. The group $\pi_1(U_{\bar{k}})$ is indeed nonabelian by Theorem 4.9.1; in characteristic 0 it is even a free profinite group.

The most important result is the following.

Theorem 4.9.6 (Tamagawa, Mochizuki) *Let k be a field that may be embedded in a finitely generated extension of the field $\mathbf{Q}_p$ of p-adic numbers for some prime p. Then every hyperbolic curve U over k is determined up to k-isomorphism by the outer Galois representation ρ_U.*

The theorem was proven by Tamagawa [102] in the important special case when k is a number field and U is affine. He first proved a result over finite fields (see Theorem 4.9.8 below), and then used a specialization argument to obtain the statement in the number field case. Mochizuki first extended Tamagawa's theorem to proper curves over number fields in [61] using techniques from logarithmic geometry, and then in [62] proved the general theorem stated above in a completely different way, exploiting p-adic Hodge theory. For a survey of this second method, see the Bourbaki lecture [22] by Faltings.

In fact, Mochizuki proved much more. In order to formulate the general result of his paper [62] we need to set up some notation. First, for U and p as in the theorem above consider the maximal pro-p-quotient $\pi_1(U_{\bar{k}})^{(p)}$ of the geometric fundamental group $\pi_1(U_{\bar{k}})$. The kernel N of the projection $\pi_1(U_{\bar{k}}) \to \pi_1(U_{\bar{k}})^{(p)}$ is preserved by all automorphisms of $\pi_1(U_{\bar{k}})$, hence it is normal in $\pi_1(U)$. We denote the quotient $\pi_1(U)/N$ by $\pi_1^{(p)}(U)$; it is an extension of $\mathrm{Gal}\,(\bar{k}|k)$ by the pro-p-group $\pi_1(U_{\bar{k}})^{(p)}$. Given profinite groups G_1, G_2 equipped with continuous surjections $G_i \twoheadrightarrow G$ onto a third profinite group G, we denote by $\mathrm{Hom}^*_G(G_1, G_2)$ the set of continuous homomorphisms $G_1 \to G_2$ that are compatible with the projections onto G up to an inner automorphism of G. Composition with inner automorphisms of G_2 equips this set with a left action by G_2; we denote the quotient by $\mathrm{Hom}^{\mathrm{Out}}_G(G_1, G_2)$. We shall see in the next chapter (Remark 5.5.3 and the subsequent discussion) that a morphism $\phi : U \to U'$ of curves over a perfect field k induces a homomorphism on fundamental groups that is well-defined up to an inner automorphism of $\pi_1(U')$, whence a well-defined element in $\mathrm{Hom}^{\mathrm{Out}}_{\mathrm{Gal}\,(\bar{k}|k)}(\pi_1^{(p)}(U), \pi_1^{(p)}(U'))$. If moreover ϕ is *dominating*, i.e. has Zariski dense image, then the induced homomorphism of fundamental groups is known to have open image. Accordingly, given curves U, U' over a perfect field k, denote by $\mathrm{Hom}^{\mathrm{dom}}_k(U, U')$ the set of k-morphisms $U \to U'$ with dense image, and by $\mathrm{Hom}^{\mathrm{Out,\,open}}_{\mathrm{Gal}\,(\bar{k}|k)}(\pi_1^{(p)}(U), \pi_1^{(p)}(U'))$ the subset of $\mathrm{Hom}^{\mathrm{Out}}_{\mathrm{Gal}\,(\bar{k}|k)}(\pi_1^{(p)}(U), \pi_1^{(p)}(U'))$ coming from homomorphisms with open image. We may now state:

Theorem 4.9.7 (Mochizuki) *Let k be a field that may be embedded in a finitely generated extension of $\mathbf{Q}_p$, and let U, U' be hyperbolic curves over k.*

The natural map

$$\mathrm{Hom}_k^{\mathrm{dom}}(U, U') \to \mathrm{Hom}_{\mathrm{Gal}\,(\bar{k}|k)}^{\mathrm{Out,\,open}}(\pi_1^{(p)}(U), \pi_1^{(p)}(U'))$$

is a bijection.

As mentioned above, there are also results for curves over finite fields. Here they are:

Theorem 4.9.8 (Tamagawa, Mochizuki) *Let* $\mathbf{F}$ *be a finite field, and let* U, U' *be hyperbolic curves over* $\mathbf{F}$*. The natural map*

$$\mathrm{Isom}(U, U') \to \mathrm{Isom}(\pi_1(U), \pi_1(U'))$$

is a bijection.

Here on the left-hand side we have the set of isomorphisms between U and U' as schemes, *regardless of the* $\mathbf{F}$*-structure*, and on the right-hand side the set of continuous isomorphisms between $\pi_1(U)$ and $\pi_1(U')$. Note that here we are not considering $\mathrm{Gal}\,(\bar{k}|k)$-isomorphisms between $\pi_1(U_{\overline{\mathbf{F}}})$ and $\pi_1(U'_{\overline{\mathbf{F}}})$ as before; in fact, over a finite field $\mathbf{F}$ the outer Galois action on $\pi_1(U_{\overline{\mathbf{F}}})$ is 'encoded' in $\pi_1(U)$.

The theorem was proven in the affine case by Tamagawa in [102] using class field theory, and in the proper case by Mochizuki [63] as a consequence of his theory of cuspidalizations. Using Tamagawa's affine result and a specialization technique, Stix [98] proved an analogue of Theorem 4.9.6 for non-constant hyperbolic curves over a finitely generated field of positive characteristic.

We conclude this survey by stating the main open question in the area, the famous Section Conjecture of Grothendieck. Let k be a perfect field, and X an integral *proper* normal curve over k. Recall from Proposition 4.7.1 that there is an exact sequence

$$1 \to \pi_1(X_{\bar{k}}) \to \pi_1(X) \xrightarrow{p} \mathrm{Gal}\,(\bar{k}|k) \to 1$$

of profinite groups. Assume $P \in X$ is a k-rational point, and let $\widetilde{P}$ be a point lying above P on the profinite cover $\widetilde{X} \to X$ introduced in Remark 4.6.6. By Corollary 4.7.3 the stabilizer of $\widetilde{P}$ in $\pi_1(X)$ maps isomorphically onto $\mathrm{Gal}\,(\bar{k}|k)$ by the projection p, and hence yields a *section* of p. The different points $\widetilde{P}$ above P are conjugate by the action of $\pi_1(X)$ (in fact, already by the action of $\pi_1(X_{\bar{k}})$, as P was assumed to be k-rational). Hence P gives rise to a conjugacy class of sections (see Remark 5.6.3 (2) for another approach). In our previous notation we may write that we have constructed a map

$$X(k) \to \mathrm{Hom}_{\mathrm{Gal}\,(\bar{k}|k)}^{\mathrm{Out}}(\mathrm{Gal}\,(\bar{k}|k), \pi_1(X)).$$

Conjecture 4.9.9 (Section Conjecture) *If* k *is finitely generated over* $\mathbf{Q}$ *and* X *has genus* ≥ 2*, then the above map is a bijection.*

In other words, each conjugacy class of sections should come from a unique k-rational point of X. The proof of injectivity is not hard and was already known to Grothendieck, but as of today, the issue of surjectivity is widely open.

Remark 4.9.10 In his letter to Faltings Grothendieck also formulated a variant of the Section Conjecture for an affine hyperbolic curve U. As the profinite branched cover $\widetilde{X}$ of Remark 4.6.6 carries an action of $\pi_1(U)$, each section $\mathrm{Gal}(\bar{k}|k) \to \pi_1(U)$ of the natural projection $\pi_1(U) \to \mathrm{Gal}(\bar{k}|k)$ induces an action of $\mathrm{Gal}(\bar{k}|k)$ on $\widetilde{X}$. Grothendieck then conjectured that this action should have a unique fixed point. As the $\mathrm{Gal}(\bar{k}|k)$-actions on $\widetilde{X}$ and $X_{\bar{k}}$ are compatible by construction, the fixed point lies above a k-rational point P of X. If $P \in U$, then the section should come from the stabilizer of a point of $\widetilde{X}$ above P. If $P \in (X \setminus U)$, then the cardinality of the sections inducing a fixed point above P should be the continuum. This latter assertion was recently verified by Esnault and Hai [20] as well as Stix [99].

Exercises

1. Let k be an algebraically closed field of characteristic not 2, $f \in k[x_1]$ a nonconstant polynomial, $Y = V(x_2^2 - f) \subset \mathbf{A}_k^2$ and $\phi : Y \to \mathbf{A}_k^1$ the morphism given by $(x_1, x_2) \mapsto x_1$. Show that ϕ is a finite morphism that is étale over the point of $\mathbf{A}_k^1$ corresponding to $a \in k$ if and only if $f(a) \neq 0$.
2. Let k be an algebraically closed field of characteristic 0, and $\phi : Y \to \mathbf{A}_k^1$ a finite surjective morphism over k such that $\mathcal{O}(Y) \cong \mathcal{O}(\mathbf{A}_k^1)[f]$ with some $f \in \mathcal{O}(Y)$. Verify directly that ϕ cannot be étale above the whole of $\mathbf{A}_k^1$.
3. Let k be an algebraically closed field, and n an integer prime to the characteristic of k. Prove in a purely algebraic way that up to isomorphism there is a unique cyclic Galois cover of $\mathbf{A}_k^1$ with group $\mathbf{Z}/n\mathbf{Z}$ étale outside 0, as in Example 4.6.12 (2). [*Hint:* Use Kummer theory (Remark 1.2.10.)]
4. Let k be an algebraically closed field of characteristic $p > 0$, and consider the rational function field $k(t)$. Choose $f \in k[t]$ so that the polynomial $x^p - x - f$ in $k(t)[x]$ is irreducible over $k(t)$.
   (a) Show that the normalization of $\mathbf{A}_k^1$ in the finite extension of $k(t)$ defined by $x^p - x - f$ is an étale Galois cover of $\mathbf{A}_k^1$ whose Galois group is $\mathbf{Z}/p\mathbf{Z}$; it is called an *Artin–Schreier cover* of $\mathbf{A}_k^1$.
   (b) Conclude that the maximal abelian quotient of exponent p of $\pi_1(\mathbf{A}_k^1)$ is an infinite dimensional $\mathbf{F}_p$-vector space of dimension equal to the cardinality of k.
5. Give an example showing that the converse statement in Lemma 4.5.10 fails if one does not assume that Y is normal.
   [*Hint:* Take $X = \mathbf{A}_k^1$, $Y = V(x_2^2 - x_1^3 + x_1^2) \subset \mathbf{A}_k^2$, Z the normalization of Y and $\phi : (x_1, x_2) \mapsto x_1$.]
6. Let k be an algebraically closed field, X an integral proper normal curve over k, and $f \in K(X)$ a rational function.

(a) Verify that the morphism $\phi_f : X \to \mathbf{P}^1_k$ defined in Section 4.4 satisfies $\phi_f(P) = f(P)$ for all closed points $P \in X$ where $f \in \mathcal{O}_{X,P}$, and $\phi_f(P) = \infty$ otherwise.

(b) Show that ϕ_f is étale exactly above those points whose preimages satisfy $(f'/f)(P) \neq 0$.

7. Prove the following sharpening of Belyi's theorem: an integral proper normal curve over an algebraically closed field k of characteristic 0 can be defined over $\overline{\mathbf{Q}}$ if and only if there is a morphism $X \to \mathbf{P}^1_k$ étale over $\mathbf{P}^1_k \setminus \{0, 1, \infty\}$ such that all ramification indices at points lying above 1 are equal to 2. [*Hint:* Compose with an appropriate Belyi function.]
8. Prove that when G is a finite group with trivial centre having a rigid system $(C_1, \dots, C_n)$ of not necessarily rational conjugacy classes, the conclusion of Theorem 4.8.7 holds with $\mathbf{Q}$ replaced by $\mathbf{Q}(\mu)$, where μ is the group of all complex roots of unity.
9. Verify that for $n \geq 3$ the three conjugacy classes in the symmetric group S_n consisting of 2-cycles, $(n-1)$-cycles and n-cycles, respectively, form a rigid system of rational conjugacy classes. [*Remark:* It then follows from Theorem 4.8.7 that S_n occurs as a Galois group over $\mathbf{Q}(t)$. For a more elementary proof, see [93], Section 4.4.]
10. Let G be a finite group with trivial centre possessing a rigid system of three rational conjugacy classes, and let $H \subset G$ be a subgroup of index 2. (Example: $G = S_n$, $H = A_n$, $n \geq 3$.) Show there exists a Galois extension of $\mathbf{Q}(t)$ regular over $\mathbf{Q}$ with group H.
[*Hint:* Apply Theorem 4.8.7 to get a Galois extension $K|\mathbf{Q}(t)$ with group G corresponding to a morphism étale outside three $\mathbf{Q}$-rational points, and set $L := K^H$. Use the Riemann–Hurwitz formula (Corollary 3.6.12) to show that the morphism $X \to \mathbf{P}^1_{\mathbf{Q}}$ corresponding to $L|\mathbf{Q}(t)$ is étale outside exactly two points, and conclude that $L \cong \mathbf{Q}(u)$ for some $u \in K(X)$.]
11. (Tamagawa, Koenigsmann) Let X be a proper normal curve over a perfect field k. Fix a separable closure K_s of the function field K of X, let $K_X \subset K_s$ be the Galois subextension with Galois group $\pi_1(X)$ over K, and $\bar{k}$ the algebraic closure of k in K_X.

(a) Let s be a section of the natural projection $\pi_1(X) \to \mathrm{Gal}(\bar{k}|k)$, and denote by D the image of s in $\pi_1(X)$; it has a natural action on K_X. Prove the equivalence of the following statements about a k-rational point P of X:

- D is the decomposition group of a closed point of the pro-étale cover $\widetilde{X} \to X$ of Remark 4.6.6 lying above P.
- For all finite extensions $L|K$ contained in the fixed field of D the corresponding proper normal curve has a k-rational point lying above P.

(b) Show that the surjectivity part of Grothendieck's Section Conjecture (Conjecture 4.9.9) is equivalent to the following statement: if k is finitely generated over $\mathbf{Q}$, then for *every* proper normal curve X of genus at least 2 the projection $\pi_1(X) \to \mathrm{Gal}(\bar{k}|k)$ has a section if and only if X has a k-rational point. [*Hint*: Use the generalization of the theorem of Faltings ([49], Chapter I, Corollary 2.2) according to which such an X has finitely many k-rational points.]

5

Fundamental groups of schemes

Though the theory of the previous chapter is sufficient for many applications, a genuine understanding of the algebraic fundamental group only comes from Grothendieck's definition of the fundamental group for schemes. His theory encompasses the classification of finite covers of complex algebraic varieties of any dimension, Galois theory for extensions of arbitrary fields and even notions coming from arithmetic such as specialization modulo a prime. Moreover, it is completely parallel to the topological situation and clarifies the role of base points and universal covers – these have been somewhat swept under the carpet in the last chapter. In his original account in [29] Grothendieck adopted an axiomatic viewpoint and presented his constructions within the context of 'Galois categories'. Here we choose a more direct approach, emphasizing the parallelism with topology. The background from algebraic geometry that is necessary for the basic constructions will be summarized in the first section. However, the proofs of some of the deeper results discussed towards the end of the chapter will require more refined techniques.

5.1 The vocabulary of schemes

In this section we collect the basic notions from the language of schemes that we shall need for the development of Grothendieck's theory of the fundamental group. Our intention is to summarize for the reader what will be needed; this concise overview can certainly not replace the study of standard references such as [34] or [64], let alone Grothendieck's *magnum opus* EGA.

We first define affine schemes. Let A be a ring (commutative with unit, as usual). The affine scheme associated with A will be a certain ringed space (recall from Section 4.4 that this means a topological space together with a sheaf of rings on it). Here is the definition of the underlying topological space.

Definition 5.1.1 The *prime spectrum* $\operatorname{Spec}(A)$ of A is the topological space whose points are prime ideals of A, and a basis of open sets is given by the sets

$$D(f) := \{P : P \text{ is a prime ideal with } f \notin P\}$$

for all $f \in A$.

For this definition to be correct, one must verify that the system of the sets $D(f)$ is closed under finite intersections. This holds because it follows from the definition that $D(f) \cap D(g) = D(fg)$ for all $f, g \in A$.

We next define a sheaf of rings on $X = \mathrm{Spec}\,(A)$ as follows. Recall that for a nonzero element $f \in A$ the notation A_f stands for the ring of fractions of A with denominators in the multiplicatively closed subset $\{1, f, f^2, \dots\} \subset A$. See ([2], Chapter 3) for the definition of rings of fractions in the case when A may contain zero-divisors.

Lemma 5.1.2 *There is a unique sheaf of rings $\mathcal{O}_X$ on $X = \mathrm{Spec}\,(A)$ satisfying $\mathcal{O}_X(D(f)) = A_f$ for all nonzero $f \in A$.*

Proof See [64], Section II.1 (or [34], Proposition II.2.2 for a direct construction). □

Definition 5.1.3 The ringed space $(X, \mathcal{O}_X)$ defined above is the *affine scheme* associated with A. The sheaf $\mathcal{O}_X$ is called its *structure sheaf.*

The structure sheaf $\mathcal{O}_X$ has the special property that its stalk at a point P is a local ring, namely the localization A_P. This follows from the fact that A_P is the direct limit of the fraction rings A_f for all $f \notin P$. A ringed space satisfying the above property is called a *locally ringed space*.

Recall that a morphism $(X, \mathcal{F}) \to (Y, \mathcal{G})$ of ringed spaces is a pair $(\phi, \phi^\sharp)$, where $\phi : X \to Y$ is a continuous map and $\phi^\sharp : \mathcal{G} \to \phi_*\mathcal{F}$ is a morphism of sheaves. Given a point $P \in X$, there is an induced map $\phi_P : \mathcal{G}_{\phi(P)} \to \mathcal{F}_P$ on stalks given by passing to the direct limit of the maps $\mathcal{G}(U) \to \phi_*\mathcal{F}(U)$ over all open subsets U containing $\phi(P)$. If $(X, \mathcal{F})$ and $(Y, \mathcal{G})$ are locally ringed spaces, we say that ϕ is a *local* homomorphism if the preimage of the maximal ideal of $\mathcal{F}_P$ is the maximal ideal of $\mathcal{G}_{\phi(P)}$. (In the case of proper normal curves considered in Section 4.4 this condition is automatically fulfilled by the morphisms considered there.)

The category of locally ringed spaces is defined to be the category whose objects are locally ringed spaces and whose morphisms are local homomorphisms.

Definition 5.1.4 A *scheme* is a locally ringed space $(X, \mathcal{O}_X)$ having an open covering $\{U_i : i \in I\}$ such that for all i the locally ringed spaces $(U_i, \mathcal{O}_X|_{U_i})$ are isomorphic to affine schemes.

We define the category of schemes (resp. affine schemes) as the *full* subcategory of that of locally ringed spaces spanned by schemes (resp. affine schemes). This means that a morphism of schemes is a morphism of locally ringed spaces.

Given a ring homomorphism $\phi : A \to B$, there is a canonically defined morphism $\mathrm{Spec}\,(\phi) : \mathrm{Spec}\,(B) \to \mathrm{Spec}\,(A)$ of schemes ([34], II.2.3 (b) or [64], II.2). The continuous map $\mathrm{Spec}\,(\phi)$ sends a prime ideal $P \in \mathrm{Spec}\,(B)$ to $\phi^{-1}(P)$, and the morphism $\mathrm{Spec}\,(\phi)^\sharp$ of sheaves is the unique one that is given over a basic open set $D(f) \subset \mathrm{Spec}\,(A)$ by the natural ring homomorphism $A_f \to B_{\phi(f)}$. Thus the rule $A \mapsto (\mathrm{Spec}\,(A), \mathcal{O}_{\mathrm{Spec}\,(A)})$ is a contravariant functor from the category of rings to that of schemes.

Proposition 5.1.5 *The above contravariant functor induces an isomorphism of the category of affine schemes with the opposite category of commutative rings with unit. The inverse functor is given by $(X, \mathcal{O}_X) \to \mathcal{O}_X(X)$.*

Proof See [34], Proposition II.2.3 or [64], II.2, Corollary 1. □

Examples 5.1.6 Here are some examples of schemes.

1. If k is a field, then the underlying space of $\mathrm{Spec}\,(k)$ consists of a single point, and the stalk of the structure sheaf at this point is k.
2. More generally, if $A \cong \bigoplus L_i$ is a finite étale k-algebra with some finite separable extensions $L_i|k$, then $\mathrm{Spec}\,(A)$ is the disjoint union of the one-point schemes $\mathrm{Spec}\,(L_i)$.
3. If A is a discrete valuation ring with fraction field K, then $X = \mathrm{Spec}\,(A)$ has one closed point corresponding to the maximal ideal M, and one non-closed point corresponding to the ideal (0). There are two open subsets, namely (0) and X. The rings of sections of $\mathcal{O}_X$ over these open subsets are K and A, respectively.
4. The affine scheme $\mathrm{Spec}\,(\mathbf{Z})$ has one closed point corresponding to each prime number p, and a non-closed point corresponding to (0). The nonempty open subsets all contain (0), and their complement in $\mathrm{Spec}\,(\mathbf{Z})$ is finite (possibly empty). The ring of sections of the structure sheaf over an open subset U is the ring of rational numbers with denominator divisible only by the primes lying outside U. This example has a direct generalization to affine schemes of the form $\mathrm{Spec}\,(A)$ with A a Dedekind ring.
5. If k is a field and A is a finitely generated k-algebra, then the closed points of $X = \mathrm{Spec}\,(A)$ correspond bijectively to the closed points of an affine variety with coordinate ring A as defined in Section 4.2. Moreover, the stalk of $\mathcal{O}_X$ at a closed point P is exactly the local ring $\mathcal{O}_{X,P}$ as defined in the previous chapter, and the topology induced on the subset $X_0 \subset X$ of closed points is exactly the Zariski topology as defined there. However, X has non-closed points as well, corresponding to non-maximal prime ideals.

6. A proper normal curve X_K as defined in Section 4.4 gives rise to a non-affine scheme. The only difference with the definition given there is that one has to add an extra point to the underlying topological space that is contained in every open subset. It corresponds to the ideals (0) in the local rings of X, and is called the *generic point.* The structure sheaf is defined in the same way. Its stalk at the generic point is isomorphic to K.

We now generalize a few other notions from the previous chapter to schemes.

Definition 5.1.7 A scheme X is called *reduced* (resp. *integral*) if for all open subsets $U \subset X$ the ring $\mathcal{O}_X(U)$ has no nilpotent elements (resp. no zero-divisors).

Recall that a topological space is irreducible if it cannot be expressed as a union of two proper closed subsets.

Lemma 5.1.8 *A scheme X is integral if and only if it is reduced and its underlying space is irreducible.*

Proof See [34], Proposition II.3.1. □

An integral scheme always has a unique point whose closure is the whole underlying space of the scheme (see [64], §II.2), its *generic point.* In the affine case it corresponds to the ideal (0).

Definition 5.1.9 The *dimension* $\dim X$ of a scheme X is the supremum of the integers n for which there exists a strictly increasing chain $Z_0 \subset Z_1 \subset \cdots \subset Z_n$ of irreducible closed subsets properly contained in X. The dimension $\dim A$ of a ring is the dimension of $\operatorname{Spec}(A)$.

There are topological properties of schemes coming from the topology of the underlying space. Thus for instance we say that a scheme is *connected* or that it is *quasi-compact* (meaning that from each open covering we may extract a finite subcovering) if the underlying space is. Similarly, properties of local rings define algebraic restrictions. A scheme X is thus said to be *locally Noetherian* if the stalks $\mathcal{O}_{X,P}$ of its structure sheaf are Noetherian local rings; if moreover X is quasi-compact, then we say that X is Noetherian. The scheme X is *normal* if the stalks $\mathcal{O}_{X,P}$ are integrally closed domains, and *regular* if the $\mathcal{O}_{X,P}$ are regular local rings. (Recall that a Noetherian local ring A with maximal ideal M and residue field $\kappa = A/M$ is said to be regular if $\dim_\kappa M/M^2 = \dim A$.) A regular scheme is integral if and only if it is connected.

Next we mention some important examples of morphisms of not necessarily affine schemes.

Examples 5.1.10

1. Given a scheme X and an open subset $U \subset X$, the ringed space given by U and $\mathcal{O}_U := (\mathcal{O}_X)|_U)$ is also a scheme, the *open subscheme* associated with U. The morphism of schemes defined by the topological inclusion $j : U \to X$ and the morphism of sheaves $\mathcal{O}_X \to j_*\mathcal{O}_U$ is called an *open immersion*.
2. A morphism $Z \to X$ of affine schemes is a closed immersion if it corresponds via Proposition 5.1.5 to a quotient map $A \to A/I$ for some ideal $I \subset A$. A general morphism is a *closed immersion* if it is injective with closed image and its restrictions to elements of an affine open covering yield closed immersions in the above sense.

We now come to an important construction for schemes that will in particular enable us to define the fibres of a morphism. Before Construction 2.4.13 we introduced the fibre product $Y \times_X Z$ of two topological spaces Y and Z equipped with maps to a third space X as the subspace of those points $(y, z) \in Y \times Z$ where y and z map to the same point of X. We also remarked that the fibre product is characterized by a universal property. It is this definition via the universal property that carries over to the category of schemes. Before we state it precisely, we define, as in topology, *a scheme over* X to be a morphism $Y \to X$ of schemes. A morphism of schemes over X is a morphism $Y \to Z$ compatible with the projections to X. In the case when $X = \mathrm{Spec}\,(k)$ for a field k, we shall abusively speak of a k-scheme.

Proposition 5.1.11 *Given a scheme X and two morphisms $p : Y \to X$, $q : Z \to X$ of schemes, the contravariant functor*

$$S \mapsto \{(\phi, \psi) \in \mathrm{Hom}(S, Y) \times \mathrm{Hom}(S, Z) : p \circ \phi = q \circ \psi\}$$

on the category of schemes over X is representable by a scheme $Y \times_X Z$ over X.

Proof See [34], Theorem II.3.3. In the case when $X = \mathrm{Spec}\,(A)$, $Y = \mathrm{Spec}\,(B)$ and $Z = \mathrm{Spec}\,(C)$ are all affine, then $Y \times_X Z$ is $\mathrm{Spec}\,(B \otimes_A C)$. The general case is handled by a patching procedure. □

The scheme $Y \times_X Z$ is called the *fibre product of Y and Z over X*. It is equipped with two canonical morphisms to Y and Z making the diagram

$$\begin{array}{ccc} Y \times_X Z & \xrightarrow{\pi_2} & Z \\ \downarrow{\scriptstyle \pi_1} & & \downarrow{\scriptstyle q} \\ Y & \xrightarrow{p} & X \end{array}$$

commute (they correspond to the identity morphism of $Y \times_X Z$ via Proposition 5.1.11). We say that the morphism π_1 is the *base change* of q via p, and similarly π_2 is the base change of p via q.

We now define the fibre of a morphism $Y \to X$ at a point P of X. For this we first need to define the inclusion morphism $i_P : \mathrm{Spec}\,(\kappa(P)) \to X$ of the point P in X. If $U = \mathrm{Spec}\,A$ is an affine open subset of X containing P, then P is identified with a prime ideal of A and we dispose of a morphism $A \to A_P$ that we may compose with the natural projection $A_P \to A_P/PA_P = \kappa(P)$. By Proposition 5.1.5 it corresponds to a morphism $\mathrm{Spec}\,(\kappa(P)) \to U$, whence i_P by composition with the inclusion map $U \to X$. It is readily verified that i_P does not depend on the choice of U.

Definition 5.1.12 Given a morphism $\phi : Y \to X$ and a point P of X, the *fibre of ϕ at P* is the scheme $Y_P := Y \times_X \mathrm{Spec}\,(\kappa(P))$, the fibre product being taken with respect to the maps ϕ and i_P.

Remark 5.1.13 The underlying topological space of a fibre product of schemes is not the topological fibre product of the underlying spaces in general. For instance, if k is a field, k_s a separable closure and $L|k$ a separable extension, then $\mathrm{Spec}\,(L) \times_{\mathrm{Spec}\,(k)} \mathrm{Spec}\,(k_s) = \mathrm{Spec}\,(L \otimes_k k_s)$ is a finite disjoint union of copies of $\mathrm{Spec}\,(k_s)$, whereas the topological fibre product is just a point.

However, given a morphism $\phi : Y \to X$ of schemes and a point P of X, the underlying topological space of the fibre Y_P is homeomorphic to the subspace $\phi^{-1}(P)$ of the underlying space of Y ([34], Ex. II.3.10).

Example 5.1.14 When $X = \mathrm{Spec}\,(A)$ with A a discrete valuation ring, a morphism $Y \to X$ has two fibres. One is over the generic point of X called the *generic fibre*; it is an open subscheme in X and may be empty. The other fibre is the one over the closed point; it is closed in X and is usually called the *special fibre*.

The notion of fibre product also allows us to define the diagonal map $\Delta : Y \to Y \times_X Y$ coming from a morphism of schemes $Y \to X$; it is induced by the identity map of Y in both coordinates. In the affine case $X = \mathrm{Spec}\,(A)$, $Y = \mathrm{Spec}\,(B)$ it is a closed immersion coming from the surjection $B \otimes_A B \to B$ induced by the multiplication map $(b_1, b_2) \mapsto b_1 b_2$. However, this is not always so in the general case ([34], Example 4.0.1), so we record it as a definition:

Definition 5.1.15 A morphism $Y \to X$ of schemes is *separated* if the diagonal map $\Delta : Y \to Y \times_X Y$ is a closed immersion.

The separatedness property of schemes is an analogue of the Hausdorff property in topology: one checks that the topological diagonal map $Y \to Y \times Y$

has closed image if and only if Y is a Hausdorff space. Separatedness thus intuitively corresponds to the Hausdorff property for all fibres. The next important definition should be thought of as the scheme-theoretic analogue of a morphism with compact fibres. First a general notion: a morphism $\phi : Y \to X$ is *locally of finite type* if X has an affine open covering by subsets $U_i = \mathrm{Spec}\,(A_i)$ so that $\phi^{-1}(U_i)$ has an open covering $V_{ij} = \mathrm{Spec}\,(B_{ij})$ with finitely generated A_i-algebras B_{ij}. It is of *finite type* if there is such an open covering with finitely many V_{ij} for each i.

Definition 5.1.16 A separated morphism $Y \to X$ of schemes is *proper* if it is of finite type and for every morphism $Z \to X$ the base change map $Y \times_X Z \to Z$ is a closed map (i.e. maps closed subsets onto closed subsets).

Examples 5.1.17

1. A closed immersion is proper ([34], Corollary II.4.8 (a)).
2. A finite morphism $Y \to X$ is proper ([64], §II.7). Here a finite morphism is defined as in the previous chapter: there is an affine open covering of X by subsets $U_i = \mathrm{Spec}\,(A_i)$ so that $\phi^{-1}(U_i) = \mathrm{Spec}\,(B_i)$ is affine, and B_i is finitely generated as an A_i-module. In fact, under a finite morphism the inverse image of every affine open subset of X has this property ([64], §II.7, Proposition 5). It follows that the class of finite morphisms is stable by composition and base change.
3. An important example of a proper morphism is the projective line $\mathbf{P}^1_X$ over a scheme X. For $X = \mathrm{Spec}\,(A)$ affine, it can be defined by patching the ringed spaces $\mathrm{Spec}\,(A[x])$ and $\mathrm{Spec}\,(A[x^{-1}])$ together along the isomorphic open subschemes $\mathrm{Spec}\,(A[x, x^{-1}])$. For general X one takes an open covering of X by open affine subschemes U_i and patches the $\mathbf{P}^1_{U_i}$ together in a straightforward way.

   More generally, one defines $\mathbf{P}^n_X$ for affine X by patching together $n+1$ copies of $\mathrm{Spec}\,(A[x_1, \dots, x_n])$ along the open subsets defined by $x_i \neq 0$ in the usual way. One then defines a morphism $Y \to X$ to be *projective* if it factors as a closed immersion $Y \to \mathbf{P}^n_X$ for some n, followed by the natural projection $\mathbf{P}^n_X \to X$. All these morphisms are proper ([34], Theorem II.4.9).

Next a general notion for schemes that generalizes the concept of modules over a ring.

Definition 5.1.18 Let X be a scheme. A *sheaf of $\mathcal{O}_X$-modules* or an *$\mathcal{O}_X$-module* for short is a sheaf of abelian groups $\mathcal{F}$ on X such that for each open $U \subset X$ the group $\mathcal{F}(U)$ is equipped with an $\mathcal{O}_X(U)$-module structure

$\mathcal{O}_X(U) \times \mathcal{F}(U) \to \mathcal{F}(U)$ making the diagram

$$\begin{array}{ccc} \mathcal{O}_X(U) \times \mathcal{F}(U) & \longrightarrow & \mathcal{F}(U) \\ \downarrow & & \downarrow \\ \mathcal{O}_X(V) \times \mathcal{F}(V) & \longrightarrow & \mathcal{F}(V) \end{array}$$

commute for each inclusion of open sets $V \subset U$. In the special case when $\mathcal{F}(U)$ is an ideal in $\mathcal{O}_X(U)$ for all U we speak of a *sheaf of ideals* on X.

Examples 5.1.19 Here are two natural situations where $\mathcal{O}_X$-modules arise.

1. Let $\phi : X \to Y$ be a morphism of schemes. We know that at the level of structure sheaves ϕ is given by a morphism $\phi^\sharp : \mathcal{O}_Y \to \phi_*\mathcal{O}_X$, whence an $\mathcal{O}_Y$-module structure on $\phi_*\mathcal{O}_X$.
2. In the previous situation the kernel $\mathcal{I}$ of the morphism $\phi^\sharp : \mathcal{O}_Y \to \phi_*\mathcal{O}_X$ (defined by $\mathcal{I}(U) = \ker(\mathcal{O}_Y(U) \to \phi_*\mathcal{O}_X(U)))$ is a sheaf of ideals on Y.

The next proposition gives a means for constructing $\mathcal{O}_X$-modules over affine schemes out of modules over the ring of global sections.

Lemma 5.1.20 *Let $X = \operatorname{Spec}(A)$ be an affine scheme, and M an A-module. There is a unique $\mathcal{O}_X$-module $\widetilde{M}$ satisfying $\widetilde{M}(D(f)) = M \otimes_A A_f$ over each basic open set $D(f) \subset X$.*

Proof The proof is similar to that of Lemma 5.1.2. See [64], Section III.1, or [34], Proposition II.5.1 for a direct approach. □

Definition 5.1.21 Let X be a scheme. A *quasi-coherent sheaf* on X is an $\mathcal{O}_X$-module $\mathcal{F}$ for which there is an open affine cover $\{U_i : i \in I\}$ of X such that the restriction of $\mathcal{F}$ to each $U_i = \operatorname{Spec}(A_i)$ is isomorphic to an $\mathcal{O}_{U_i}$-module of the form $\widetilde{M}_i$ with some A_i-module M_i. If moreover each M_i is finitely generated over A_i, then $\mathcal{F}$ is a *coherent sheaf*. The sheaf $\mathcal{F}$ is *locally free* if we may choose the above data in such a way that the M_i are free A-modules. Locally free sheaves of rank 1 are called *invertible sheaves*.

Remark 5.1.22 It can be shown that for an affine scheme $X = \operatorname{Spec}(A)$ the functor $M \to \widetilde{M}$ establishes an equivalence between the category of A-modules and that of quasi-coherent sheaves on X. See [34], Corollary II.5.5 or [64], III.1, Corollary to Proposition 1.

We now return to the first example in 5.1.19 and investigate the question of determining whether a morphism $\phi : X \to Y$ yields a quasi-coherent sheaf $\phi_*\mathcal{O}_X$ on Y. This is not true in general, but imposing restrictions on ϕ yields sufficient conditions.

Here is such a condition: a morphism $\phi : X \to Y$ of schemes is *affine* if Y has a covering by affine open subsets $U_i = \operatorname{Spec} A_i$ such that for each i the open subscheme $\phi^{-1}(U_i)$ of X is affine as well. By definition, finite morphisms (Example 5.1.10 (2)) are affine. The following lemma is then almost tautological for $\mathcal{F} = \mathcal{O}_X$, and the general case can be reduced to it; we leave it as an exercise to the readers.

Lemma 5.1.23 *If $\phi : X \to Y$ is an affine morphism and $\mathcal{F}$ is a quasi-coherent sheaf on X, then $\phi_*\mathcal{F}$ is a quasi-coherent sheaf on Y. If moreover $\mathcal{F}$ is coherent and ϕ is finite, then $\phi_*\mathcal{F}$ is coherent.*

Remark 5.1.24 If $\phi : X \to Y$ is an affine morphism, then $\phi_*\mathcal{O}_X$ is actually a sheaf of $\mathcal{O}_Y$*-algebras*. It can be shown that the contravariant functor $X \mapsto \phi_*\mathcal{O}_X$ induces an anti-equivalence of categories between affine morphisms $X \to Y$ and quasi-coherent sheaves of $\mathcal{O}_X$-algebras. The inverse functor is obvious when X and Y are affine, and the general case is handled by patching; see [34], Exercise II.5.17.

Remark 5.1.25 Nakayama's lemma ([48], Chapter X, Lemma 4.3) is a key tool in the study of coherent sheaves. For instance, it can be used to show that if the stalk of a coherent sheaf is 0 at a point P, then the sheaf restricts to 0 in an open neighbourhood of P. Similarly, if the stalk at P is a free $\mathcal{O}_{X,P}$-module, then the sheaf is locally free in a neighbourhood of P. See the discussion at the end of §III.2 of [64].

We now introduce an important class of quasi-coherent sheaves, the sheaves of relative differentials. First a construction for rings.

Definition 5.1.26 Given a morphism $A \to B$ of rings, we define the B-module $\Omega^1_{B|A}$ of *relative differential forms* as follows: if $F(B)$ denotes the free B-module generated by symbols db for all $b \in B$, let $\Omega^1_{B|A}$ be the quotient of $F(B)$ by the submodule generated by elements of the form da, $d(b_1 + b_2) - db_1 - db_2$ or $d(b_1b_2) - b_1db_2 - b_2db_1$ for some $a \in A$ or $b_1, b_2 \in B$.

Example 5.1.27 Assume that B arises as the quotient of the polynomial ring $A[x_1, \ldots, x_n]$ by an ideal $(f_1, \ldots, f_m)$. Then $\Omega^1_{B|A}$ is the quotient of the free B-module generated by the dx_i modulo the submodule generated by the elements $\sum_i \partial_i f_j dx_i$ $(1 \le j \le m)$, where ∂_i denotes the partial derivative with respect to x_i. This follows immediately from the above construction.

It follows from the construction that for an A-algebra A' one has an isomorphism $\Omega^1_{B\otimes_A A'|A'} \cong \Omega^1_{B|A} \otimes_A A'$; this is the *base change property* of differentials. Also, for a multiplicative subset S of B one has $\Omega^1_{B_S|A} \cong \Omega^1_{B|A} \otimes_B B_S$; this is the *localization property* of differentials. It follows from the latter property that the quasi-coherent sheaf $\widetilde{\Omega}^1_{B|A}$ on Spec (B) satisfies $\widetilde{\Omega}^1_{B|A}(D(f)) = \Omega^1_{B_f|A}$ for all basic open sets $D(f) \subset$ Spec (B).

This defines the sheaf of relative differential forms for morphisms of affine schemes. To extend the definition to general morphisms we need an algebraic lemma. Let $A \to B$ be a ring homomorphism, and let I be the kernel of the multiplication map $m : B \otimes_A B \to B$ sending $b_1 \otimes b_2$ to $b_1 b_2$. Then since multiplication by elements of I is trivial on I/I^2, there is a natural $(B \otimes_A B)/I$-module structure on I/I^2. This is in fact a B-module structure since $(B \otimes_A B)/I \cong B$ by surjectivity of m.

Lemma 5.1.28 *There is a canonical isomorphism of B-modules $\Omega^1_{B|A} \cong I/I^2$.*

Proof See [64], III.1, Theorem 4. □

The lemma motivates the following construction. Let $\phi : X \to Y$ be a separated morphism of schemes, and let $\Delta : X \to X \times_Y X$ be the associated diagonal morphism. Let $\mathcal{I} \subset \mathcal{O}_{X \times_Y X}$ be the kernel of the morphism of structure sheaves $\Delta^\sharp : \mathcal{O}_{X \times_Y X} \to \Delta_* \mathcal{O}_X$. It defines the closed subscheme $\Delta(X) \subset X \times_Y X$. As in the affine case, we see that $\mathcal{I}/\mathcal{I}^2$ is a $\mathcal{O}_{X \times_Y X}/\mathcal{I}$-module. But the latter sheaf is zero outside $\Delta(X)$, and may be identified with $\mathcal{O}_{\Delta(X)}$ (more precisely, with the extension of this sheaf by 0).

Definition 5.1.29 The sheaf of relative differentials $\Omega_{X|Y}$ is the $\mathcal{O}_X$-module defined by pulling back the $\mathcal{O}_{\Delta(X)}$-module $(\mathcal{I}/\mathcal{I}^2)$ via the isomorphism $X \xrightarrow{\sim} \Delta(X)$.

We finally discuss two important applications of differentials. The first is the definition of smoothness for a scheme of finite type over a field k (i.e. for a morphism $X \to$ Spec (k) of finite type).

Definition 5.1.30 Let k be a field, and X a separated k-scheme of finite type whose irreducible components are of the same dimension d. We say that X is *smooth* over k if the sheaf $\Omega_{X|\mathrm{Spec}\,(k)}$ is locally free of rank d.

Using Example 5.1.27 one shows that for k algebraically closed the condition on $\Omega_{X|\mathrm{Spec}\,(k)}$ is equivalent to saying that for an affine open subset of the form $U =$ Spec $(k[x_1, \ldots, x_n]/(f_1, \ldots, f_m))$ the Jacobian determinant of the f_i has rank $n - d$ at all closed points of U; see [34], Theorem II.8.15. Thus we have defined an algebraic analogue of the concept of a complex submanifold of $\mathbf{C}^n$. For k perfect there is also a purely algebraic characterization going back to Zariski: X is smooth over k if and only if it is a regular scheme (see [34], Theorem II.8.15 for k algebraically closed; the proof of the general case is similar).

The other application is the differential characterization of finite étale algebras.

Proposition 5.1.31 *A finite dimensional algebra A over a field is étale if and only if $\Omega^1_{A|k} = 0$.*

Proof The 'only if' part is an easy application of Example 5.1.27 and the primitive element theorem. The 'if' part is a bit more difficult; see e.g. [18], Corollary 16.16. □

5.2 Finite étale covers of schemes

We now come to the general definition of finite étale morphisms of schemes.

Definition 5.2.1 A finite morphism $\phi : X \to S$ of schemes is *locally free* if the direct image sheaf $\phi_* \mathcal{O}_X$ is locally free (of finite rank). If moreover each fibre X_P of ϕ is the spectrum of a finite étale $\kappa(P)$-algebra, then we speak of a *finite étale* morphism. A *finite étale cover* is a surjective finite étale morphism.

Remarks 5.2.2

1. In view of Remark 5.1.24, to define a finite locally free morphism is the same as defining a quasi-coherent $\mathcal{O}_X$-algebra $\mathcal{A}$ that is locally free of finite rank as an $\mathcal{O}_X$-module. Finite étale morphisms correspond to $\mathcal{O}_X$-algebras such that $\mathcal{A}_P \otimes_{\mathcal{O}_{X,P}} \kappa(P)$ is a finite étale $\kappa(P)$-algebra for all $P \in X$.
2. A finitely generated module M over a local ring A is free if and only if it is *flat*, i.e. tensoring with M takes short exact sequences of A-modules to short exact sequences (see e.g. [48], Chapter XVI, Theorem 3.8). Thus a finite morphism of schemes is locally free if and only if $\phi_* \mathcal{O}_X$ is a sheaf of flat $\mathcal{O}_S$-modules. In the latter case one calls ϕ a flat morphism.

   In general, a not necessarily finite morphism is defined to be étale if it is flat, locally of finite type and the fibre over each point P has an open covering by spectra of finite étale $\kappa(P)$-algebras. In this book we shall not consider non-finite étale morphisms.
3. The image of a finite and locally free morphism ϕ is both open and closed. Indeed, it is closed because ϕ is finite, hence proper, and it is open because by local freeness $\phi_* \mathcal{O}_X$ has nonzero stalks over an open subset of S.

There is an important reformulation of the definition of a finite étale morphism that brings it closer to the notion of a finite cover in topology. To state it, define first a *geometric point* of a scheme S as a morphism $\bar{s} : \mathrm{Spec}\,(\Omega) \to S$, where Ω is an algebraically closed field. The image of $\bar{s}$ is a point s of S such that

Ω is an algebraically closed extension of $\kappa(s)$. Given a morphism $\phi : X \to S$ and a geometric point $\bar{s} : \mathrm{Spec}(\Omega) \to S$ of S, the *geometric fibre* $X_{\bar{s}}$ of ϕ over $\bar{s}$ is defined to be the fibre product $X \times_S \mathrm{Spec}(\Omega)$ induced by $\bar{s} : \mathrm{Spec}(\Omega) \to S$. Now Proposition 1.5.6 implies that the fibres of ϕ are spectra of finite étale algebras if and only if its geometric fibres are of the form $\mathrm{Spec}(\Omega \times \cdots \times \Omega)$, i.e. they are finite disjoint unions of points defined over Ω. (The latter is to be understood in the scheme-theoretic sense, because for instance in Example 4.5.4 the fibre over 0 is topologically a point, but not a $\mathbf{C}$-point as a scheme.) We shall see that by working with the geometric fibres of a finite étale morphism one obtains a nice theory analogous to that of topological covers.

Remarks 5.2.3 Here are some properties of finite étale morphisms that are more or less immediate from the definition.

1. If $\phi : X \to S$ and $\psi : Y \to X$ are finite étale morphisms, then so is the composite $\phi \circ \psi : Y \to S$. Local freeness is immediate to check, and the property of fibres follows as in Lemma 4.5.10.
2. If $\phi : X \to S$ is a finite étale morphism and $Z \to S$ is any morphism, then $X \times_S Z \to Z$ is a finite étale morphism. This is again immediate from the definition.

We next check that the definition of finite étale covers given above generalizes the one for normal curves used in the previous chapter. To see this, note first that the separability assumption made in Definition 4.5.1 corresponds to étaleness of the fibre over the generic point. Thus it suffices to prove the following lemma.

Lemma 5.2.4 *Let $\phi : X \to S$ be a finite surjective morphism of integral schemes, where S is Noetherian, normal and of dimension one. Then ϕ is locally free.*

Proof We may assume that $X = \mathrm{Spec}(B)$ and $S = \mathrm{Spec}(A)$ are affine, and ϕ comes from an injective ring homomorphism $\lambda : A \to B$. At a point $P \in S$ the stalk of $\phi_*\mathcal{O}_X$ is the spectrum of the localization $B_{\lambda(P)}$, which is a finitely generated A_P-module. Choose elements $t_1, \ldots, t_n \in B_P$ whose images modulo PB_P form a basis of the $\kappa(P)$-vector space B_P/PB_P. By Nakayama's lemma ([48], Chapter X, Lemma 4.3) they generate B_P over A_P. It then suffices to see that the t_i are linearly independent over the fraction field K of A_P. If not, there is a nontrivial relation $\sum a_i t_i = 0$ with $a_i \in K$. As S is Noetherian and normal of dimension one, the local ring A_P is a discrete valuation ring (Proposition 4.1.9), so PA_P is a principal ideal. By multiplying with a suitable power of a generator of PA_P we may assume that all the a_i lie in A_P and not all of them are in PA_P. But then reducing modulo P we obtain a nontrivial relation among the t_i in B_P/PB_P, a contradiction. □

Remark 5.2.5 There are other important examples of finite étale covers $X \to S$ with S Noetherian and normal of dimension one. For instance, when $X = \mathrm{Spec}\,(B)$ and $S = \mathrm{Spec}\,(A)$ are affine and the morphism comes from an inclusion $A \subset B$, then A and B are Dedekind rings and the points in the fibre over $P \in S$ correspond to the factors P_i in the decomposition $PB = P_1^{e_1} \cdots P_r^{e_r}$ (see Facts 4.1.5). In particular, X_P is étale if and only if all e_i are equal to 1.

This is particularly interesting when $K \subset L$ are finite extensions of $\mathbf{Q}$, and A (resp. B) is the integral closure of $\mathbf{Z}$ in K (resp. B). The morphism $\mathrm{Spec}\,(B) \to \mathrm{Spec}\,(A)$ is finite (Fact 4.1.4) and surjective (Fact 4.1.1 (4)). The case when it is a finite étale cover corresponds in classical parlance to an *unramified extension* of number fields $K \subset L$. According to a famous theorem of Minkowski (see e.g. [69], Chapter III, Theorem 2.18) the scheme $\mathrm{Spec}\,(\mathbf{Z})$ has no nontrivial finite étale covers. This is the arithmetic analogue of the simply connectedness of the complex affine line.

We now give a simple example of a finite étale cover $X \to S$ where S is not necessarily one-dimensional.

Example 5.2.6 Let $S = \mathrm{Spec}\,(A)$ and $X = \mathrm{Spec}\,(B)$ be affine schemes, where $B = A[x]/(f)$ with a monic polynomial $f \in A[x]$ of degree d. As B is freely generated as an A-module by the images of $1, x, x^2, \dots x^{d-1}$ in B, the morphism $\phi : \mathrm{Spec}\,(B) \to \mathrm{Spec}\,(A)$ is finite and locally free. If moreover $(f, f') = (1)$ in $A[x]$, then ϕ is finite étale. Indeed, if $P \in S$, then the fibre X_P is the spectrum of $B \otimes_A \kappa(P) \cong \kappa(P)[x]/(\bar{f})$, where $\bar{f}$ is the image of f in $\kappa(P)[x]$. It has only simple roots by our assumption on f', so this is a finite étale $\kappa(P)$-algebra.

In the previous example we have $\Omega^1_{X/S} = 0$. This is a general fact, as the following proposition shows.

Proposition 5.2.7 *Let $\phi : X \to S$ be a finite and locally free morphism. The following are equivalent.*

1. *The morphism ϕ is étale.*
2. *The sheaf of relative differentials $\Omega^1_{X/S}$ is 0.*
3. *The diagonal morphism $\Delta : X \to X \times_S X$ coming from ϕ is an isomorphism of X onto an open and closed subscheme of $X \times_S X$.*

See [59], Proposition I.3.5 for a generalization without the assumption that ϕ is finite and locally free.

Proof To show that (1) $\Leftrightarrow$ (2) we may assume that X and S are affine, and then by the localization and base change properties of differentials and Nakayama's lemma we are reduced to showing that for all $P \in S$ the fibre X_P is the spectrum of a finite étale $\kappa(P)$-algebra if and only if $\Omega^1_{X_P|\mathrm{Spec}\,(\kappa(P))} = 0$. This is the differential characterization of finite étale algebras (Proposition 5.1.31).

Next we show (2) $\Rightarrow$ (3). As ϕ is finite, hence separated, the diagonal morphism $\Delta : X \to X \times_S X$ is a closed immersion. It corresponds to a coherent sheaf of ideals $\mathcal{I}$ on $X \times_S X$. The restriction of the quotient $\mathcal{I}/\mathcal{I}^2$ to $\Delta(X)$ is isomorphic to $\Omega^1_{X/S}$, so it is 0 by assumption (2). It follows using Nakayama's lemma that the stalk of $\mathcal{I}$ is trivial at all points of $\Delta(X)$. As $\mathcal{I}$ is coherent, we obtain that $\mathcal{I} = 0$ on an open subset of $X \times_S X$, so $\Delta(X)$ is both open and closed in $X \times_S X$.

Finally, for (3) $\Rightarrow$ (1) let $\bar{s} : \mathrm{Spec}\,(\Omega) \to S$ be a geometric point. Taking the base change of the morphism $\Delta : X \to X \times_S X$ over S via $\bar{s}$, we obtain a morphism $\Delta_{\bar{s}}$ from the geometric fibre $X_{\bar{s}} = \mathrm{Spec}\,(\Omega) \times_S X$ to

$$(X_{\bar{s}}) \times_S X = (\mathrm{Spec}\,(\Omega) \times_S X) \times_{\mathrm{Spec}\,(\Omega)} \mathrm{Spec}\,(\Omega) \times_S X = X_{\bar{s}} \times_{\mathrm{Spec}\,(\Omega)} X_{\bar{s}}.$$

As property (3) behaves well with respect to base change, $\Delta_{\bar{s}}$ is an isomorphism of $X_{\bar{s}}$ onto an open and closed subscheme of $X_{\bar{s}} \times_{\mathrm{Spec}\,(\Omega)} X_{\bar{s}}$. The geometric fibre $X_{\bar{s}}$ is the spectrum of a finite dimensional Ω-algebra, and as such it has finitely many points, all having residue field Ω because Ω is algebraically closed. If $\bar{t} : \mathrm{Spec}\,(\Omega) \to X_{\bar{s}}$ is such a point, then by taking yet another base change, this time by $\bar{t}$, we obtain a morphism $\mathrm{Spec}\,(\Omega) \to X_{\bar{s}}$, which is again an isomorphism onto an open and closed subscheme of $X_{\bar{s}}$. As $\mathrm{Spec}\,(\Omega)$ is connected, this must be a connected component. We conclude that $X_{\bar{s}}$ as a scheme is a finite disjoint union of points, as required. □

Remark 5.2.8 The equivalence (1) $\Leftrightarrow$ (2) of the proposition gives a criterion for checking étaleness in practice. Assume $\phi : X \to S$ is finite and locally free (as we have seen, the latter property is automatic for S Noetherian and normal of dimension one), and moreover each point of S has an affine open neighbourhood $U = \mathrm{Spec}\,(A)$ such that $\phi^{-1}(U) = \mathrm{Spec}\,(B)$, with $B = A[x_1, \dots, x_n]/(f_1, \dots, f_n)$ for some monic polynomials f_i. If over all these open subsets the Jacobian determinant $\det(\partial_j f_i)_{i,j}$ maps to a unit in B, then $\Omega^1_{X|S} = 0$, and so ϕ is a finite étale morphism. This is the *Jacobian criterion* for étaleness. (The converse is also true; see [59], Corollary I.3.16.)

Using the previous proposition we now prove that finite étale covers are locally trivial in an appropriate sense. Call a finite étale cover $\phi : X \to S$ *trivial* if X as a scheme over S is isomorphic to a finite disjoint union of copies of S, and the map ϕ restricts to the identity on each component.

Proposition 5.2.9 *Let S be a connected scheme, and $\phi : X \to S$ an affine surjective morphism. Then ϕ is a finite étale cover if and only if there is a finite,* *locally free and surjective morphism* $\psi : Y \to S$ *such that* $X \times_S Y$ *is a trivial cover of* Y.

Proof To prove the 'if' part, we first show that ϕ must be finite and locally free. As ψ is such a map, each point of S is contained in an affine open subset $U = \mathrm{Spec}\,(A)$ over which ψ restricts to a morphism $\mathrm{Spec}\,(C) \to \mathrm{Spec}\,(A)$ with an A-algebra C that is finitely generated and free as an A-module. If ϕ restricts to $\mathrm{Spec}\,(B) \to \mathrm{Spec}\,(A)$ over U, then the base change over $\mathrm{Spec}\,(B)$ is of the form $\mathrm{Spec}\,(B \otimes_A C) \to \mathrm{Spec}\,(B)$. Here $B \otimes_A C$ is a finitely generated free C-module by assumption, so it is also a finitely generated and free A-module. As on the other hand it is isomorphic to a finite direct sum of copies of B, this is only possible if B is finitely generated and free over A.

Next let $\bar{s} : \mathrm{Spec}\,(\Omega) \to Y$ be a geometric point of Y; by composition with ψ it yields a geometric point of S. The geometric fibres $X_{(\bar{s} \circ \psi)}$ and $(X \times_S Y)_{\bar{s}}$ are isomorphic, and the latter is a finite disjoint union of copies of $\mathrm{Spec}\,(\Omega)$ by assumption. As ψ is surjective, this shows that all fibres of ϕ are étale.

Now to the 'only if' part. As S is connected, we see as in topology that all fibres of ϕ have the same cardinality n. Following Lenstra [52], we use induction on n, the case $n = 1$ being trivial. For $n > 1$ we consider the base change $X \times_S X \to X$. By part (3) of the previous proposition the diagonal map Δ induces a section of this map, and in fact $X \times_S X$ is the disjoint union of $\Delta(X)$ with some open and closed subscheme X'. As the inclusion map $X' \to X \times_S X$ and the projection $X \times_S X \to X$ are both finite and étale (the latter by Remark 5.2.3 (2)), we see from Remark 5.2.3 (1) that so is their composite $X' \to X$. By construction, it has fibres of cardinality $n - 1$, so the inductive hypothesis yields a finite, locally free and surjective morphism $\psi' : Y \to X$ such that $X' \times_X Y$ is isomorphic to the disjoint union of $(n-1)$ copies of Y. But then $(X \times_S X) \times_X Y \cong X \times_S Y$ is the disjoint union of n copies of Y. It remains to notice that the composite map $\psi := \psi' \circ \phi$ is also finite, locally free and surjective, being the composite of two such maps. □

Remark 5.2.10 In topology covers are characterized by the property that they become trivial after restricting to sufficiently small open subsets. Notice that the restriction of a cover $Y \to X$ above an open subset $U \subset X$ is none but the fibre product $Y \times_X U \to U$ in the category of topological spaces. This explains the analogy of the above proposition with the topological situation.

In fact the proposition says that finite étale covers are locally trivial for the *Grothendieck topology* where covering families are given by surjective finite locally free morphisms. For a discussion of Grothendieck topologies, see e.g. [59] or [109].

We conclude this section with some facts related to special schemes. Among these normal schemes will play a prominent role, so we first summarize a few known facts about them.

Facts 5.2.11

1. Recall that we defined a scheme to be normal if its local rings are integrally closed domains. In the locally Noetherian case there is a useful algebraic criterion for this:

   Serre's Normality Criterion *A reduced Noetherian local ring is normal if and only if for each prime ideal P of height 1 the localization A_P is a discrete valuation ring, and each prime divisor of a principal ideal is of height 1.*

   Here the height of a prime ideal P is by definition the dimension of the localization A_P. For a proof of the criterion, see [57], Theorem 23.8.
2. Given an integral scheme S with function field K and a finite separable extension $L|K$, the *normalization* of S in L is defined to be a normal integral scheme $\widetilde{S}$ with function field L together with a finite surjective morphism $\nu : \widetilde{S} \to S$. The normalization exists and is unique ([64], §III.8, Theorem 3). In the affine case $S = \mathrm{Spec}\,(A)$ it is the spectrum of the integral closure of A in L (compare with the proof of Theorem 4.3.10); in the general case it is obtained by patching normalizations of affine open subschemes together.

This being said, we return to finite étale covers.

Proposition 5.2.12 *Let S be a connected scheme, and $\phi : X \to S$ a finite étale cover. Any of the following properties of S implies the same property for X.*

1. *S is reduced.*
2. *S is regular.*
3. *S is normal and locally Noetherian.*

Proof In all three cases we may assume $X = \mathrm{Spec}\,(B)$ and $S = \mathrm{Spec}\,(A)$ are affine.

Assume first S is reduced. Let P be a point of S, which we view as a prime ideal of A. By assumption the localization A_P has no nilpotents, and we shall prove that the same holds for $C_P = B \otimes_A A_P$. This will suffice, as the local rings of points of X lying above P are localizations of C_P, and ϕ is surjective. Let $P_1, \ldots, P_r$ be the minimal prime ideals of A_P. Their intersection is the ideal of nilpotents in A_P, so by assumption the natural map $A_P \to \prod(A_P/P_i)$ is injective. But C_P is a free A_P-module (because ϕ is locally free), hence by tensoring with C_P we again obtain an injective map $C_P \to \prod(C_P/P_iC_P)$. It thus suffices to show that the C_P/P_iC_P have no nilpotents. But each Spec (C_P/P_iC_P) is finite étale over Spec (A_P/P_i) by Remark 5.2.3 (2), so by replacing A_P with A_P/P_i and C_P with C_P/P_iC_P we reduce to the case where A_P is an integral domain. Denoting by K_P its fraction field, we obtain that Spec $(C_P \otimes_{A_P} K_P) \to$ Spec (K_P) is a finite étale morphism (again by Remark 5.2.3 (2)), and that the natural map $C_P \to C_P \otimes_{A_P} K_P$ induced by the inclusion $A_P \to K_P$ is injective (again because ϕ is locally free). But then C_P has no nilpotents because $C_P \otimes_{A_P} K_P$ is a direct product of fields.

Now return to the case where $X =$ Spec (B) and $S =$ Spec (A) are affine. As ϕ is a finite surjective morphism, they have the same dimension, and if Q is a prime ideal of B with $Q \cap A = P$, then P and Q must have the same height. Moreover, it follows from the definition of étaleness that the image of the maximal ideal M_P of $\mathcal{O}_{S,P}$ generates the maximal ideal M_Q of $\mathcal{O}_{X,Q}$. From this we conclude that the regularity of S implies the regularity of X, and also that if M_P is principal, then so is M_Q. This shows that both conditions of Serre's normality criterion recalled above are preserved, whence the proposition in the last two cases. □

The strongest result about finite étale covers of a regular scheme is the following.

Theorem 5.2.13 (Zariski–Nagata purity theorem) *Let $\phi : X \to S$ be a finite surjective morphism of integral schemes, with X normal and S regular. Assume that the fibre X_P of ϕ above each codimension 1 point P of S is étale over $\kappa(P)$. Then ϕ is a finite étale cover.*

Here the codimension of P means the dimension of the local ring $\mathcal{O}_{X,P}$. The name 'purity' comes from the following reformulation: for $\phi : X \to S$ a finite surjective morphism of integral schemes with X normal and S regular, the closed subscheme $Z \subset S$ over which ϕ is *not* étale (the 'branch locus') must be of *pure* codimension 1, i.e. each irreducible component of Z must have codimension 1.

The statement is of local nature, therefore it is enough to prove it when X is the spectrum of a regular local ring. In this case it is a difficult piece of commutative algebra. See [67], Theorem 41.1 for Nagata's original proof, and [30], X.3.4 for a proof by Grothendieck.

Corollary 5.2.14 *Let S be a regular integral scheme, and $U \subset S$ an open subscheme whose complement consists of points of codimension ≥ 2. The functor $X \mapsto X \times_S U$ induces an equivalence between the category of finite étale covers of S and the category of finite étale covers of U.*

Proof To begin with, the morphism $X \times_S U \to U$ is a finite étale cover of U by Remark 5.2.3 (2). Fully faithfulness holds because if X and Y are finite and locally free over S, then any morphism $X \times_S U \to Y \times_S U$ over a dense open subset $U \subset S$ extends uniquely to a morphism $X \to Y$ over S by local freeness (to see this, look at the corresponding morphism of $\mathcal{O}_U$-algebras). For essential surjectivity take a finite étale cover $Z \to U$. By Proposition 5.2.12 (2) it is a regular scheme, so it is a finite disjoint union of regular integral schemes Z_i finite and étale over U. Let X_i be the normalization of S in the function field of Z_i. It is normal and finite étale over each codimension 1 point of S because these all lie in U by assumption. Hence $X_i \to S$ is a finite étale cover by the theorem. Moreover, the disjoint union X of the X_i satisfies $X \times_S U \cong Z$ by construction. □

5.3 Galois theory for finite étale covers

In this section we develop an analogue of the basic Galois theory of topological covers. We begin with a characterization of sections of finite étale covers.

Proposition 5.3.1 *Let $\phi : X \to S$ be a finite étale cover, and let $s : S \to X$ be a section of ϕ (i.e. a morphism satisfying $\phi \circ s = \mathrm{id}_S$). Then s induces an isomorphism of S with an open and closed subscheme of X. In particular, if S is connected, then s maps S isomorphically onto a whole connected component of X.*

For the proof we need the following useful lemma.

Lemma 5.3.2 *Let $\phi : X \to S$ and $\psi : Y \to X$ be morphisms of schemes.*

1. *If $\phi \circ \psi$ is finite and ϕ is separated, then ψ is finite.*
2. *If moreover $\phi \circ \psi$ and ϕ are finite étale, then so is ψ.*

Proof The diagonal morphism $X \to X \times_S X$ coming from ϕ is a closed immersion as ϕ is separated. In particular, it is a finite morphism. Taking fibre products over X with Y via ψ we obtain a morphism $\Gamma_\psi : Y \to Y \times_S X$ (the 'graph' of ψ) that is again finite. The second projection $p_2 : Y \times_S X \to X$ is also finite, being the base change of $\phi \circ \psi : Y \to S$ by $\phi : X \to S$. Thus $p_2 \circ \Gamma_\psi = \psi$ is finite.

If ϕ is moreover étale, then by Proposition 5.2.7 the diagonal morphism $X \to X \times_S X$ is an isomorphism of X onto an open and closed subscheme (a union of connected components) of $X \times_S X$. As such, it is certainly a finite étale morphism, hence so is $\Gamma_\psi : Y \to Y \times_S X$ by Remark 5.2.3 (2). As above, $p_2 : Y \times_S X \to X$ is finite étale as well, and hence so is $p_2 \circ \Gamma_\psi = \psi$ by Remark 5.2.3 (1). □

Proof of Proposition 5.3.1 By the lemma s is a finite étale morphism, hence its image is both open and closed in X by Remark 5.2.2 (3). As it is injective, the proposition follows. □

The above property of sections enables us to prove the following analogue of Proposition 2.2.2 which played a key role in the topological theory.

Corollary 5.3.3 *If $Z \to S$ is a connected S-scheme and $\phi_1, \phi_2 : Z \to X$ are two S-morphisms to a finite étale S-scheme X with $\phi_1 \circ \bar{z} = \phi_2 \circ \bar{z}$ for some geometric point $\bar{z} : \mathrm{Spec}\,(\Omega) \to Z$, then $\phi_1 = \phi_2$.*

Proof By passing to the fibre product $Z \times_S X \to Z$ and using the base change property of étale morphisms (Remark 5.2.3 (2)) we may assume $S = Z$. Then we have to prove that if two sections of a finite étale cover $X \to S$ of a connected scheme S coincide at a geometric point, then they are equal. This follows from the proposition because each such section, being an isomorphism of S onto a connected component of X, is determined by the image of a geometric point. □

Given a morphism of schemes $\phi : X \to S$, define $\mathrm{Aut}(X|S)$ to be the group of scheme automorphisms of X preserving ϕ. By convention, automorphisms act from the left. For a geometric point $\bar{s} : \mathrm{Spec}\,(\Omega) \to S$ there is a natural left action of $\mathrm{Aut}(X|S)$ on the geometric fibre $X_{\bar{s}} = X \times_S \mathrm{Spec}\,(\Omega)$ coming by base change from its action on X. We have the following property analogous to Proposition 2.2.4:

Corollary 5.3.4 *If $\phi : X \to S$ is a connected finite étale cover, the nontrivial elements of $\mathrm{Aut}(X|S)$ act without fixed points on each geometric fibre. Hence $\mathrm{Aut}(X|S)$ is finite.*

Proof Applying the previous corollary with $\phi_1 = \phi$, $\phi_2 = \phi \circ \lambda$ for some automorphism $\lambda \in \mathrm{Aut}(X|S)$ yields the first statement. It then follows that the permutation representation of $\mathrm{Aut}(X|S)$ on the underlying sets of geometric fibres is faithful. But these sets are finite, whence the second statement. □

To continue the parallelism with the topological situation, we consider quotients by group actions. By the corollary just proven we need to restrict to quotients by finite groups.

Construction 5.3.5 Let $\phi : X \to S$ be an affine surjective morphism of schemes, and $G \subset \mathrm{Aut}(X|S)$ a finite subgroup. Define a ringed space $G\backslash X$ and a morphism $\pi : X \to G\backslash X$ of ringed spaces as follows. The underlying topological space of $G\backslash X$ is to be the quotient of X by the action of G, and the underlying continuous map of π the natural projection. Then define the structure sheaf of $G\backslash X$ as the subsheaf $(\pi_*\mathcal{O}_X)^G$ of G-invariant elements in $\pi_*\mathcal{O}_X$.

Proposition 5.3.6 *The ringed space $G\backslash X$ constructed above is a scheme, the morphism π is affine and surjective, and ϕ factors as $\phi = \psi \circ \pi$ with an affine morphism $\psi : G\backslash X \to S$.*

Proof We may assume, using the affineness assumption on ϕ, that $X = \mathrm{Spec}\,(B)$ and $S = \mathrm{Spec}\,(A)$ are affine, and ϕ comes from a ring homomorphism $\lambda : A \to B$. Then it suffices to show that the ringed space $G\backslash X$ is isomorphic to the spectrum of B^G, the ring of G-invariants of B. This will imply the claim including the assertions on π, because B is integral over B^G (as every $b \in B$ is a root of the monic polynomial $\prod(x - \sigma(b)) \in B^G[x]$, where σ runs over the elements of G), and therefore Fact 4.1.1 (4) implies the surjectivity of π.

To identify the underlying space of $G\backslash X$ with $X_G := \mathrm{Spec}\,(B^G)$ it is enough to identify them as sets, as a closed subset $V(I) \subset X$ induces the closed subset $V(I^G) \subset X_G$. As we have just seen the surjectivity of $\pi : X \to X_G$, we have to show that the fibres of the map $X \to X_G$ coming from the inclusion $B^G \to B$ are the G-orbits of $\mathrm{Spec}\,(B)$. Assume that two G-orbits $\{\sigma(P) : \sigma \in G\}$ and $\{\sigma(Q) : \sigma \in G\}$ of points of B lie above the same point $P^G \in \mathrm{Spec}\,(B^G)$. As the fibre $X_{\kappa(P^G)}$ is zero-dimensional, the $\sigma(P)$ and $\sigma(Q)$ induce maximal ideals $\sigma(\overline{P})$ and $\sigma(\overline{Q})$ of the ring $\overline{B} = B \otimes_{B^G} \kappa(P^G)$ with $\bigcap \sigma(\overline{P}) = \bigcap \sigma(\overline{Q}) = 0$. But using the Chinese Remainder Theorem we find $\bar{b} \in \overline{B}$ with $\bar{b} \in \sigma(\overline{P})$ and $\bar{b} \notin \sigma(\overline{Q})$ for all $\sigma \in G$, which is a contradiction.

Finally, to show $(\pi_*\mathcal{O}_X)^G \cong \mathcal{O}_{X_G}$, notice that the first sheaf is quasi-coherent, being the kernel of the morphism of quasi-coherent sheaves

$$\pi_*\mathcal{O}_X \to \bigoplus_{\sigma \in G} \pi_*\mathcal{O}_X, \quad s \mapsto (\ldots, \sigma(s) - s, \ldots).$$

Thus it is enough to check the isomorphism on the rings of sections over X_G, which are B^G in both cases. □

The scheme $G\backslash X$ over S is the *quotient of X by G*. It can be shown that it is characterized by a universal property: every morphism $\lambda : X \to Y$ in the category of schemes affine and surjective over S such that λ is constant on the orbits of G factors uniquely through $G\backslash X$.

Proposition 5.3.7 *Let $\phi : X \to S$ be a connected finite étale cover, and $G \subset \mathrm{Aut}(X|S)$ a finite group of S-automorphisms of X. Then $X \to G\backslash X$ is a finite étale cover of $G\backslash X$, and $G\backslash X$ is a finite étale cover of S.*

In [29] Grothendieck proved this under the additional assumption that the schemes are locally Noetherian. We give a proof due to Lenstra [52] that works in general.

Proof Thanks to Lemma 5.3.2 (2) it suffices to prove that $G\backslash X \to S$ is a finite étale cover. Apply Proposition 5.2.9 to obtain a base change $X \times_S Y \to Y$ such that $X \times_S Y$ is a finite disjoint union of copies of Y, i.e. $X \times_S Y \cong F \times Y$ for a finite set F. There is a natural action of G on $X \times_S Y$ coming by base change from its action of X, which yields an isomorphism $G\backslash(X \times_S Y) \cong (G\backslash F) \times Y$. Now observe that $G\backslash(X \times_S Y) \cong (G\backslash X) \times_S Y$. Indeed, there is a natural map $X \times_S Y \to (G\backslash X) \times_S Y$ that is constant on G-orbits, whence a map $G\backslash(X \times_S Y) \to (G\backslash X) \times_S Y$. To see that it induces an isomorphism we may argue over a small affine neighbourhood $U = \mathrm{Spec}\,(A)$ of each point of S, with preimages $\mathrm{Spec}\,(B)$ and $\mathrm{Spec}\,(C)$ in X and Y, respectively. There the isomorphism to be proven translates to $B^G \otimes_A C \xrightarrow{\sim} (B \otimes_A C)^G$, which holds for U sufficiently small, as C is then a free A-module with trivial G-action. We thus obtain that $(G\backslash X) \times_S Y \cong (G\backslash F) \times Y$, which is again a finite disjoint union of copies of Y. We may then conclude by applying Proposition 5.2.9 in the other direction. □

As in topology, we define a *connected* finite étale cover $X \to S$ to be *Galois* if its S-automorphism group acts transitively on geometric fibres. The following analogue of Theorem 2.2.10 is now proven in the same way as in topology, using Corollaries 5.3.1 and 5.3.3 as well as Proposition 5.3.7 instead of the corresponding topological facts.

Proposition 5.3.8 *Let $\phi : X \to S$ be a finite étale Galois cover. If $Z \to S$ is a connected finite étale cover fitting into a commutative diagram*

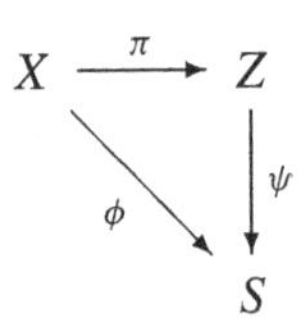

then $\pi : X \to Z$ is a finite étale Galois cover, and actually $Z \cong H\backslash X$ with some subgroup H of $G=\mathrm{Aut}(X|S)$. In this way we get a bijection between subgroups of G and intermediate covers Z as above. The cover $\psi : Z \to S$ is Galois if and only if H is a normal subgroup of G, in which case $\mathrm{Aut}(Z|S) \cong G/H$.

We next prove a key proposition that generalizes not so much a well-known topological fact, but rather a basic statement in field theory. Namely, in Lemma 1.3.1 we proved that every finite separable field extension can be embedded in a finite Galois extension and there is a smallest such extension, the Galois closure.

Proposition 5.3.9 *Let $\phi : X \to S$ be a connected finite étale cover. There is a morphism $\pi : P \to X$ such that $\phi \circ \pi : P \to S$ is a finite étale Galois cover, and moreover every S-morphism from a Galois cover to X factors through p.*

The following proof is due to Serre.

Proof Fix a geometric point $\bar{s} : \mathrm{Spec}\,(\Omega) \to S$, and let $F = \{\bar{x}_1, \dots, \bar{x}_n\}$ be the finite set of $\mathrm{Spec}\,(\Omega)$-points of the geometric fibre $X_{\bar{s}}$. An ordering of the points $\bar{x}_i$ induces a canonical geometric point $\bar{x}$ of the n-fold fibre product $X^n = X \times_S \cdots \times_S X$, giving $\bar{x}_i$ in the i-th component. Let P be the connected component of X^n containing the image of $\bar{x}$, and let $\pi : P \to X$ be the map induced by the first projection of X^n to X. Using Remarks 5.2.3 we see that P is a finite étale cover of S via $\phi \circ \pi$.

Next we show that each point in the geometric fibre $P_{\bar{s}}$ can be represented by an n-tuple $(\bar{x}_{\sigma(1)}, \dots, \bar{x}_{\sigma(n)})$ for some permutation σ of the $\bar{x}_i$. Indeed, each point of $X^n_{\bar{s}}$ corresponds to an element of F^n, so we only have to show that the points concentrated on P have distinct coordinates. But by Proposition 5.2.7 (3) the diagonal image $\Delta(X)$ of X in $X \times_S X$ is open and closed, and therefore so is its inverse image by a projection π_{ij} mapping X^n to the (i, j)-components. As P is connected, $\pi_{ij}^{-1}(\Delta(X)) \cap P \neq \emptyset$ would imply $\pi_{ij}^{-1}(\Delta(X)) \supset P$, which is impossible because $\bar{x}$ hits P away from any of the $\pi_{ij}^{-1}(\Delta)$.

Now to show that P is Galois over S, remark that each permutation σ of the $\bar{x}_i$ induces an S-automorphism ϕ_σ of X^n by permuting the components. If $\phi_\sigma \circ \bar{x} \in P_{\bar{s}}$, then $\phi_\sigma(P) \cap P \neq \emptyset$ and so $\phi_\sigma \in \mathrm{Aut}(P|S)$ by connectedness of P. Thus $\mathrm{Aut}(P|S)$ acts transitively on one geometric fibre, from which we conclude as in Remark 2.2.8 that P is Galois.

Finally, if $q : Q \to X$ is an S-morphism with Q Galois, choose a preimage $\bar{y}$ of $\bar{x}_1$. As q is a surjective morphism of covers by Proposition 5.3.8, after composing with appropriate elements of $\mathrm{Aut}(Q|S)$ we get n maps $q = q_1, \dots, q_n : Q \to X$ such that $q_i \circ \bar{y} = \bar{x}_i$. Whence an S-morphism $Q \to X^n$ that factors through P, for it maps $\bar{y}$ to $\bar{x}$ and Q is connected. This concludes the proof. □

To close this section, we briefly discuss another point of view on Galois covers that is often useful. First some definitions.

Definition 5.3.10 Let S be a scheme. A *group scheme* over S is a morphism of schemes $p : G \to S$ that has a section $e : S \to G$, together with S-morphisms $\mathrm{mult} : G \times_S G \to G$ ('multiplication') and $i : G \to G$ ('inverse'), subject to

the commutative diagrams

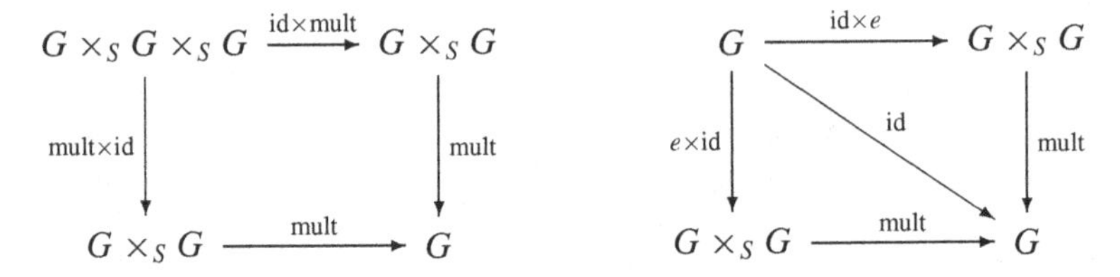

and

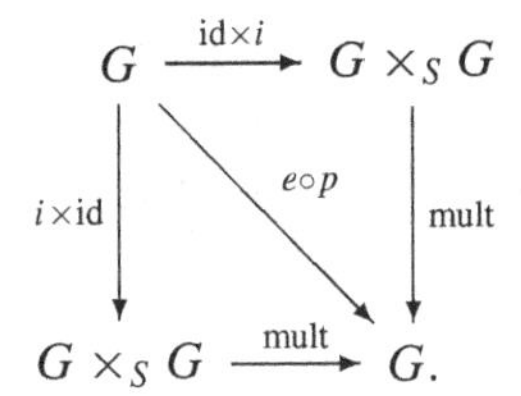

We say that G is *finite* if the structure morphism $p : G \to S$ is finite. Similarly, G is a *finite flat* (resp. *finite étale*) group scheme if $p : G \to S$ is finite locally free (resp. finite étale).

Examples 5.3.11

1. Let S be a scheme, Γ a finite group of order n. We define the *constant group scheme* $\Gamma_S \to S$ corresponding to Γ over S to be the disjoint union of n copies of S indexed by Γ and equipped with the projection map given by the identity on each component. The group operation is induced by that on Γ. (Explicitly, a point (P_1, P_2) of $\Gamma_S \times_S \Gamma_S$ lies in some component indexed by a pair $(g_1, g_2) \in \Gamma \times \Gamma$, and on that component may be identified with a point $P \in S$. We define $P_1 \cdot P_2$ to be the point P on the component of Γ_S indexed by $g_1 g_2$. The inverse is given similarly.) This is a finite étale group scheme over S.
2. Assume $S = \mathrm{Spec}\,(k)$ for a field k. In this case a finite étale group scheme $G \to S$ is of the form $G = \mathrm{Spec}\,(A)$ with a finite étale k-algebra A. Hence it corresponds via Theorem 1.5.2 to a finite set equipped with a continuous action of $\mathrm{Gal}\,(k_s|k)$. The fibre of G over the geometric point given by an algebraic closure of k_s carries a group structure coming from that of G; it is compatible with the Galois action. Conversely, Galois-equivariant group operations on a finite continuous $\mathrm{Gal}\,(k_s|k)$-set come from a group scheme structure on the corresponding finite étale k-scheme, by a similar reasoning as in the proof of Theorem 2.5.15. Thus the category of finite étale k-group schemes is equivalent to the category of finite groups Γ carrying a continuous $\mathrm{Gal}\,(k_s|k)$-action. Here the constant group scheme $\Gamma_{\mathrm{Spec}\,(k)}$ corresponds to Γ with trivial Galois action. For a general finite

étale group scheme G there is always a finite separable field extension $L|k$ such that $G \times_S \mathrm{Spec}\,(L)$ is isomorphic to the constant group scheme $\Gamma_{\mathrm{Spec}\,(L)}$.

Definition 5.3.12 Let S be a connected scheme, and $G \to S$ a finite flat group scheme. A *(left) G-torsor* or *principal homogeneous space* over S is a finite locally free surjective morphism $X \to S$ together with a group action $\rho : G \times_S X \to X$ (defined by diagrams similar to those in the above definition) such that the map $(\rho, \mathrm{id}) : G \times_S X \to X \times_S X$ is an isomorphism.

Another characterization of G-torsors is the following.

Lemma 5.3.13 *Let S be a connected scheme, $G \to S$ a finite flat group scheme, and $X \to S$ a scheme over S equipped with a left action $\rho : G \times_S X \to X$.*

These data define a G-torsor over S if and only if there exists a finite locally free surjective morphism $Y \to S$ such that $X \times_S Y \to Y$ is isomorphic, as a Y-scheme with $G \times_S Y$-action, to $G \times_S Y$ acting on itself by left translations.

Proof For the 'only if' part, take $Y = X$. For the 'if' part, note first that since $G \times_S Y \to Y$ and $Y \to S$ are finite, locally free and surjective, the same must be true of $X \to S$. Also, the map $G \times_S X \to X \times_S X$ corresponds via Remark 5.1.24 to a morphism $\lambda : (\phi_* \mathcal{O}_X) \otimes_{\mathcal{O}_S} (\phi_* \mathcal{O}_X) \to (\phi_* \mathcal{O}_X) \otimes_{\mathcal{O}_S} (\phi_* \mathcal{O}_G)$. We have $G \times_S Y \cong X \times_S Y$ by assumption, and therefore the map $(\rho, \mathrm{id}_X, \mathrm{id}_Y)$: $G \times_S X \times_S Y \to X \times_S X \times_S Y$ is an isomorphism. This means that λ becomes an isomorphism after tensoring with $\phi_* \mathcal{O}_Y$. But $\phi_* \mathcal{O}_Y$ is locally free, so λ must be an isomorphism. □

Remark 5.3.14 One defines torsors under arbitrary flat group schemes and states a lemma exactly as above, except that 'finite, locally free and surjective' must be replaced by 'flat, locally of finite presentation and surjective' everywhere. However, the proof of the general lemma requires flat descent theory. See [59], Section III.4.

Example 5.3.15 Let k be a field, $m > 1$ an integer invertible in k. The spectrum of the finite étale k-algebra $k[\mu_m] := k[x]/(x^m - 1)$ is a finite étale group scheme because it corresponds to the $\mathrm{Gal}\,(k_s|k)$-set given by the group μ_m of m-th roots of unity. It is the *group scheme $(\mu_m)_k$ of m-th roots of unity* over k.

Given an element $a \in k^\times \setminus k^{\times m}$, the spectrum of the field extension $k(\sqrt[m]{a})$ is a μ_m-torsor. The action $\mathrm{Spec}\,(k[\mu_m]) \times_{\mathrm{Spec}\,(k)} \mathrm{Spec}\,(k(\sqrt[m]{a})) \to \mathrm{Spec}\,(k(\sqrt[m]{a}))$ is induced by multiplication with roots of unity; explicitly, it comes from the k-morphism $k(\sqrt[m]{a}) \to k(\sqrt[m]{a}) \otimes_k k[\mu_m]$ sending $\sqrt[m]{a}$ to $\sqrt[m]{a} \otimes \zeta$, where ζ is a primitive m-th root of unity. When $\zeta \in k$, the group scheme $(\mu_m)_k$ is isomorphic to the constant group scheme $\mathbf{Z}/m\mathbf{Z}$ over k, and $k(\sqrt[m]{a})$ is a Galois extension with group $\mathbf{Z}/m\mathbf{Z}$; therefore $k(\sqrt[m]{a}) \otimes_k k(\sqrt[m]{a})$ is a finite direct product of copies of $k(\sqrt[m]{a})$ carrying a permutation action of $\mathbf{Z}/n\mathbf{Z}$. Thus in this case we see that $Y = \mathrm{Spec}\,(k(\sqrt[m]{a}))$ satisfies the requirement of the above lemma. In the general case $Y = \mathrm{Spec}\,(k(\zeta, \sqrt[m]{a}))$ will do.

A generalization of Kummer theory states that actually all μ_m-torsors over $\mathrm{Spec}\,(k)$ arise in this way. See e.g. [111], Sections 18.2 and 18.4.

Proposition 5.3.16 *Let S be a connected scheme, and G a finite étale group scheme over S.*

1. *If G is a constant group scheme Γ_S, then a G-torsor is the same as a finite étale Galois cover with group Γ.*
2. *If there is a morphism $S \to \mathrm{Spec}\,(k)$ with a field k, and G arises from an étale k-group scheme G_k by base change to S, then every G-torsor $Y \to S$ is a finite étale cover of S. Moreover, there is a finite separable extension $L|k$ such that $Y \times_{\mathrm{Spec}\,(k)} \mathrm{Spec}\,(L) \to S \times_{\mathrm{Spec}\,(k)} \mathrm{Spec}\,(L)$ is a Galois étale cover.*

Proof Statement (1) follows from Proposition 5.2.9 and Lemma 5.3.13. To prove (2), one first takes a separable extension $L|k$ for which the fibre product $G_k \times_{\mathrm{Spec}\,(k)} \mathrm{Spec}\,(L)$ is a constant group scheme, and then applies statement (1) to $Y \times_{\mathrm{Spec}\,(k)} \mathrm{Spec}\,(L) \to S \times_{\mathrm{Spec}\,(k)} \mathrm{Spec}\,(L)$; the conclusion follows by another application of Proposition 5.2.9. □

5.4 The algebraic fundamental group in the general case

We now come to the construction of the algebraic fundamental group of a scheme. The discussion will be parallel to the classification of covers in topology, but there are two important differences: there is no a priori definition of the monodromy action, and the fibre functor is not representable. Grothendieck overcame these difficulties by using categorical constructions.

We begin with some notation and terminology. For a scheme S denote by Fet_S the category whose objects are finite étale covers of S, and the morphisms are morphisms of schemes over S. Fix a geometric point $\bar{s} : \mathrm{Spec}\,(\Omega) \to S$. For an object $X \to S$ of Fet_S we consider the geometric fibre $X \times_S \mathrm{Spec}\,(\Omega)$ over $\bar{s}$, and denote by $\mathrm{Fib}_{\bar{s}}(X)$ its underlying set. Given a morphism $X \to Y$ in Fet_S, there is an induced morphism of schemes $X \times_S \mathrm{Spec}\,(\Omega) \to Y \times_S \mathrm{Spec}\,(\Omega)$, whence a set-theoretic map $\mathrm{Fib}_{\bar{s}}(X) \to \mathrm{Fib}_{\bar{s}}(Y)$. We have thus defined a set-valued functor $\mathrm{Fib}_{\bar{s}}$ on Fet_S; we call it the *fibre functor* at the geometric point $\bar{s}$.

We now define the monodromy action on the fibres in an abstract way. Quite generally, given a functor F between two categories $\mathcal{C}_1$ and $\mathcal{C}_2$, an *automorphism* of F is a morphism of functors $F \to F$ that has a two-sided inverse.

Composition of morphisms then equips the set $\mathrm{Aut}(F)$ of automorphisms of F with a group structure, and we call the resulting group the *automorphism group* of F. Notice that for every object C of $\mathcal{C}_1$ and automorphism $\phi \in \mathrm{Aut}(F)$ there is by definition a morphism $F(C) \to F(C)$ induced by ϕ. For a set-valued functor this gives a natural left action of $\mathrm{Aut}(F)$ on $F(C)$.

Definition 5.4.1 Given a scheme S and a geometric point $\bar{s} : \mathrm{Spec}\,(\Omega) \to S$, we define the *algebraic fundamental group* $\pi_1(S, \bar{s})$ as the automorphism group of the fibre functor $\mathrm{Fib}_{\bar{s}}$ on Fet_S.

By the preceding discussion, there is a natural left action of $\pi_1(S, \bar{s})$ on $\mathrm{Fib}_{\bar{s}}(X)$ for each finite étale S-scheme X, and therefore $\mathrm{Fib}_{\bar{s}}$ takes its values in the category of left $\pi_1(S, \bar{s})$-sets. The main theorem is now the following.

Theorem 5.4.2 (Grothendieck) *Let S be a connected scheme, and let $\bar{s} : \mathrm{Spec}\,(\Omega) \to S$ be a geometric point.*

1. *The group $\pi_1(S, \bar{s})$ is profinite, and its action on $\mathrm{Fib}_{\bar{s}}(X)$ is continuous for every X in Fet_S.*
2. *The functor $\mathrm{Fib}_{\bar{s}}$ induces an equivalence of Fet_S with the category of finite continuous left $\pi_1(S, \bar{s})$-sets. Here connected covers correspond to sets with transitive $\pi_1(S, \bar{s})$-action, and Galois covers to finite quotients of $\pi_1(S, \bar{s})$.*

In [29] this was proven under the additional assumption that S is locally Noetherian, because Proposition 5.3.7 was only proven in that case.

Example 5.4.3 The theorem contains as a special case the case $S = \mathrm{Spec}\,(k)$ for a field k. Here a finite étale S-scheme X is the spectrum of a finite étale k-algebra. For a geometric point $\bar{s}$ the fibre functor maps a connected cover $X = \mathrm{Spec}\,(L)$ to the underlying set of $\mathrm{Spec}\,(L \otimes_k \Omega)$, which is a finite set indexed by the k-algebra homomorphisms $L \to \Omega$. The image of each such homomorphism lies in the separable closure k_s of k in Ω via the embedding given by $\bar{s}$. So finally we obtain that $\mathrm{Fib}_{\bar{s}}(X) \cong \mathrm{Hom}_k(L, k_s)$ for all X of the form $\mathrm{Spec}\,(L)$, and therefore $\pi_1(S, \bar{s}) \cong \mathrm{Gal}\,(k_s|k)$. Thus in this case the theorem is equivalent to Theorem 1.5.2.

Note that although in the above example the fibre functor is identified with the functor $X \mapsto \mathrm{Hom}(\mathrm{Spec}\,(k_s), X)$, this does *not* mean that the fibre functor is representable, for k_s is not a finite étale k-algebra. However, it is the union of its finite Galois subextensions which already are. Passing to the associated affine schemes, this motivates the following definition.

Definition 5.4.4 Let $\mathcal{C}$ be a category, and F a set-valued functor on $\mathcal{C}$. We say that F is *pro-representable* if there exists an inverse system $P = (P_\alpha, \phi_{\alpha\beta})$

of objects of $\mathcal{C}$ indexed by a directed partially ordered set Λ and a functorial isomorphism

$$\varinjlim \operatorname{Hom}(P_\alpha, X) \cong F(X) \tag{5.1}$$

for each object X in $\mathcal{C}$.

Notice that in the above definition the inverse limit of the P_α may not exist in $\mathcal{C}$, but the direct limit of the $\operatorname{Hom}(P_\alpha, X)$ does in the category of sets. Recall that by definition the direct limit of a direct system $(S_\alpha, \phi_{\alpha\beta})$ of sets is the disjoint union of the sets S_α modulo the equivalence relation where $s_\alpha \in S_\alpha$ is equivalent to $s_\beta \in S_\beta$ if $\phi_{\alpha\gamma}(s_\alpha) = \phi_{\alpha\gamma}(s_\beta)$ for some $\gamma \geq \alpha, \beta$.

Remark 5.4.5 If F is pro-representable by an inverse system $P = (P_\alpha, \phi_{\alpha\beta})$, then for each α the identity map of P_α gives rise to a class in $\operatorname{Hom}(P_\alpha, P_\alpha)$, and hence to a class in the direct limit on the left-hand side of (5.1) with $X = P_\alpha$. It thus gives rise to an element $p_\alpha \in F(P_\alpha)$, and for $\alpha \leq \beta$ the morphism $F(\phi_{\alpha\beta})$ maps p_β to p_α. The p_α thus define an element (p_α) of the inverse limit $\varprojlim F(P_\alpha)$. The isomorphism (5.1) is then induced by (p_α) in the sense that a morphism $\phi : P_\alpha \to X$ is mapped to the image of p_α by $F(\phi)$.

In Example 5.4.3 the fibre functor was pro-representable by the inverse system of spectra of finite (Galois) extensions contained in k_s. We now prove that this holds in general.

Proposition 5.4.6 *Under the assumptions of the theorem the fibre functor* $\operatorname{Fib}_{\bar{s}}$ *is pro-representable.*

Proof Take the index set Λ to be the set of all finite étale Galois covers $P_\alpha \to S$, and define $P_\alpha \leq P_\beta$ if there is a morphism $P_\beta \to P_\alpha$. This partially ordered set is directed, because if $P_\alpha, P_\beta \in \Lambda$, we may apply Proposition 5.3.9 to a connected component Z of the fibre product $P_\alpha \times_S P_\beta$ to obtain $P_\gamma \in \Lambda$ together with maps $P_\gamma \to Z \to P_\alpha$, $P_\gamma \to Z \to P_\beta$.

The objects of the inverse system will be the P_α themselves, so we next have to define the morphisms $\phi_{\alpha\beta}$. If $P_\alpha \leq P_\beta$ in the above ordering, then by definition there exists a morphism $\phi : P_\beta \to P_\alpha$ over S. This ϕ is in general not unique, so we 'rigidify' the situation as follows. For each $P_\alpha \in \Lambda$ we fix an arbitrary element $p_\alpha \in \operatorname{Fib}_{\bar{s}}(P_\alpha)$. Since $P_\beta \to S$ is a Galois cover, we find by Corollary 5.3.3 a unique S-automorphism λ of P_β so that $\phi \circ \lambda$ maps p_β to p_α. Defining $\phi_{\alpha\beta} := \phi \circ \lambda$ we obtain an inverse system $(P_\alpha, \phi_{\alpha\beta})$ such that moreover for each $\alpha \leq \beta$ we have $\operatorname{Fib}_{\bar{s}}(\phi_{\alpha\beta})(p_\beta) = p_\alpha$.

As in Remark 5.4.5 above, for every X in Fet_S and every $P_\alpha \in \Lambda$ there is a natural map $\operatorname{Hom}(P_\alpha, X) \to \operatorname{Fib}_{\bar{s}}(X)$ sending $\phi \in \operatorname{Hom}(P_\alpha, X)$ to $\operatorname{Fib}_{\bar{s}}(\phi)(p_\alpha)$. These maps are compatible with the transition maps in the inverse system defined above, whence a functorial map $\varinjlim \mathrm{Hom}(P_\alpha, X) \to \mathrm{Fib}_{\bar{s}}(X)$. To conclude the proof we have to construct a functorial inverse to this map. To do so, we may assume X is connected (otherwise take disjoint unions), and consider the Galois closure $\pi : P \to X$ given by Proposition 5.3.9. Here P is one of the $P_\alpha \in \Lambda$ by definition, and since it is Galois, for each $\bar{x} \in \mathrm{Fib}_{\bar{s}}(X)$ we find a unique S-automorphism λ as above so that $\mathrm{Fib}_{\bar{s}}(\pi \circ \lambda)$ maps the distinguished element $p_\alpha \in \mathrm{Fib}_{\bar{s}}(P_\alpha)$ to $\bar{x}$. Now sending $\bar{x}$ to the class of $\pi \circ \lambda$ in $\varinjlim \mathrm{Hom}(P_\alpha, X)$ yields the required inverse. □

An inspection of the above proof shows that the maps $\phi_{\alpha\beta}$ in the system pro-representing the fibre functor $\mathrm{Fib}_{\bar{s}}$ depend on the choice of the system of geometric points (p_α); however, once such a system (p_α) is fixed, the pro-representing system becomes unique. This fact will be crucial for the proof of the next corollary.

Corollary 5.4.7 *Every automorphism of the functor* $\mathrm{Fib}_{\bar{s}}$ *comes from a unique automorphism of the inverse system* $(P_\alpha, \phi_{\alpha\beta})$ *constructed in the proof above.*

Here by definition an automorphism of $(P_\alpha, \phi_{\alpha\beta})$ is a collection of automorphisms $\lambda_\alpha \in \mathrm{Aut}(P_\alpha|S)$ compatible with the transition maps $\phi_{\alpha\beta}$.

Proof An automorphism of $\mathrm{Fib}_{\bar{s}}$ maps the system (p_α) of distinguished elements to another system (p'_α). As the P_α are Galois, for each α there is a unique automorphism $\lambda_\alpha \in \mathrm{Aut}(P_\alpha|S)$ sending p_α to p'_α. Since the p_α form a compatible system, so do the λ_α, whence the corollary. □

Before stating the next corollary, recall from Chapter 2 that the opposite group of a group G is the group G^{op} with the same underlying set but multiplication defined by $(x, y) \mapsto yx$.

Corollary 5.4.8 *The automorphism groups* $\mathrm{Aut}(P_\alpha)^{op}$ *form an inverse system whose inverse limit is isomorphic to* $\pi_1(S, \bar{s})$.

Proof The inverse system comes from Proposition 5.3.8: if $P_\alpha \leq P_\beta$ in the partial order of the proof above, then since the covers are Galois, there is a natural surjective group homomorphism $\mathrm{Aut}(P_\beta|S) \twoheadrightarrow \mathrm{Aut}(P_\alpha|S)$. The elements of the inverse limit are exactly the automorphisms of the inverse system $(P_\alpha, \phi_{\alpha\beta})$, which in turn correspond bijectively to automorphisms of the fibre functor by the previous corollary. The isomorphism with the opposite group then comes from the contravariance of the Hom-functor. □

Proof of Theorem 5.4.2 Apply Proposition 5.4.6 to find an inverse system $(P_\alpha, \phi_{\alpha\beta})$ of Galois covers pro-representing the functor $\mathrm{Fib}_{\bar{s}}$. By Corollary 5.3.4 the groups $\mathrm{Aut}(P_\alpha|S)$ are finite for all α, hence $\pi_1(S, \bar{s})$ is a profinite group by the previous corollary. An automorphism of the inverse system $(P_\alpha, \phi_{\alpha\beta})$ induces an automorphism of $\mathrm{Fib}_{\bar{s}}(X) = \varinjlim(P_\alpha, X)$, whence a left action of $\pi_1(S, \bar{s})$ on $\mathrm{Fib}_{\bar{s}}(X)$ for each X in Fet_S. This action is continuous, because if $\bar{x} \in \mathrm{Fib}_{\bar{s}}(X)$ comes from a class in $\mathrm{Hom}(P_\alpha, X)$, then the action of $\pi_1(S, \bar{s})$ factors through $\mathrm{Aut}(P_\alpha|S)^{op}$.

The proof of the second statement of the theorem is parallel to that in the special case of fields done in Theorem 1.5.2. We indicate the proof of essential surjectivity, leaving the details for fully faithfulness to the readers. Given a finite continuous left $\pi_1(S, \bar{s})$-set E, we may assume that the $\pi_1(S, \bar{s})$-action is transitive by decomposing E in orbits. The stabilizer U of a point $x \in E$ is an open subgroup of $\pi_1(S, \bar{s})$, and therefore contains an open normal subgroup V_α arising as the kernel of a projection $\pi_1(S, \bar{s}) \to \mathrm{Aut}(P_\alpha|S)^{op}$ since the V_α form a basis of open neighbourhoods of 1 in $\pi_1(S, \bar{s})$. Let $\overline{U}$ be the image of U in $\mathrm{Aut}(P_\alpha|S)^{op}$, and let X be the quotient of P_α by the action of $\overline{U}^{op}$ constructed in Lemma 5.3.7. Then $E \cong \mathrm{Fib}_{\bar{s}}(X)$. □

Finally we make the link with the theory of the previous chapter. In fact, it generalizes to an arbitrary integral normal scheme S.

Proposition 5.4.9 *Let S be an integral normal scheme. Denote by K_s a fixed separable closure of the function field K of S, and by K_S the composite of all finite subextensions $L|K$ of K_s such that the normalization of S in L is étale over S. Then $K_S|K$ is a Galois extension, and* $\mathrm{Gal}\,(K_S|K)$ *is canonically isomorphic to the fundamental group $\pi_1(S, \bar{s})$ for the geometric point $\bar{s}$:* $\mathrm{Spec}\,(\overline{K}) \to S$, *where $\overline{K}$ is the algebraic closure of K containing K_s.*

Proof A finite étale cover $X \to S$ is normal by Proposition 5.3.2 (3). Its generic fibre is a finite product of finite separable extensions $L_i|K$, so by uniqueness of normalization (Fact 5.2.11 (2)) X must be the disjoint union of the normalizations of S in the L_i. This being said, one proves exactly as in Theorem 4.6.4 that $K_S|K$ is a Galois extension, and that the category of finite étale S-schemes is equivalent to that of finite continuous left $\mathrm{Gal}\,(K_S|K)$-sets, except that for the composition and base change properties of finite étale covers one applies Remarks 5.2.3 as in the proof of Lemma 3.4.2. By construction, the equivalence is induced by the fibre functor at the geometric point $\bar{s}$. □

5.5 First properties of the fundamental group

Now that we have constructed the fundamental group, we discuss some of its basic properties.

First we show that, just as in topology, fundamental groups of the same S corresponding to different base points $\bar{s}, \bar{s}'$ are (non-canonically) isomorphic. We begin by constructing isomorphisms between different fibre functors.

Proposition 5.5.1 *Let S be a connected scheme. Given two geometric points $\bar{s}: \mathrm{Spec}\,(\Omega) \to S$ and $\bar{s}': \mathrm{Spec}\,(\Omega') \to S$, there exists an isomorphism of fibre functors* $\mathrm{Fib}_{\bar{s}} \xrightarrow{\sim} \mathrm{Fib}_{\bar{s}'}$.

Proof By Proposition 5.4.6 both fibre functors are pro-representable. Moreover, the proof shows that the representing inverse systems have the same index set Λ and objects P_α, only the transition morphisms may be different. Denote them by $\phi_{\alpha\beta}$ and $\psi_{\alpha\beta}$, respectively. Proving the proposition amounts to constructing an isomorphism of the inverse system $(P_\alpha, \phi_{\alpha\beta})$ onto the system $(P_\alpha, \psi_{\alpha\beta})$, i.e. a system of automorphisms $\lambda_\alpha \in \mathrm{Aut}(P_\alpha|S)$ transforming the maps $\phi_{\alpha\beta}$ to the $\psi_{\alpha\beta}$. Assume given a pair $\alpha \leq \beta$ in Λ and an automorphism $\lambda_\beta \in \mathrm{Aut}(P_\beta|S)$. Consider the distinguished elements $p_\alpha \in \mathrm{Fib}_{\bar{s}}(P_\alpha)$ and $p_\beta \in \mathrm{Fib}_{\bar{s}}(P_\beta)$. By construction we have $\mathrm{Fib}_{\bar{s}}(\phi_{\alpha\beta})(p_\beta) = p_\alpha$. On the other hand, set $p'_\alpha := \mathrm{Fib}_{\bar{s}}(\psi_{\alpha\beta})(\lambda_\beta(p_\beta))$. As P_α is Galois, there is a unique automorphism $\lambda_\alpha \in \mathrm{Aut}(P_\alpha|S)$ mapping p_α to p'_α. Corollary 5.3.3 applied with $\bar{z} = p_\alpha$, $\phi_1 = \psi_{\alpha\beta} \circ \lambda_\beta$ and $\phi_2 = \lambda_\alpha \circ \phi_{\alpha\beta}$ then implies that the diagram of S-morphisms

$$\begin{array}{ccc} P_\beta & \xrightarrow{\lambda_\beta} & P_\beta \\ {\scriptstyle \phi_{\alpha\beta}}\downarrow & & \downarrow{\scriptstyle \psi_{\alpha\beta}} \\ P_\alpha & \xrightarrow{\lambda_\alpha} & P_\alpha \end{array}$$

commutes. Define $\rho_{\alpha\beta} : \mathrm{Aut}(P_\beta|S) \to \mathrm{Aut}(P_\alpha|S)$ to be the map sending each $\lambda_\beta \in \mathrm{Aut}(P_\beta|S)$ to the λ_α defined as above. It is a map of sets but in general not a group homomorphism. We thus obtain an inverse system $(\mathrm{Aut}(P_\alpha|S), \rho_{\alpha\beta})$ of finite sets. The inverse limit of such a system is nonempty (see Corollary 3.4.12). An element of it defines the required isomorphism of $(P_\alpha, \phi_{\alpha\beta})$ onto $(P_\alpha, \psi_{\alpha\beta})$. □

Corollary 5.5.2 *Let S be a connected scheme. For any two geometric points $\bar{s}: \mathrm{Spec}\,(\Omega) \to S$ and $\bar{s}': \mathrm{Spec}\,(\Omega') \to S$ there exists a continuous isomorphism of profinite groups $\pi_1(S, \bar{s}') \xrightarrow{\sim} \pi_1(S, \bar{s})$.*

Proof An isomorphism $\lambda : \mathrm{Fib}_{\bar{s}} \xrightarrow{\sim} \mathrm{Fib}_{\bar{s}'}$ of fibre functors induces an isomorphism of their automorphism groups via $\phi \mapsto \lambda^{-1} \circ \phi \circ \lambda$. Continuity of this isomorphism with respect to the profinite structure follows from the construction in the proof above. □

Remark 5.5.3 In analogy with the topological situation (Proposition 2.4.7) an isomorphism $\mathrm{Fib}_{\bar{s}} \xrightarrow{\sim} \mathrm{Fib}_{\bar{s}'}$ of fibre functors is called a *path* from $\bar{s}'$ to $\bar{s}$ (sometimes the French word *chemin* is used). As in Remark 2.4.8 it follows from the above statements that the isomorphism $\pi_1(S, \bar{s}') \xrightarrow{\sim} \pi_1(S, \bar{s})$ induced by a path from $\bar{s}'$ to $\bar{s}$ depends on the path but is unique up to an inner automorphism of $\pi_1(S, \bar{s})$ (or $\pi_1(S, \bar{s}')$). In particular, the maximal abelian (profinite) quotient of $\pi_1(S, \bar{s})$ does not depend on the choice of the geometric base point.

Next we investigate functoriality with respect to base point preserving morphisms. Let S and S' be connected schemes, equipped with geometric points $\bar{s} : \mathrm{Spec}\,(\Omega) \to S$ and $\bar{s}' : \mathrm{Spec}\,(\Omega) \to S'$, respectively. Assume given a morphism $\phi : S' \to S$ with $\phi \circ \bar{s}' = \bar{s}$. Then ϕ induces a base change functor $BC_{S,S'} : \mathrm{Fet}_S \to \mathrm{Fet}_{S'}$ sending an object X to the base change $X \times_S S'$ and a morphism $X \to Y$ to the induced morphism $X \times_S S' \to Y \times_S S'$. Moreover, the condition $\phi \circ \bar{s}' = \bar{s}$ implies that there is an equality of functors $\mathrm{Fib}_{\bar{s}} = \mathrm{Fib}_{\bar{s}'} \circ BC_{S,S'}$. Consequently, every automorphism of the functor $\mathrm{Fib}_{\bar{s}'}$ induces an automorphism of $\mathrm{Fib}_{\bar{s}}$ via composition with $BC_{S,S'}$, and we have a map

$$\phi_* : \pi_1(S', \bar{s}') \to \pi_1(S, s)$$

which is readily seen to be a continuous homomorphism of profinite groups.

The base change functor $X \to X \times_S S'$ corresponds via Theorem 5.4.2 to the functor sending the $\pi_1(S, \bar{s})$-set $\mathrm{Fib}_{\bar{s}}(X)$ to the $\pi_1(S', \bar{s}')$-set obtained by composing with ϕ_*. We now investigate how properties of the functor $BC_{S,S'}$ are reflected by properties of the morphism ϕ_*.

Proposition 5.5.4

1. *The map ϕ_* is trivial if and only if for every connected finite étale cover $X \to S$ the base change $X \times_S S'$ is a trivial cover (i.e. isomorphic to a finite disjoint union of copies of S').*
2. *The map ϕ_* is surjective if and only if for every connected finite étale cover $X \to S$ the base change $X \times_S S'$ is connected as well.*

Proof A finite étale cover is trivial if and only if it corresponds to a finite set with trivial action of the fundamental group. This immediately gives the 'only if' part of the first statement. For the 'if' part assume that $\mathrm{im}\,(\phi_*)$ is nontrivial, and choose an open subgroup of $\pi_1(S, \bar{s})$ not containing the whole of $\mathrm{im}\,(\phi_*)$. Then the action of $\pi_1(S', \bar{s}')$ on the coset space $\pi_1(S, \bar{s})/U$ induced by ϕ_* is nontrivial, so the finite étale cover corresponding to $U \backslash \pi_1(S, \bar{s})$ pulls back to a nontrivial cover of S'.

In the second statement the 'only if' part follows from the fact that connected covers correspond via the fibre functor to sets with transitive $\pi_1(S, \bar{s})$-action. For the 'if' part assume that $\mathrm{im}\,(\phi_*)$ is not the whole of $\pi_1(S, \bar{s})$. It is then a proper closed subgroup by compactness of $\pi_1(S', \bar{s}')$, so we find a nontrivial open subgroup $U \subset \pi_1(S, \bar{s})$ containing it (see Lemma 5.5.7 (1) below for a stronger statement). Then $\pi_1(S', \bar{s}')$ acts trivially on the coset space $U \backslash \pi_1(S, \bar{s})$

via ϕ_*, which means that the connected cover corresponding to $U\backslash\pi_1(S,\bar{s})$ pulls back to a trivial cover of S' (different from S'). □

We now prove a common generalization of both parts of the previous proposition. Observe first that identifying a finite continuous transitive left $\pi_1(S,\bar{s})$-set F with the coset space $U\backslash\pi_1(S,\bar{s})$ of some open subgroup U amounts to choosing a point $\bar{x}$ of F which will correspond to the coset of 1. In terms of the étale cover $X \to S$ corresponding to F this is the choice of a geometric point $\bar{x}$ in the fibre above $\bar{s}$. We may also view $\bar{x}$ as a geometric point of a base change $X \times_S S'$.

Proposition 5.5.5 *Let $U \subset \pi_1(S,\bar{s})$ be an open subgroup, and let $X \to S$ be the connected cover corresponding to the coset space $U\backslash\pi_1(S,\bar{s})$, together with the base point $\bar{x}$ described above.*

The subgroup U contains the image of ϕ_* if and only if the finite étale cover $X \times_S S' \to S'$ has a section $S' \to X \times_S S'$ sending $\bar{s}$ to $\bar{x}$.

Proof We have $\mathrm{im}(\phi_*) \subset U$ if and only if $\pi_1(S',\bar{s}')$ acts trivially on $\bar{x}$ via ϕ_*. This in turn implies that the whole connected component of $\bar{x}$ in $X \times_S S'$ is fixed by $\pi_1(S',\bar{s}')$. In other words, this component is a one-sheeted trivial cover of S', and hence mapped isomorphically onto S' by the map $X \times_S S' \to S'$. □

The corresponding characterization of $\ker(\phi_*)$ is the following.

Proposition 5.5.6 *Let $U' \subset \pi_1(S',\bar{s}')$ be an open subgroup, and let $X' \to S'$ be the cover corresponding to the coset space $U'\backslash\pi_1(S',\bar{s}')$.*

The subgroup U' contains the kernel of ϕ_* if and only if there exists a finite étale cover $X \to S$ and a morphism $X_i \to X'$ over S', where X_i is a connected component of $X \times_S S'$.

For the proof we need an easy group-theoretic lemma.

Lemma 5.5.7 *Let G be a profinite group, and $H \subset G$ a closed subgroup.*

1. *The intersection of the open subgroups of G containing H is exactly H.*
2. *Given an open subgroup $V' \subset H$, there is an open subgroup $V \subset G$ with $V \cap H = V'$.*

Proof For the first statement, given $g \in G \setminus H$ we shall find an open subgroup of G containing H but not g. To do so, by closedness of H we first pick an open normal subgroup $N \subset G$ with $gN \cap H = \emptyset$ (such an N exists, because the gN of this type form a basis of open neighbourhoods of g). Denoting by $p : G \to G/N$ the natural projection, the open subgroup $p^{-1}(p(H)) \subset G$ will do.

For statement (2) use the closedness of V' in G and part (1) to write both H and V' as the intersection of the open subgroups of G containing them. Since $[H : V']$ is finite, we find finitely many open subgroups $V_1, \dots, V_n$ in G containing V' but not H such that $V' = V_1 \cap \dots \cap V_n \cap H$. Thus the subgroup $V := V_1 \cap \dots \cap V_n$ will do. □

Proof of Proposition 5.5.6 For the 'only if' part let X be as in the statement of the proposition, and let $U \subset \pi_1(S, \bar{s})$ be an open subgroup with $X \cong U \backslash \pi_1(S, \bar{s})$. By choosing an appropriate geometric base point we may identify the component $X_i \subset X \times_S S'$ with the coset space $U'' \backslash \pi_1(S', \bar{s}')$, for some open subgroup $U'' \subset \pi_1(S', \bar{s}')$. Note that U'' must contain $\ker(\phi_*)$, as $\ker(\phi_*)$ stabilizes the base point by construction. To say that there is a morphism $X_i \to X'$ is then equivalent to the inclusion $U'' \subset U'$, but U'' contains $\ker(\phi_*)$ as we just said, and we are done.

Conversely, assume $U' \supset \ker(\phi_*)$. As $H = \phi_*(\pi_1(S', \bar{s}'))$ is a closed subgroup in $\pi_1(S, \bar{s})$ (being compact), and $V' = \phi_*(U')$ is open in H (being compact of finite index), we may apply the lemma to find an open subgroup $V \subset \pi_1(S, \bar{s})$ with $V \cap H = V'$, giving rise to a connected finite étale cover $X \to S$. Again a connected component X_i of $X \times_S S'$ corresponds to some coset space $U'' \subset \pi_1(S', \bar{s}')$, and there is a morphism $X_i \to X'$ if and only if $U'' \subset U'$. As both groups here contain $\ker(\phi_*)$, the required inclusion is equivalent to $\phi_*(U'') \subset \phi_*(U')$, but this holds by construction. □

Corollary 5.5.8 *The map ϕ_* is injective if and only if for every connected finite étale cover $X' \to S'$ there exists a finite étale cover $X \to S$ and a morphism $X_i \to X'$ over S', where X_i is a connected component of $X \times_S S'$.*

In particular, if every connected finite étale cover $X' \to S'$ is of the form $X \times_S S' \to S'$ for a finite étale cover $X \to S$, then ϕ is injective.

Proof This follows from the proposition since the intersection of the open subgroups of $\pi_1(S', \bar{s}')$ is trivial. □

Corollary 5.5.9 *Let $S'' \xrightarrow{\psi} S' \xrightarrow{\phi} S$ be a a sequence of morphisms of connected schemes, and let $\bar{s}, \bar{s}', \bar{s}''$ be geometric points of S, S' and S'', respectively, satisfying $\bar{s} = \phi \circ \bar{s}'$ and $\bar{s}' = \psi \circ \bar{s}''$. The sequence*

$$\pi_1(S'', \bar{s}'') \xrightarrow{\psi_*} \pi_1(S', \bar{s}') \xrightarrow{\phi_*} \pi_1(S, \bar{s})$$

is exact if and only if the following two conditions are satisfied.

1. *For every finite étale cover $X \to S$ the base change $X \times_S S'' \to S''$ is a trivial cover of S''.*

2. *Given a connected finite étale cover $X' \to S'$ such that $X' \times_{S'} S''$ has a section over S'', there exists a connected finite étale cover $X \to S$ and an S'-morphism from a connected component of $X \times_S S'$ onto X'.*

Proof This is a consequence of Propositions 5.5.4 (1), 5.5.5 and 5.5.6. □

Remark 5.5.10 In part (2) of the above criterion it suffices to consider finite *Galois* covers. This is because $\ker(\phi_*)$, being a closed normal subgroup of $\pi_1(S', \bar{s}')$, is the intersection of the open *normal* subgroups containing it.

5.6 The homotopy exact sequence and applications

We now prove a generalization of Proposition 4.7.1. Recall that given a field k, a k-scheme $X \to \mathrm{Spec}\,(k)$ is *geometrically integral* if $X \times_{\mathrm{Spec}\,(k)} \mathrm{Spec}\,(\bar{k})$ is integral for an algebraic closure $\bar{k} \supset k$.

Proposition 5.6.1 *Let X be a quasi-compact and geometrically integral scheme over a field k. Fix an algebraic closure $\bar{k}$ of k, and let $k_s|k$ be the corresponding separable closure. Write $\overline{X} := X \times_{\mathrm{Spec}\,(k)} \mathrm{Spec}\,(k_s)$, and let $\bar{x}$ be a geometric point of $\overline{X}$ with values in $\bar{k}$. The sequence of profinite groups*

$$1 \to \pi_1(\overline{X}, \bar{x}) \to \pi_1(X, \bar{x}) \to \mathrm{Gal}\,(k_s|k) \to 1$$

induced by the maps $\overline{X} \to X$ and $X \to \mathrm{Spec}\,(k)$ is exact.

For the proof we need a standard lemma.

Lemma 5.6.2 *Given a finite étale cover $\overline{Y} \to \overline{X}$, there is a finite extension $L|k$ contained in k_s and a finite étale cover Y_L of $X_L := X \times_{\mathrm{Spec}\,(k)} \mathrm{Spec}\,(L)$ so that $\overline{Y} \cong Y_L \times_{\mathrm{Spec}\,(L)} \mathrm{Spec}\,(k_s)$.*

Similarly, elements of $\mathrm{Aut}(\overline{Y}|\overline{X})$ *come from* $\mathrm{Aut}(Y_L|X_L)$ *for L sufficiently large.*

Proof We prove the first statement, the proof of the second one being similar. Since X is quasi-compact by assumption, it has a finite covering by open affine subschemes. Let $U_i = \mathrm{Spec}\,(A_i)$ $(1 \leq i \leq n)$ be a finite affine open covering of X so that the preimage in $\overline{Y}$ of each $\overline{U}_i := \mathrm{Spec}\,(A_i \otimes_k k_s)$ is an affine open subscheme of the form $\mathrm{Spec}\,(\overline{B}_i)$, where $\overline{B}_i$ is a finitely generated $A_i \otimes_k k_s$-module. Each $\overline{B}_i$ then arises as a quotient of some polynomial ring $(A_i \otimes_k k_s)[x_1, \ldots, x_m]$ by an ideal generated by finitely many polynomials $f_1, \ldots, f_r$. The finitely many coefficients involved in the f_j generate a finitely generated k_s-subalgebra $A'_i \subset A_i \otimes_k k_s$, itself a quotient of a polynomial ring over k_s by an ideal generated by finitely many polynomials $g_1, \ldots, g_s$. If $L|k$ is the finite extension generated by the coefficients of the g_l, then the coefficients of the f_j are contained in $A_i \otimes_k L$, and hence $B_i \cong ((A_i \otimes_k L)[x_1, \dots, x_m]/(f_1, \dots, f_r)) \otimes_L k_s$. Moreover, such an isomorphism holds for all i for L sufficiently large. Similarly, one sees that the isomorphisms showing the compatibility of the $\overline{Y}_{U_i}$ over the overlaps $U_i \cap U_j$ can be defined by equations involving only finitely many coefficients. Thus an extension L that is so large that it contains all the coefficients involved satisfies the requirements of the lemma. □

Proof of Proposition 5.6.1 Injectivity of the map $\pi_1(\overline{X}, \bar{x}) \to \pi_1(X, \bar{x})$ follows from the criterion of Corollary 5.5.8 in view of the lemma. Similarly, the surjectivity of $\pi_1(X, \bar{x}) \to \mathrm{Gal}\,(k_s|k)$ follows from Proposition 5.5.4 (2) and our assumption that X is geometrically connected.

It thus remains to prove exactness in the middle. For this we apply the criterion of Corollary 5.5.9, complemented by Remark 5.5.10. Condition (1) is straightforward: each finite étale cover of the form $X_L \to X$, with $L|k$ a finite extension, yields a trivial cover when pulled back to $\overline{X}$. For condition (2), assume $Y \to X$ is a finite étale Galois cover such that $Y_{k_s} \to \overline{X}$ has a section. As X is integral, the generic fibre of $Y \to X$ is the spectrum of a finite Galois extension K of the function field $k(X)$ of X that splits in a finite direct product of copies of $k_s(X)$ when tensored with k_s. This is possible if and only if $K \cong k(X) \otimes_k L$ for some finite Galois extension $L|k$. Consider the corresponding étale Galois cover $X_L \to X$. It has the same function field as Y, so they are isomorphic over the generic point of X. This means that there is a dense open subset $U \subset X$ such that $Y \times_X U$ and $(X_L) \times_X U$ are isomorphic. As Y and X_L are locally free over U, this implies $Y \cong X_L$, as required. □

Remarks 5.6.3

1. Using the proposition we may extend the definition the outer Galois representation $\rho_X : \mathrm{Gal}\,(k_s|k) \to \mathrm{Out}(\pi_1(X, \bar{x}))$ of Section 4.7 to an arbitrary quasi-compact and geometrically integral scheme over a field k.
2. The proposition also enables us to give another, slightly more precise formulation of Grothendieck's Section Conjecture discussed at the end of Section 4.9. Given a k-rational point $y : \mathrm{Spec}\, k \to X$, it induces by functoriality a map $\sigma_y : \mathrm{Gal}\,(\bar{k}|k) \to \pi_1(X, \bar{y})$ for a geometric point $\bar{y}$ lying above y. This is not quite a section of the exact sequence of the proposition because of the difference of base points. But by Remark 5.5.3 a path from $\bar{y}$ to $\bar{x}$ induces an isomorphism $\lambda : \pi_1(X, \bar{y}) \xrightarrow{\sim} \pi_1(X, \bar{x})$ that is uniquely determined up to inner automorphism of $\pi_1(X, \bar{x})$. The composite $\lambda \circ \sigma_y$ is then a section of the exact sequence uniquely determined up to conjugation by an element of $\pi_1(\bar{X}, \bar{x})$ (because the section induces the identity map of $\mathrm{Gal}\,(\bar{k}|k)$). The Section Conjecture predicts that in the

case when X is a smooth projective curve of genus at least 2 and k is finitely generated over $\mathbf{Q}$ this construction establishes a bijection between k-rational points of X and conjugacy classes of sections.

Using more algebraic geometry it is possible to prove a relative version of Proposition 5.6.1, under a properness assumption.

Proposition 5.6.4 *Let S be a Noetherian integral scheme, and $\phi : X \to S$ a proper flat morphism with geometrically integral fibres. Let $\bar{s} : \mathrm{Spec}\,(\Omega) \to S$ be a geometric point of S such that Ω is the algebraic closure of the residue field of the image of $\bar{s}$ in S, and let $\bar{x}$ be a geometric point of the geometric fibre $X_{\bar{s}}$. The sequence*

$$\pi_1(X_{\bar{s}}, \bar{x}) \to \pi_1(X, \bar{x}) \to \pi_1(S, \bar{s}) \to 1$$

is exact.

For the notion of a flat morphism, see Remark 5.2.2 (2). The main tool in the proof is the Stein factorization theorem. Let us recall the statement.

Facts 5.6.5 A proper morphism $\phi : X \to S$ of Noetherian integral schemes has a *Stein factorization* $\phi = \psi \circ \phi'$, where $\psi : S' \to S$ is a finite morphism and $\phi' : X \to S'$ satisfies $\phi'_* \mathcal{O}_X = \mathcal{O}_{S'}$. For a proof, see [34], Corollary III.11.5 (under the unnecessary assumption that ϕ is projective). The condition $\phi'_* \mathcal{O}_X = \mathcal{O}_{S'}$ implies that ϕ' has geometrically connected fibres ([34], Corollary III.11.3 and its proof); in fact, the two conditions are equivalent by the existence of the Stein factorization.

Proof of Proposition 5.6.4 We first prove surjectivity of the map $\pi_1(X, \bar{x}) \to \pi_1(S, \bar{s})$ by means of the criterion of Proposition 5.5.4 (2). Assume given a connected finite étale cover $\rho : Y \to S$. We have to show that $\rho_X : Y \times_S X \to X$ is again connected. By our assumption on ϕ we have $\phi_* \mathcal{O}_X = \mathcal{O}_S$, from which we conclude $\phi_*(\rho_{X*} \mathcal{O}_{X \times_S Y}) = \rho_* \mathcal{O}_Y$. Were $X \times_S Y$ disconnected, the sheaf $\rho_{X*} \mathcal{O}_{X \times_S Y}$ would decompose as a nontrivial direct product of sheaves of $\mathcal{O}_X$-algebras, corresponding to the regular functions on different components. But then the same would be true of $\rho_* \mathcal{O}_Y$, contradicting the connectedness of Y.

Next we show exactness in the middle by checking the conditions of Corollary 5.5.9. Condition (1) is automatic, so we turn to condition (2). Take a connected finite étale cover $X' \to X$, and consider the Stein factorization of the composite map $X' \to X \to S$. We obtain a finite morphism $\psi : S' \to S$ and a morphism $\phi' : X' \to S'$ with $\phi'_* \mathcal{O}_{X'} = \mathcal{O}_{S'}$. As we assumed ϕ flat, we may apply Corollary III.12.9 of [34]. It first shows that $\psi_* \mathcal{O}_{S'} = (\psi \circ \phi')_* \mathcal{O}_{X'}$ is locally free, i.e. ψ is a locally free morphism. Secondly, it shows that for each point P of S we have $(\psi_* \mathcal{O}_{S'})_P \otimes_{\mathcal{O}_{X,P}} \kappa(P) = ((\psi \circ \phi')_* \mathcal{O}_{X'})_P \otimes_{\mathcal{O}_{X,P}} \kappa(P) \cong \mathcal{O}_{X'_P}(X'_P)$, where $X'_P = X' \times_S \mathrm{Spec}\,(\kappa(P))$. But since ϕ had proper geometrically integral fibres and $X' \to X$ is finite étale, here $\mathcal{O}_{X'_P}(X'_P) \otimes_{\kappa(P)} \overline{\kappa(P)}$ must be a finite direct product of copies of $\overline{\kappa(P)}$. We conclude that $\psi : S' \to S$ is a finite étale cover; it is connected because so is X', and ϕ' is surjective by the condition $\phi'_* \mathcal{O}_{X'} = \mathcal{O}_{S'}$. Consider the natural map $X' \to X \times_S S'$ given by the universal property of the fibre product. It is a map of finite étale covers of X. Assuming, as in condition (2) of Corollary 5.5.9, that the base change of $X' \to X$ to $X_{\bar{s}}$ has a section implies that the base change of $X \times_S S'$ has a section as well. This shows that the finite étale cover $X' \to X \times_S S'$ must be of degree 1, i.e. an isomorphism. □

Corollary 5.6.6 *Let k be an algebraically closed field, and X, Y Noetherian connected schemes over k. Assume moreover that X is proper and geometrically integral, and choose geometric points $\bar{x} : \mathrm{Spec}\,(k) \to X$, $\bar{y} : \mathrm{Spec}\,(k) \to Y$ with values in k. The natural morphism*

$$\pi_1(X \times Y, (\bar{x}, \bar{y})) \to \pi_1(X, \bar{x}) \times \pi_1(Y, \bar{y}) \tag{5.2}$$

induced by the projections of $X \times Y$ to X and Y, respectively, is an isomorphism.

Proof The map (5.2) may be inserted in the commutative diagram of exact sequences

$$\begin{array}{ccccccccc}
1 & \longrightarrow & \pi_1(X, \bar{x}) & \longrightarrow & \pi_1(X \times Y, (\bar{x}, \bar{y})) & \longrightarrow & \pi_1(Y, \bar{y}) & \longrightarrow & 1 \\
 & & \downarrow & & \downarrow & & \downarrow & & \\
1 & \longrightarrow & \pi_1(X, \bar{x}) & \longrightarrow & \pi_1(X, \bar{x}) \times \pi_1(Y, \bar{y}) & \longrightarrow & \pi_1(Y, \bar{y}) & \longrightarrow & 1.
\end{array}$$

Here the upper row comes from the previous proposition applied to the morphism $X \times Y \to Y$ and the geometric point $\bar{y}$. Injectivity on the left comes from the fact that the projection $X \times Y \to X$ yields a section of the inclusion map $X \to X \times Y$ (where X is identified with the fibre $X_{\bar{y}}$) sending $(\bar{x}, \bar{y})$ to $\bar{x}$. The outer vertical maps are identity maps, hence the middle one must be an isomorphism. □

We conclude this section by two applications of Corollary 5.6.6, taken from the classic paper of Lang and Serre [51]. The first one is:

Proposition 5.6.7 *Let $k \subset K$ be an extension of algebraically closed fields, and let X be a proper integral scheme over k. The map $\pi_1(X_K, \bar{x}_K) \to \pi_1(X, \bar{x})$ induced by the projection $X_K \to X$ is an isomorphism for every geometric point $\bar{x}$ of X.*

Proof If $Y \to X$ is a connected finite étale cover, then Y is reduced (Proposition 5.2.12 (1)) and the generic fibre must be a finite field extension of the function field of X. Thus Y is integral as well. Since k is algebraically closed in K, the tensor product $K(Y) \otimes_k K$ is a field, where $K(Y)$ is the function field of Y. This shows that Y_K is again connected, whence the surjectivity of the map $\pi_1(X_K, \bar{x}_K) \to \pi_1(X, \bar{x})$ by Proposition 5.5.4 (2).

For the proof of injectivity we start with a connected finite étale cover $Y \to X_K$. As in the proof of Lemma 5.6.2 we find a subfield $k' \subset K$ finitely generated over k and a connected finite étale cover $\phi' : Y' \to X_{k'}$ with $Y' \times_{\mathrm{Spec}\,(k')} \mathrm{Spec}\,(K) \cong Y$. In fact, the argument shows that there is an integral affine k-scheme T with function field k' and a finite morphism $\phi_T : \mathcal{Y} \to X \times_{\mathrm{Spec}\,(k)} T$ of schemes of finite type over T such that the induced morphism $\mathcal{Y} \times_T \mathrm{Spec}\,(k') \to X_{k'}$ is isomorphic to ϕ'. Since the sheaves $\phi_{T*}\mathcal{O}_{\mathcal{Y}}$ (resp. $\Omega_{\mathcal{Y}|(S\times T)}$) are coherent and $\phi'_*\mathcal{O}_{Y'}$ is locally free (resp. $\Omega_{Y'|S_{k'}} = 0$) by Proposition 5.2.7, we see that after replacing T with an affine open subscheme and restricting ϕ_T we obtain a connected finite étale cover. We now choose a geometric point $\bar{t} : \mathrm{Spec}\,(k) \to T$ and apply the isomorphism $\pi_1(X \times T, (\bar{x}, \bar{t})) \xrightarrow{\sim} \pi_1(X, \bar{t}) \times \pi_1(T, \bar{t})$ of the previous corollary. By Corollary 5.5.8 it implies that we may find connected finite étale covers $Z \to X$, $T' \to T$ together with a morphism $Z \times T' \to \mathcal{Y}$ of schemes over $X \times T$. The function field of T' still embeds in K, so by base change with $T' \to T$ we may assume $T' = T$. Then the fibre of $Z \times T \to T$ at a k-point P of T is a finite étale cover of X equipped with a morphism to Y over X_K, and the criterion of Corollary 5.5.8 is satisfied. □

In the next section we shall give another proof of the proposition due to Pop that is more natural but works under somewhat different assumptions. The properness assumption in the proposition is essential in positive characteristic; see Exercise 8.

The second application of Corollary 5.6.6 concerns *abelian varieties*. Recall that an abelian variety over a field k means a proper geometrically integral group scheme over k; it is necessarily commutative by [65], §II.4.

Proposition 5.6.8 *Let A be an abelian variety over an algebraically closed field k. Given a finite étale cover $\phi : Y \to A$, the scheme Y can be equipped with the structure of an abelian variety over k such that the map ϕ becomes a homomorphism of k-group schemes.*

In fact, the map ϕ is an *isogeny* of abelian varieties, i.e. a surjective homomorphism with finite kernel. This follows immediately from the proposition and the assumption that ϕ is a finite étale cover.

Proof It is enough to prove the proposition in the case when $Y \to A$ is a Galois finite étale cover. Indeed, by Propositions 5.3.8 and 5.3.9 a general $Y \to A$ is a quotient of a Galois étale cover by the action of a finite group of automorphisms. But the quotient of an abelian variety by the action of a finite group is again an abelian variety ([65], §7, Theorem 4).

Denote by $m_A : A \times A \to A$ the group operation of A, and consider the fibre product square

$$\begin{array}{ccc} Y' = (A \times A) \times_A Y & \longrightarrow & Y \\ \downarrow & & \downarrow \phi \\ A \times A & \xrightarrow{m_A} & A. \end{array}$$

We contend that Y' is connected. Indeed, the base change Y'_1 of $Y' \to A \times A$ by the closed embedding $\{0\} \times A \to A$ is isomorphic to Y, hence it is connected. Thus it must be contained in a connected component of Y', but all connected components of Y' project surjectively onto $A \times A$ and therefore meet Y'_1. This is only possible if Y' is connected.

Thus $Y' \to A \times A$ is a Galois étale cover with group $G = \mathrm{Aut}(Y|A)$. Applying Corollary 5.6.6 (together with Propositions 5.3.8 and 5.3.9) we see that there exist étale Galois covers $Z_1 \to A$ and $Z_2 \to A$, with groups G_1 and G_2, respectively, such that Y' is a quotient of the direct product $Z_1 \times Z_2 \to A \times A$. In particular, there is a normal subgroup $H \subset G_1 \times G_2$ with $G \cong (G_1 \times G_2)/H$. Replacing the Z_i by their quotients by the actions of the $G_i \cap H$ we may assume $H \cap G_i = \{1\}$ for $i = 1, 2$. Thus the G_i may be identified with normal subgroups of G that generate G and centralize each other. But by construction $Y'_1 \to \{0\} \times A$ must be an étale Galois cover with group G_2, and similarly for the base change $Y'_2 \to A \times \{0\}$. It follows that $G = G_1 = G_2$, and G is commutative. Moreover, we have isomorphisms $Y_i \cong Y'_i \cong Y$ for $i = 1, 2$. This yields a map $Y \times Y \to Y'$, whence also a map $m_Y : Y \times Y \to Y$ by composing with the projection $Y' \to Y$. Fix a point 0_Y of Y in the fibre above 0. Modifying m_Y by an automorphism of Y if necessary we may assume $m_Y(0_Y, 0_Y) = 0_Y$.

We contend that m_Y is a commutative group operation on Y with 0_Y as null element. By construction this will also imply that Y is an abelian variety (being proper over k) and that ϕ is a group homomorphism. To see that m_Y is commutative, consider the automorphism $\sigma_Y : Y \times Y$ given by permuting the factors. The maps m_Y and $m_Y \circ \sigma_Y : Y \times Y \to Y$ coincide at the point $(0_Y, 0_Y)$, so they are the same by Corollary 5.3.3. Similar reasonings show that m_Y is associative with null element 0_Y. The existence of an inverse is equivalent to the property that the product map $(\mathrm{id}_Y, m_Y) : Y \times Y \to Y \times Y$ is an isomorphism. We know that (id_A, m_A) is an isomorphism. Therefore from the commutative diagram

$$\begin{array}{ccc} Y \times Y & \xrightarrow{(\mathrm{id}_Y, m_Y)} & Y \times Y \\ {\scriptstyle \phi\times\phi}\downarrow & & \downarrow{\scriptstyle \phi\times\phi} \\ A \times A & \xrightarrow{(\mathrm{id}_A, m_A)} & A \times A \end{array}$$

we infer that the composite $(\phi \times \phi) \circ (\mathrm{id}_Y, m_Y) : Y \times Y \to A \times A$ is an étale Galois cover with group $G \times G$. But so is $\phi \times \phi$, so (id_Y, m_Y) must indeed be an isomorphism. □

Corollary 5.6.9 *Let $n > 0$ be an integer divisible by the degree of ϕ. There is a map $\psi : A \to Y$ and a commutative diagram*

$$\begin{array}{ccc} A & \xrightarrow{\psi} & Y \\ & {\scriptstyle n_A}\searrow & \downarrow{\scriptstyle \phi} \\ & & A \end{array} \qquad (5.3)$$

where $n_A : A \to A$ is the multiplication-by-n map of A.

Proof We have just proven that ϕ is an isogeny, and by assumption on n we have $n\ker(\phi) = 0$. Consider the multiplication-by-n map $n_Y : Y \to Y$. Since $\ker(\phi) \subset \ker(n_Y)$ and $A \cong Y/\ker(\phi)$ as an abelian variety (compare with Proposition 5.3.6), the map n_Y induces a homomorphism of abelian varieties $\psi : A \to Y$ satisfying $n_Y = \psi \circ \phi$. But we also have $n_A = \phi \circ \psi$ because for $P \in A$ we find $Q \in Y$ with $\phi(Q) = P$ by surjectivity of ϕ, and so

$$\phi(\psi(P)) = \phi(\psi(\phi(Q)) = \phi(nQ) = n\phi(Q) = nP.$$

This proves the corollary. □

We can now completely determine the fundamental group of the abelian variety A. Recall that given a prime number ℓ, the *ℓ-adic Tate module $T_\ell(A)$* of A is defined as the inverse limit of the inverse system formed by the ℓ^r-torsion subgroups ${}_{\ell^r}A(k)$ for all $r > 0$. Here $A(k)$ denotes the group of k-rational points of A (i.e. the group of sections of the structure morphism $A \to \mathrm{Spec}\,(k)$). The direct product of the $T_\ell(A)$ for all ℓ is the *full Tate module $T(A)$* of A; it is also the inverse limit of the inverse system of n-torsion subgroups ${}_nA(k)$ partially ordered by divisibility.

Theorem 5.6.10 *Let A be an abelian variety over an algebraically closed field. The fundamental group of A is commutative, and there are natural* *isomorphisms*

$$\pi_1(A) \cong T(A) \cong \prod_{\ell} T_\ell(A).$$

We have omitted the base point from the notation as it plays no role in the commutative case (Remark 5.5.3).

Proof When n is prime to the characteristic of k, the map $n_A : A \to A$ is a finite étale Galois cover with group $\ker(n_A)(k) \cong {}_nA(k)$. Indeed, the fibres of n_A are all given by the constant group scheme ${}_nA(k)$. By the previous corollary each finite étale cover $Y \to A$ of degree dividing n arises as a quotient of this cover.

When n is divisible by the characteristic of k, the finite map n_A is not étale (the group scheme $\ker(n_A)$ is not reduced). Still, we can make the finite group ${}_nA(k)$ act on A by translation; the quotient is A because $A(k)$ is divisible ([65], §6, Application 2). We thus get a finite étale Galois cover $n' : A \to A$ with group ${}_nA(k)$, and moreover the map n_A factors as $n_A = n' \circ \rho$ for some $\rho : A \to A$ as in the previous proof. The map ρ must induce a purely inseparable extension of function fields for degree reasons (see [95], Section II.6, Theorems 3 and 4). Therefore the image of the embedding of function fields $\psi^* : K(Y) \to K(A)$, with ψ as in Corollary 5.6.9, must lie in $n'^*K(A)$. As Y is the normalization of A in the extension $K(Y)|\phi^*K(A)$ and similarly for A and $K(A)|n'^*K(A)$, we obtain that a diagram of the form (5.3) exists with n' in place of n_A.

Thus in both cases we conclude $\pi_1(A)/n\pi_1(A) \cong {}_nA(k)$, and the theorem follows by taking the inverse limit over all n. □

Remark 5.6.11 From the theory of abelian varieties it is known (see [65], §6) that in the case when ℓ is prime to the characteristic of k one has $T_\ell(A) \cong \mathbf{Z}_\ell^{2g}$, where g is the dimension of A. When k has characteristic $p > 0$, one has $T_p(A) \cong \mathbf{Z}_p^\rho$ with an integer $0 \leq \rho \leq g$, the *p-rank* of A. This determines the structure of $\pi_1(A)$ completely.

5.7 Structure theorems for the fundamental group

Already in the case of curves the more precise structural results for the fundamental group relied on comparison with the theory over $\mathbf{C}$. In higher dimension the situation is similar. We begin with a basic finiteness result.

Proposition 5.7.1 *Let X be a smooth projective integral scheme over* $\mathbf{C}$. *For every geometric base point $\bar{x}$ the fundamental group $\pi_1(X, \bar{x})$ is topologically finitely generated.*

Recall that a profinite group is topologically finitely generated if it contains a dense finitely generated subgroup. For the proof we need a Bertini type lemma.

Lemma 5.7.2 *Let k be an algebraically closed field, $X \subset \mathbf{P}^n_k$ a smooth connected closed subscheme of dimension at least 2, and $Y \to X$ a connected finite étale cover. There exists a hyperplane $H \subset \mathbf{P}^n_k$ not containing X so that $X \cap H$ is again smooth and connected, and moreover the fibre product $Y \times_X (H \cap X)$ is connected as well.*

Proof By Theorem II.8.18 in [34] one may find H with $X \cap H$ smooth. Then Remark III.7.9.1 of the same reference shows the connectedness of $X \cap H$ for this H. The same argument carries over *mutatis mutandis* to to the closed immersion $Y \times_X (H \cap X) \hookrightarrow Y$ and shows the connectedness of $Y \times_X (H \cap X)$. □

Proof of Proposition 5.7.1 Since by Remark 5.2.3 (2) the fibre product $Y \times_X (H \cap X)$ in the lemma is a connected finite étale cover of $H \cap X$, we obtain from Proposition 5.5.4 (2) a surjection $\pi_1(H \cap X, \bar{x}) \twoheadrightarrow \pi_1(X, \bar{x})$ for a geometric point $\bar{x}$ of $H \cap X$. As $H \cap X$ is smooth and projective of strictly smaller dimension than X, we may apply induction on dimension to obtain a surjection $\pi_1(C, \bar{x}) \twoheadrightarrow \pi_1(X, \bar{x})$, where $C \subset X$ is a smooth projective curve obtained by cutting X with a linear subspace in $\mathbf{P}^n$, and $\bar{x}$ is a geometric point of C. Over $k = \mathbf{C}$ we know from Theorem 4.6.7 that $\pi_1(C, \bar{x})$ is topologically finitely generated, hence so is $\pi_1(X, \bar{x})$. □

Remark 5.7.3 The key point in the above proof was the surjectivity of the map $\pi_1(H \cap X, \bar{x}) \to \pi_1(X, \bar{x})$. In fact, when X has dimension at least 3, the *Lefschetz hyperplane theorem* states that this map is actually an isomorphism for *any* hyperplane H (and $X \cap H$ is connected).

Over $k = \mathbf{C}$ there is a simple topological proof by Andreotti and Frenkel (see [60], §7). In the general case there is a proof by Grothendieck in [30]. He constructs an equivalence of categories between Fet_X and $\mathrm{Fet}_{X \cap H}$ in three steps. First he takes an open subscheme $U \subset X$ containing $X \cap H$. As X has dimension at least 3, each codimension 1 closed subset meets $X \cap H$ and hence the complement of U must be of codimension ≥ 2. Corollary 5.2.14 then implies an equivalence between Fet_X and Fet_U. Next he considers the formal completion $\widehat{X}$ of X along $X \cap H$ ([34], Section II.9) and defines finite étale covers for formal schemes. He then proves a nontrivial algebraization theorem asserting the equivalence of Fet_U and $\mathrm{Fet}_{\widehat{X}}$. Finally, the equivalence between $\mathrm{Fet}_{\widehat{X}}$ and $\mathrm{Fet}_{X \cap H}$ results by a generalization of Exercise 7. For a simplified account of Grothendieck's proof, see [33], Chapter IV.

Now let us return to Proposition 5.7.1. A similar argument involving reduction to the case of curves via cutting with hyperplanes shows that the topological fundamental group of the complex manifold associated with X is finitely generated (in the usual sense for groups). This hints at the possibility of generalizing the comparison between algebraic and topological fundamental groups to higher dimension. Such a generalization involves a higher dimensional Riemann Existence Theorem that we now briefly explain.

Given a scheme X of finite type over $\mathbf{C}$, one may define the *complex analytic space* X^{an} associated with X (see [34], Appendix B, §1). When X is smooth, the space X^{an} is a complex manifold as defined in Remark 3.1.2. A morphism $\phi : Y \to X$ of schemes of finite type over $\mathbf{C}$ induces a holomorphic map $\phi^{\mathrm{an}} : Y^{\mathrm{an}} \to X^{\mathrm{an}}$ of analytic spaces. We then have the following vast generalization of Theorem 4.6.7 (see [29], Exposé XII, Corollaire 5.2).

Theorem 5.7.4 *Let X be a connected scheme of finite type over $\mathbf{C}$. The functor $(Y \to X) \mapsto (Y^{\mathrm{an}} \to X^{\mathrm{an}})$ induces an equivalence of the category of finite étale covers of X with that of finite topological covers of X^{an}. Consequently, for every $\mathbf{C}$-point $\bar{x} : \mathrm{Spec}\,(\mathbf{C}) \to X$ this functor induces an isomorphism*

$$\widehat{\pi_1^{\mathrm{top}}(X^{\mathrm{an}}, \bar{x})} \xrightarrow{\sim} \pi_1(X, \bar{x})$$

where on the left-hand side we have the profinite completion of the topological fundamental group of X with base point $\mathrm{im}\,(\bar{x})$.

The hard part of the theorem is the essential surjectivity of the functor. As in the theory of Riemann surfaces, one first proves that every finite topological cover $p : Y \to X^{\mathrm{an}}$ can be equipped with a canonical structure of analytic space for which p becomes a proper holomorphic map. Next there is a reduction to the case where X is normal. One then has to use a deep theorem, first proven by Grauert and Remmert, which says that for normal X every proper analytic map $Z \to X^{\mathrm{an}}$ with finite fibres is isomorphic to $Y^{\mathrm{an}} \to X^{\mathrm{an}}$ for a finite surjective morphism of schemes $Y \to X$. In the case when X is a projective variety over $\mathbf{C}$ this is part of Serre's famous GAGA theorems. The proof given in [29], Exposé XII does not use the theorem of Grauert and Remmert, but needs Hironaka's resolution of singularities.

Remark 5.7.5 Despite the above theorem, there is in general a big difference between the topological and algebraic fundamental groups. Toledo [104] constructed a smooth projective variety X over $\mathbf{C}$ such that the intersection of all subgroups of finite index in $\pi_1^{\mathrm{top}}(X, \bar{x})$ is a free group of infinite rank. In other words, the completion map $\pi_1^{\mathrm{top}}(X, \bar{x}) \to \pi_1(X, \bar{x})$ has a free kernel of infinite rank. It is an open question of Serre whether there exist examples with $\pi_1(X, \bar{x}) = \{1\}$ but $\pi_1^{\mathrm{top}}(X, \bar{x}) \neq \{1\}$.

Now let us return to Proposition 5.7.1. Combined with Proposition 5.6.7 it yields:

Corollary 5.7.6 *Let k be an algebraically closed field of characteristic 0, and let X be a smooth connected projective scheme over k. For every geometric base point $\bar{x}$ the fundamental group $\pi_1(X, \bar{x})$ is topologically finitely generated.*

Proof Applying Proposition 5.6.7 to the inclusion $\overline{\mathbf{Q}} \hookrightarrow \mathbf{C}$ we obtain the result for $k = \overline{\mathbf{Q}}$. Then we deduce the result for general k using the inclusion $\overline{\mathbf{Q}} \hookrightarrow k$. □

Before leaving the case of characteristic 0, we give an alternate approach to the above corollary due to Pop. It uses the following well-known lemma on profinite groups.

Lemma 5.7.7 *Let G_1 be a profinite group that has only finitely many open normal subgroups of index N for every integer $N > 0$. Assume that G_2 is another profinite group such that for every open normal subgroup $U \subset G_1$ there is an open normal subgroup $V \subset G_2$ with $G_1/U \cong G_2/V$, and vice versa. Then:*

1. *There exists an isomorphism of profinite groups $G_1 \cong G_2$.*
2. *Every continuous surjection $G_1 \twoheadrightarrow G_2$ (or $G_2 \twoheadrightarrow G_1$) is an isomorphism.*

Proof For part (1) denote by U_N (resp. V_N) the intersection of all open normal subgroups of index at most N in G_1 (resp. G_2). By assumption $[G_1 : U_N]$ is finite. If $V \subset G_2$ is an open normal subgroup that is an intersection of open normal subgroups of index at most N, there is an open normal subgroup $U \supset U_N$ in G_1 with $G_1/U \cong G_2/V$. Thus G_2/V is a quotient of G_1/U_N. Now let $W \subset G_2$ be an open normal subgroup of maximal index such that G_2/W arises as a quotient of G_1/U_N. By the previous discussion $G_2/(W \cap V)$ is again a quotient of G_1/U_N for every open normal $V \subset G_2$ of index at most N. It follows from the maximality assumption that the inclusion $W \cap V \subset W$ must be an equality, and hence $V_N = W$. In particular, $[G_2 : V_N]$ is finite. As the finite groups G_1/U_N and G_2/V_N have the same quotients, the natural surjection $G_1/U_N \twoheadrightarrow G_2/V_N$ is an isomorphism.

Now consider the finite set X_N of isomorphisms $G_1/U_N \xrightarrow{\sim} G_2/V_N$ for each $N > 0$. The X_N form a natural inverse system indexed by the positive integers, because composing a given isomorphism $G_1/U_{N+1} \xrightarrow{\sim} G_2/V_{N+1}$ with the projection $G_2/V_{N+1} \to G_2/V_N$ induces an isomorphism $G_1/U_N \to G_2/V_N$. Thus we obtain an inverse system of nonempty finite sets, and the inverse limit is nonempty by Lemma 3.4.12. An element of $\varprojlim X_N$ is a compatible system of isomorphisms $G_1/U_N \xrightarrow{\sim} G_2/V_N$, so it induces a continuous isomorphism

$\varprojlim G_1/U_N \xrightarrow{\sim} \varprojlim G_2/V_N$. But these inverse limits are G_1 and G_2, respectively, because the U_N are cofinal in the system of open normal subgroups of G_1, and similarly for the V_N.

For part (2) let $\phi : G_1 \twoheadrightarrow G_2$ be a continuous surjection. For each $N > 0$ it induces a surjection $\phi_N : G_1/U_N \twoheadrightarrow G_2/V_N$. But we have just seen that these are finite groups of the same order, so ϕ_N must be an isomorphism. Passing to the inverse limit we obtain that ϕ is an isomorphism. The argument for a surjection $G_2 \to G_1$ is similar. □

Second proof of Corollary 5.7.6 Assume first k may be embedded in $\mathbf{C}$. Then the surjectivity part of the proof of Proposition 5.6.7 (which was elementary) shows that the map $\pi_1(X_{\mathbf{C}}, \bar{x}_{\mathbf{C}}) \to \pi_1(X, \bar{x})$ is surjective. What therefore remains to be shown is the injectivity part of Proposition 5.6.7 for an extension $k \subset K$ of algebraically closed fields of characteristic 0 and X smooth and projective, *knowing already that $\pi_1(X, \bar{x})$ is topologically finitely generated.* This enables us to apply the lemma, which reduces us to showing that $\pi_1(X, \bar{x})$ and $\pi_1(X_K, \bar{x}_K)$ have the same finite quotients. By surjectivity we already know that every finite quotient of $\pi_1(X, \bar{x})$ is also a finite quotient of $\pi_1(X_K, \bar{x}_K)$. For the converse, assume given a finite connected Galois cover $\phi : Y \to X_K$ with group G. As in the proof of Proposition 5.6.7 we find a subfield $k' \subset K$ finitely generated over k, an integral affine k-scheme T with function field k' and a finite étale cover $\phi_T : \mathcal{Y} \to X \times_{\operatorname{Spec}(k)} T$ of schemes of finite type over T such that the induced morphism $\mathcal{Y} \times_T \operatorname{Spec}(k') \to X_{k'}$ is isomorphic to ϕ'. Moreover, by construction ϕ_T is a Galois cover with group G. Now a Bertini type lemma ([95], II.6.1, Theorem 1) allows us to find a k-point P of T such that the fibre $\mathcal{Y}_P$ of the composite map $\mathcal{Y} \to X \times_{\operatorname{Spec}(k)} T \to T$ over P is connected. We obtain a finite étale Galois cover $\mathcal{Y}_P \to X$ with group G. □

Remark 5.7.8 Though it works under more restrictive assumptions than that of Proposition 5.6.7, the above proof of the isomorphism $\pi_1(X_K, \bar{x}) \xrightarrow{\sim} \pi_1(X, \bar{x})$ has several advantages. To begin with, it avoids the use of the difficult Proposition 5.6.4, and in particular the properness assumption that was crucial in its proof. Instead, all we need here is the fact that $\pi_1(X_{\mathbf{C}}, \bar{x}_{\mathbf{C}})$ is topologically finitely generated. In particular, the above proof also works for a normal *affine* curve X over an algebraically closed field of characteristic 0 (by virtue of Theorem 4.6.7), and we obtain a result used in the previous chapter.

On the other hand, if one is only interested in proving Proposition 5.6.7 for proper normal schemes but not Corollary 5.7.6, even the transcendental input is superfluous. In fact, in order to be able to apply the lemma, all one needs is the fact that $\pi_1(X, \bar{x})$ has only finitely many open normal subgroups of index N for every integer $N > 0$. But, as already noted in the previous chapter, this was proven in a purely algebraic way for proper normal schemes (in arbitrary characteristic) by Lang and Serre ([51], th. 4).

We now extend the previous results to positive characteristic. The extension is based on Grothendieck's specialization theory for the fundamental group which we now explain. We shall work over the spectrum of a complete discrete valuation ring A. (Recall that a local ring A with maximal ideal M is said to be complete if the natural map $A \to \varprojlim A/M^i$ is an isomorphism.) Examples of complete discrete valuation rings are the ring $\mathbf{Z}_p$ of p-adic integers for a prime p (Example 1.3.4 (4)) and the formal power series ring $k[[t]]$ over a field k. We shall tacitly use the key property that the integral closure of a complete discrete valuation ring in a finite extension of its fraction field is again a complete discrete valuation ring (see [69], Chapter II, Theorem 4.8 and its proof).

Let us introduce some notation. Denote by K the fraction field and by κ the residue field of the complete discrete valuation ring A. The affine scheme Spec (A) will be denoted by S, and $\eta : \mathrm{Spec}\,(K) \to S$ (resp. $s : \mathrm{Spec}\,(\kappa) \to S$) will stand for the generic (resp. closed) points of S. Fix geometric points $\bar{\eta}$ (resp. $\bar{s}$) lying above η (resp. s). Given a morphism of schemes $X \to S$, we denote by X_η, X_s, $X_{\bar{\eta}}$, $X_{\bar{s}}$ its base changes by the corresponding maps.

Theorem 5.7.9 (Grothendieck) *Let S be as above, and let $\phi : X \to S$ be a proper morphism. Fix geometric points $\bar{x}$ and $\bar{y}$ of $X_{\bar{\eta}}$ and X_s, respectively.*

1. *The natural map $\pi_1(X_s, \bar{y}) \to \pi_1(X, \bar{y})$ induced by the map $X_s \to X$ is an isomorphism.*
2. *Assume moreover that κ is algebraically closed, ϕ is flat, and the geometric fibres $X_{\bar{\eta}}$, $X_{\bar{s}}$ are reduced. Then the natural map $\pi_1(X_{\bar{\eta}}, \bar{x}) \to \pi_1(X, \bar{x})$ is surjective.*

This is proven in [29], Exposé X using deep algebraization techniques for formal schemes. We refer to Section 8.5.C of [36] for an excellent exposition.

If k is algebraically closed and $s = \bar{s}$, we may consider the composite map

$$sp : \pi_1(X_{\bar{\eta}}, \bar{x}) \to \pi_1(X, \bar{x}) \xrightarrow{\sim} \pi_1(X, \bar{y}) \xrightarrow{\sim} \pi_1(X_{\bar{s}}, \bar{y}) \qquad (5.4)$$

where the middle isomorphism is a non-canonical one coming from Corollary 5.5.2, the first map is the one in part (2) of the theorem, and the last one is the inverse of the isomorphism of part (1). It is called a *specialization map* for the fundamental group associated with ϕ. As it depends on the choice of a path from $\bar{x}$ to $\bar{y}$, it is unique up to inner automorphism.

Concerning the specialization map the main result is the following.

Theorem 5.7.10 (Grothendieck) *Keep notations and assumptions as above, and assume moreover that ϕ is proper and smooth with geometrically connected* *fibres. Then the specialization map sp induces an isomorphism*

$$\pi_1(X_{\bar{\eta}}, \bar{x})^{(p')} \xrightarrow{\sim} \pi_1(X_{\bar{s}}, \bar{y})^{(p')},$$

where the superscripts (p') denote the maximal prime-to-p quotients of the profinite groups involved.

The proof uses a lemma on discrete valuation rings. It is often referred to by the slogan 'ramification kills ramification'. For background and terminology concerning extensions of discrete valuation rings, see Section 4.1.

Lemma 5.7.11 (Abhyankar) *Let A be a discrete valuation ring with maximal ideal P, fraction field K, and perfect residue field κ. Let $K_1|K$, $K_2|K$ be two finite Galois extensions. Denote by A_i the integral closure of A in K_i, and fix maximal ideals P_i lying above P for $i = 1, 2$. Assume that the orders e_i of the inertia subgroups I_i of the P_i are prime to the characteristic of κ, and that moreover e_1 divides e_2.*

Then the finite morphism $\mathrm{Spec}\,(C) \to \mathrm{Spec}\,(A_2)$ *is étale, where C denotes the integral closure of A in the composite field K_1K_2.*

Proof Let Q be a maximal ideal of C lying above P. As the Galois groups $\mathrm{Gal}\,(K_i|K)$ act transitively on the set of maximal ideals of A_i lying above P for $i = 1, 2$ (Fact 4.1.3 (1)), we may assume without loss of generality that $Q \cap A_i = P_i$ for $i = 1, 2$. Restriction to the subfields K_i yields an injective homomorphism $\mathrm{Gal}\,(K_1K_2|K) \to \mathrm{Gal}\,(K_1|K) \times \mathrm{Gal}\,(K_2|K)$. Moreover, if we denote by I_Q the inertia subgroup of Q in $\mathrm{Gal}\,(K_1K_2|K)$, then the above map induces an injection $I_Q \hookrightarrow I_1 \times I_2$ such that the projections $I_Q \to I_i$ are surjective for $i = 1, 2$. Here all three groups have order prime to the residue characteristic by assumption, hence all of them are cyclic by Fact 4.1.3 (4). By the assumption $e_1|e_2$ the elements of $I_1 \times I_2$, and hence of I_Q, have order dividing e_2. But I_Q surjects onto I_2, so it must have order e_2 and the projection $I_Q \to I_2$ must be an isomorphism. This implies that the inertia group of Q over P_2 is trivial. As Q here was arbitrary, the conclusion follows from Corollary 4.1.7 (and the isomorphism (4.1) preceding it). □

Proof of Theorem 5.7.10 In view of Theorem 5.7.9 (1) we have to show that the map $\pi_1(X_{\bar{\eta}}, \bar{x}) \to \pi_1(X, \bar{x})$ induces an isomorphism on maximal prime-to-p quotients.

We first show that the map $\pi_1(X_{\bar{\eta}}, \bar{x}) \to \pi_1(X, \bar{x})$ itself is surjective. By Proposition 5.5.4 (2) we have to check that for each connected finite étale cover $\pi : Y \to X$ the base change $Y_{\bar{\eta}} := Y \times_X \bar{\eta}$ remains connected. The scheme $Y_{\bar{\eta}}$ is the inverse limit of the schemes $Y_{\eta'} := Y \times_S \mathrm{Spec}\,(K')$, where $K'|K$ is a finite extension contained in the algebraic closure of K corresponding to $\bar{\eta}'$. As an inverse limit of connected topological spaces is again connected, it suffices to check the connectedness of the $Y_{\eta'}$. But $Y_{\eta'}$ is the generic fibre (so in particular an open subscheme) of $Y' = Y \times_S S'$, where S' is the spectrum of the normalization A' of A in K', so therefore it is enough to check the connectedness of Y'. If Y' were the disjoint union of two nonempty closed subsets Z_1 and Z_2, one of them, say Z_1, would contain the special fibre Y_s. Indeed, Y_s is the same as X_s (since $s = \bar{s}$) and X_s is connected by assumption. But then the image of Z_2 by the map $Y' \to S'$ induced by $\phi \circ \pi$ via base change would be the open point η' of S'. This is impossible, because $Y' \to S'$ is a proper, hence closed map.

For injectivity on maximal prime-to-p quotients we have to show by virtue of Corollary 5.5.8 (taking the argument of Remark 5.5.10 into account) that every finite étale Galois cover $\overline{Y} \to X_{\bar{\eta}}$ of degree prime to p comes via base change from a Galois cover $Y \to X$ of degree prime to p. As before, we know that $\overline{Y}$ comes via base change from a finite étale cover of $X_{\eta'}$, with η' corresponding to a finite extension $K'|K$. On the other hand, for each finite extension $K'|K$ we have, in the above notation, an isomorphism $\pi_1(X \times_S S', \bar{y}) \xrightarrow{\sim} \pi_1(X, \bar{y})$, because both groups here are isomorphic to $\pi_1(X_{\bar{s}}, \bar{y})$ by Theorem 5.7.9 (1). Thus after replacing S by S' we may assume that $\overline{Y}$ comes from a finite Galois cover $Y_\eta \to X_\eta$, and we have to show that there is a finite extension $K''|K$ such that, with obvious notation, $\overline{Y}$ comes by base change from an étale Galois cover $Y'' \to X \times_S S''$ of degree prime to p. To see this, write $Z \to X$ for the normalization of X in the function field $K(Y_\eta)$ of Y_η. Note that $K(Y_\eta)|K(X)$ is a finite Galois extension; denote by d its degree. By Zariski–Nagata purity (Theorem 5.2.13) it suffices to find $K''|K$ such that $Z \times_S S'' \to X \times_S S''$ is étale over all codimension 1 points of $X \times_S S''$, for then we may set $Y'' := Z \times_S S''$. Now all codimension 1 points of X lie in X_η except for the generic point ξ of X_s. So taking Remark 5.2.3 (2) into account it is enough to find S'' such that $Z \times_S S''$ is étale over the points of $X \times_S S''$ lying above ξ.

Now if we denote by π a generator of the maximal ideal of A, then by construction π also generates the maximal ideal of the discrete valuation ring $\mathcal{O}_{X,\xi}$. The extension $K(X)(\sqrt[d]{\pi})|K(X)$ is a finite Galois extension of degree d by Kummer theory. Moreover, Corollary 4.1.7 implies that there is a single point ξ' on the normalization of X in $K(X)(\sqrt[d]{\pi})$ lying above ξ, with cyclic inertia group of order d. Hence we may apply Lemma 5.7.11 to the extensions $K(Y_\eta)$ and $K(X)(\sqrt[d]{\pi})$ of $K(X)$ to conclude that the fibre over ξ' of the normalization Y'' of X in the composite $K(Y_\eta) \cdot K(X)(\sqrt[d]{\pi})$ is étale. It remains to notice that by construction $Y'' \cong X \times_S S''$, where S'' is the spectrum of the normalization of A in the extension $K(\sqrt[d]{\pi})|K$. □

Remark 5.7.12 In fact, one may define the specialization map in the following more general situation: S is a locally Noetherian scheme, $X \to S$ is a proper morphism with connected geometric fibres, and s_0, s_1 are scheme-theoretic points of S such that s_0 lies in the closure of s_1 in S. Fix geometric points $\bar{s}_i$ lying above the s_i and geometric points $\bar{x}_i$ of the corresponding geometric fibres. We then have a specialization homomorphism $\pi_1(X_{\bar{s}_1}, \bar{x}_1) \to \pi_1(X_{\bar{s}_0}, \bar{x}_0)$ defined as follows. Replace S by the spectrum of the local ring $\mathcal{O}_{Z,s_0}$ of s_0 on the closure Z of s_1 in S; this does not affect the geometric fibres $X_{\bar{s}_i}$. Let A be the completion of the localization of the integral closure of $\mathcal{O}_{Z,s_0}$ in its fraction field K_0 by a prime ideal of height 1. It is a complete discrete valuation ring containing $\mathcal{O}_{Z,s_0}$, hence the specialization map (5.4) is defined for $X \times_S \operatorname{Spec}(A) \to \operatorname{Spec}(A)$. The geometric special fibre $X_{\bar{s}_0}$ is preserved, and the natural map $X_{\bar{\eta}} \to X_{\bar{s}_1}$ induces an isomorphism on fundamental groups by Proposition 5.6.7. Composing with this isomorphism (and possibly changing base points) we obtain the required map $\pi_1(X_{\bar{s}_1}, \bar{x}_1) \to \pi_1(X_{\bar{s}_0}, \bar{x}_0)$. The statements of Theorems 5.7.9 (2) and 5.7.10 carry over immediately to this more general specialization map.

Grothendieck's main motivation for developing the above theory was to prove Theorem 4.9.1 in positive characteristic. Let us restate it in the proper case.

Theorem 5.7.13 *Let k be an algebraically closed field of characteristic $p \geq 0$, and let X be an integral proper normal curve of genus g over k. For every geometric point $\bar{x}$ of X the group $\pi_1(X, \bar{x})$ is topologically finitely generated, and its maximal prime to p-quotient $\pi_1(X, \bar{x})^{(p')}$ is isomorphic to the profinite p'-completion of the group*

$$\Pi_{g,0} := \langle a_1, b_1, \dots, a_g, b_g \mid [a_1, b_1] \dots [a_g, b_g] = 1 \rangle.$$

As already remarked in the previous chapter, for $k = \mathbf{C}$ this follows from the topological theory and the Riemann existence theorem, and then for general k of characteristic 0 it follows from Proposition 5.6.7. When k has positive characteristic, the proof proceeds in three steps.

1. First one uses the fact that there exists a discrete valuation ring A with fraction field K of characteristic 0 and residue field κ isomorphic to k ([57], Theorem 29.1). Here A may be assumed complete by taking its completion.

2. Then one uses the existence of a smooth proper scheme $\mathcal{X} \to \operatorname{Spec}(A)$ with $\mathcal{X}_s \cong X$ and $\mathcal{X}_\eta$ a smooth proper curve over K. This can be proven in several ways. The original approach of Grothendieck was to extend X first to a formal $\operatorname{Spec}(A)$-scheme and use an algebraization theorem; see [36], Theorem 8.5.19.

Another approach is to use the classification of curves. For instance, one may embed X in projective space using the third power of the canonical sheaf, and then identify its isomorphism class with a k-point $[X]$ of the Hilbert scheme denoted by H_g^0 in [15]. According to Corollaries 1.7 and 1.9 of that paper the scheme H_g^0 is smooth over $\mathrm{Spec}\,(\mathbf{Z})$, and therefore a form of Hensel's lemma (see e.g. [64], §III.5) can be applied to extend $[X]$ to an A-valued point. (This argument works for $g \geq 2$, but the cases $g = 0, 1$ are easily handled directly.)

3. Finally, one applies Theorem 5.7.10.

Corollary 5.7.14 *Let X be a smooth connected projective scheme over an algebraically closed field. For each geometric base point $\bar{x}$ the fundamental group $\pi_1(X, \bar{x})$ is topologically finitely generated.*

Proof Like Proposition 5.7.1, this is proven by reduction to the case of curves treated in the theorem above. □

Combining the arguments involved in the above proof with techniques of flat descent theory, Grothendieck proved in ([29], X.2.10) that the corollary holds for an arbitrary proper connected scheme over an algebraically closed field. Without the properness assumption the corollary is false in positive characteristic even for smooth curves: see Theorem 4.9.5.

In the case of non-proper schemes it is therefore often useful to work with a fundamental group whose p-part is also topologically finitely generated in characteristic $p > 0$. Such a fundamental group is defined by means of finite étale covers tamely ramified at infinity.

Definition 5.7.15 Let X be a normal integral scheme, and $U \subset X$ an open subscheme. Assume given a connected finite étale cover $Y \to U$. We say that Y is *tamely ramified along* $X \setminus U$ if for each codimension 1 point P of X not lying in U the closed points of the normalization of $\mathrm{Spec}\,(\mathcal{O}_{X,P})$ in the function field $K(Y)$ have ramification indices prime to the characteristic of $\kappa(P)$.

Assume now that U is a regular integral scheme that is separated of finite type over a base scheme S which is the spectrum of a field, a complete discrete valuation ring or $\mathbf{Z}$. We say that a connected finite étale cover $Y \to U$ is tamely ramified if for all normal integral schemes X proper over S that contain U as a dense open subscheme the cover Y is tamely ramified along $X \setminus U$. This definition naturally extends to non-connected covers by considering the function fields of various components of Y.

In the above situation fix moreover a geometric point $\bar{u}$ of U. We define the *tame fundamental group* $\pi_1^t(U, \bar{u})$ as the automorphism group of the restriction of the fibre functor $\mathrm{Fib}_{\bar{u}}$ to the full subcategory of Fet_U spanned by tamely ramified finite étale covers of U.

Remarks 5.7.16

1. It follows from the definition that $\pi_1^t(U, \bar{u})$ is a quotient of $\pi_1(U, \bar{u})$; in particular, it is a profinite group. When $\bar{u}$ is given by an algebraic closure $\overline{K(U)}$ of the function field $K(U)$, we may identify $\pi_1^t(U, \bar{u})$ with the Galois group $\mathrm{Gal}\,(K_U^t|K(Y))$, where K_U^t is the compositum in $\overline{K(U)}$ of the function fields of tamely ramified connected finite étale covers of U. This makes sense, because composita of tamely ramified extensions of discrete valuation rings are again tamely ramified ([69], Chapter II, Corollary 7.9).
2. In the case when U has dimension 1, there is a unique regular proper X containing U as an open subscheme. We have seen this in the case of normal curves in the previous chapter; the general case is similar. Thus in this case the tameness condition is to be checked for a single compactification X.
3. The above definition of tameness is taken from Kerz and Schmidt [43]. They show that if there exists a regular proper X containing U such that the complement $X \setminus U$ is a so-called normal crossing divisor, then $Y \to U$ is tamely ramified if and only if it is tamely ramified along $X \setminus U$. The latter condition is the definition of tameness adopted in [29]. In [43] it is also shown that a finite étale cover $Y \to U$ is tamely ramified if and only if the base change $X \times_S C \to C$ is a tamely ramified cover of C for every regular integral separated scheme of finite type over S that has dimension 1. This implies in particular that the notion of tameness is preserved by base change.
4. Assume that $X \to S$ is a proper smooth morphism with geometrically integral fibres of dimension 1, where S is the spectrum of a complete discrete valuation ring with algebraically closed residue field. Assume moreover given a closed subscheme $Y \subset X$ such that the composite $Y \to X \to S$ is a finite étale cover. Set $U = X \setminus Y$, fix geometric points $\bar{\eta}$ and $\bar{s}$ lying above the generic and closed points of S, and choose geometric points $\bar{u}$ and $\bar{y}$ of the the geometric fibres $U_{\bar{\eta}}$ and $U_{\bar{s}}$, respectively. In this situation there is a specialization map $\pi_1^t(U_{\bar{\eta}}, \bar{u}) \to \pi_1^t(U_{\bar{s}}, \bar{y})$ which is surjective and induces an isomorphism on maximal prime-to-p quotients. For a proof, see the chapter by Orgogozo and Vidal in [8].

   As a consequence, one obtains by a similar argument as in Theorem 5.7.13 that when U is an integral normal curve over an algebraically closed field, the tame fundamental group $\pi_1^t(U, \bar{u})$ is topologically finitely generated ([29], exposé XIII, corollaire 2.12). Of course, in characteristic 0 this uses the topological result of Remark 3.6.4. By means of a hyperplane section argument as in Proposition 5.7.1 one gets the more general statement that for an open subscheme U of a smooth projective integral scheme over an algebraically closed field the group $\pi_1^t(U, \bar{u})$ is topologically finitely generated.

5.8 The abelianized fundamental group

Abelian covers of a scheme are easier to describe than general finite étale covers, and they already capture a lot of information. They can be very conveniently studied using étale cohomology; see [59], especially Section III.4. Here we content ourselves with an elementary exposition which, however, heavily uses the theory of abelian varieties.

Let X be a regular integral Noetherian separated scheme with function field K. Recall that the group $\mathrm{Div}(X)$ of divisors on X is by definition the free abelian group with basis consisting of the codimension 1 points of X. A *principal divisor* is a divisor of the form $\sum v_P(f)P$, where f is a nonzero element of the function field of X and v_P is the discrete valuation associated with the local ring $\mathcal{O}_{X,P}$ of the codimension 1 point P (see Proposition 4.1.9 and the subsequent discussion). This is indeed a divisor because there are only finitely many codimension 1 points P that correspond to a prime ideal on an open affine subscheme that contains f. Principal divisors form a subgroup in $\mathrm{Div}(X)$, and the *Picard group* $\mathrm{Pic}\,(X)$ of X may be defined in this case as the quotient of $\mathrm{Div}(X)$ by the subgroup of principal divisors (compare with [34], Corollary II.6.16).

Torsion elements in $\mathrm{Pic}\,(X)$ of order prime to the characteristic of K give rise to finite étale covers of X via the following construction.

Construction 5.8.1 Assume moreover that the residue fields of codimension 1 points of X are perfect. Fix an integer m prime to the characteristic of K, and assume K contains a primitive m-th root of unity. Let $D \in \mathrm{Div}(X)$ be a divisor whose class in $\mathrm{Pic}\,(X)$ has order dividing m. Then by definition there exists a function $f \in K$ with $\mathrm{div}(f) = mD$; such an f is unique up to multiplication with an everywhere regular function on X. Choose an m-th root $\sqrt[m]{f}$ in a fixed algebraic closure of K, and let Y be the normalization of X in the extension $K(\sqrt[m]{f})|K$. Note that this field extension is Galois with group $\mathbf{Z}/m\mathbf{Z}$ (Example 1.2.9 (2)); in particular it is separable.

Lemma 5.8.2 *The morphism $\phi : Y \to X$ constructed above is a finite étale Galois cover with group $\mathbf{Z}/m\mathbf{Z}$.*

Proof Pick a codimension 1 point P of X. If f is a unit in $\mathcal{O}_{X,P}$, then the fibre Y_P of ϕ above P is isomorphic to the spectrum of the $\kappa(P)$-algebra $\kappa(P)[x]/(x^m - \bar{f})$, where $\bar{f}$ is the image of f in $\kappa(P)$. As we assumed $\kappa(P)$

to be perfect, this is a finite étale $\kappa(P)$-algebra. If f is not a unit in $\mathcal{O}_{X,P}$, pick a point Q of Y lying above P (necessarily of codimension 1). In the discrete valuation v_Q associated with $\mathcal{O}_{Y,Q}$ we have $v_Q(f) = m\, v_Q(\sqrt[m]{f})$, and $v_Q(f) \neq 0$ because either f or its inverse lies in the maximal ideal of $\mathcal{O}_{X,P}$ by assumption. By Proposition 4.1.6 and Fact 4.1.3 (3) this is only possible if Q is the only point of Y lying above P, and the residue field extension $\kappa(Q)|\kappa(P)$ is Galois with group $\mathbf{Z}/m\mathbf{Z}$. Hence Y_P is again étale. Applying Zariski–Nagata purity (Theorem 5.2.13) we obtain that $Y \to X$ is a finite étale cover; it is moreover Galois with group $\mathbf{Z}/m\mathbf{Z}$ by construction. □

From now on we write $\pi_1^{\mathrm{ab}}(X)$ for the maximal abelian profinite quotient of the fundamental group $\pi_1(X, \bar{x})$ with respect to some geometric base point $\bar{x}$. It can also be described as the quotient of $\pi_1(X, \bar{x})$ by the closure of its commutator subgroup. According to Remark 5.5.3 it does not depend on the choice of the geometric base point.

Proposition 5.8.3 *For X and m as in Construction 5.8.1 there is an exact sequence*

$$1 \to \mathcal{O}_X(X)^\times/\mathcal{O}_X(X)^{\times m} \to \mathrm{Hom}(\pi_1^{\mathrm{ab}}(X), \mathbf{Z}/m\mathbf{Z}) \to {}_m\mathrm{Pic}\,(X) \to 0.$$

Here for an abelian group A the notation ${}_mA$ stands for the m-torsion subgroup, and continuous homomorphisms are considered.

Proof Each continuous homomorphism $\pi_1^{\mathrm{ab}}(X) \to \mathbf{Z}/m\mathbf{Z}$ factors through a cyclic quotient of $\pi_1^{\mathrm{ab}}(X)$ of order dividing m, which in turn corresponds to a cyclic Galois étale cover $Y \to X$. In this way we obtain a one-to-one correspondence between $\mathrm{Hom}(\pi_1^{\mathrm{ab}}(X), \mathbf{Z}/m\mathbf{Z})$ and cyclic Galois étale covers of degree dividing m.

To define the map $\mathcal{O}_X(X)^\times/\mathcal{O}_X(X)^{\times m} \to \mathrm{Hom}(\pi_1^{\mathrm{ab}}(X), \mathbf{Z}/m\mathbf{Z})$, pick a function $f \in \mathcal{O}_X(X)^\times$, and normalize X in the extension $K(\sqrt[m]{f})|K$. The beginning of the above proof shows that we obtain a finite étale cover of X. To define the next map in the sequence, start with a cyclic Galois étale cover $Y \to X$ with group $\mathbf{Z}/m\mathbf{Z}$. Replacing Y by a connected component (and lowering m) if necessary, we may assume Y is integral. The function field extension $K(Y)|K$ is then Galois with group $\mathbf{Z}/m\mathbf{Z}$, so by Kummer theory (Remark 1.2.10) $K(Y)$ is of the form $K(\sqrt[m]{f})$ for some $f \in K$. As Y is normal (Proposition 5.2.12 (3)) and finite over X, it must be isomorphic to the normalization of X in the extension $K(\sqrt[m]{f})|K$. Since $Y \to X$ is étale, we must have $v_P(f) = v_Q(f)$ for all points Q lying above a P of codimension 1. But $v_Q(f) = m v_Q(\sqrt[m]{f})$, whence $\mathrm{div}(f) = mD$ for some $D \in \mathrm{Div}(X)$. As $\mathrm{Div}(X)$ is a free abelian group, the said D is uniquely defined, so by passing to the class of D we get a map $\mathrm{Hom}(\pi_1^{\mathrm{ab}}(X), \mathbf{Z}/m\mathbf{Z}) \to {}_m\mathrm{Pic}\,(X)$. This map is in fact surjective, as it has a retraction by the lemma above.

The two maps in the previous paragraph are homomorphisms. Indeed, the sum $\lambda + \lambda'$ of two homomorphisms $\lambda, \lambda' : \pi_1^{\mathrm{ab}}(X) \to \mathbf{Z}/m\mathbf{Z}$ induces via composition with the map $\pi_1^{\mathrm{ab}}(\mathrm{Spec}\,(K)) \to \pi_1^{\mathrm{ab}}(X)$ the sum $\lambda_K + \lambda'_K$ of the two restrictions $\lambda_K, \lambda'_K : \mathrm{Gal}\,(\overline{K}|K)^{\mathrm{ab}} \to \mathbf{Z}/m\mathbf{Z}$. This in turn corresponds via Kummer theory to the product ff' of the corresponding functions $f, f' \in K^\times$, and finally we have $\mathrm{div}(ff') = \mathrm{div}(f) + \mathrm{div}(f')$. Exactness of the sequence at the other terms then follows from the construction. □

Remark 5.8.4 It is possible to construct the exact sequence of the proposition for an arbitrary Noetherian connected scheme such that m is prime to the residue characteristic of each point; see [59], Proposition III.4.14. (Here of course one has to work with the more general definition of the Picard group via isomorphism classes of invertible sheaves.)

In characteristic $p > 0$ we can use Artin–Schreier theory to construct étale Galois covers with group $\mathbf{Z}/p\mathbf{Z}$.

Construction 5.8.5 Assume K is of characteristic $p > 0$, and let $f \in \mathcal{O}_X(X)$ be an everywhere regular function. In a fixed algebraic closure of K choose a function g with $g^p - g = f$, and let Y be the normalization of X in the extension $K(g)|K$. The field extension here is Galois with group $\mathbf{Z}/p\mathbf{Z}$ by Example 1.2.9 (5). The fibre Y_P above a point P is the spectrum of the $\kappa(P)$-algebra $\kappa(P)[x]/(x^p - x - \bar{f})$, where $\bar{f}$ is the image of f in $\kappa(P)$. This is a product of copies of $\kappa(P)$ or a Galois field extension with group $\mathbf{Z}/p\mathbf{Z}$. It follows as above that $Y \to X$ is indeed a finite étale Galois cover with group $\mathbf{Z}/p\mathbf{Z}$.

Remark 5.8.6 The above construction yields an injective map

$$\mathcal{O}_X(X)/\wp(\mathcal{O}_X(X)) \to \mathrm{Hom}(\pi_1^{\mathrm{ab}}(X), \mathbf{Z}/p\mathbf{Z})$$

where $\wp : \mathcal{O}_X(X) \to \mathcal{O}_X(X)$ is the map $f \mapsto f^p - f$. Serre has identified the cokernel with the invariants of the cohomology group $H^1(X, \mathcal{O}_X)$ under the action of the Frobenius morphism. Nowadays this result is viewed as a consequence of the Artin–Schreier exact sequence for étale cohomology (see [59], Proposition III.4.12).

In the special case where X is affine we have $H^1(X, \mathcal{O}_X) = 0$ by Serre's vanishing theorem ([34], Theorem II.3.7), so the above map is an isomorphism. On the other hand, when X is proper over an algebraically closed field, it is the term $\mathcal{O}_X(X)/\wp(\mathcal{O}_X(X))$ that is trivial, and $\mathrm{Hom}(\pi_1^{\mathrm{ab}}(X), \mathbf{Z}/p\mathbf{Z})$ is isomorphic to the subgroup of Frobenius-invariant elements in $H^1(X, \mathcal{O}_X)$.

Proposition 5.8.3 relates prime-to-p quotients of the abelianized fundamental group to torsion in the Picard group. In order to obtain more precise information we need deep theorems concerning the Picard variety.

Facts 5.8.7 Let X be a smooth, projective integral scheme over an algebraically closed field k. The group $\mathrm{Pic}\,(X)$ can be identified with the group of k-points of a commutative k-group scheme Pic_X. The identity component $\mathrm{Pic}^0_X \subset \mathrm{Pic}_X$ is an abelian variety called the *Picard variety of* X. If X is defined over a subfield $F \subset k$ (i.e. arises via base change to $\mathrm{Spec}\,(k)$ from a smooth projective scheme over $\mathrm{Spec}\,(F)$), then so does Pic^0_X. For these facts, see [44], Theorem 9.5.4.

The abelian group $\mathrm{Pic}^0_X(k)$ is divisible (a general fact about abelian varieties; see [65], §6, Application 2). The quotient $NS(X) := \mathrm{Pic}_X(k)/\mathrm{Pic}^0_X(k)$ is a finitely generated abelian group ([59], Theorem VI.11.7). It is called the *Néron–Severi group* of X. When X is a curve, one has $NS(X) \cong \mathbf{Z}$, and when X is an abelian variety, then $NS(X) \cong \mathbf{Z}^r$ for some $r > 0$ ([65], §19, Corollary 2).

Keeping the assumptions under which the above facts hold, we have for each $m > 0$ a commutative diagram

$$\begin{array}{ccccccccc} 0 & \longrightarrow & \mathrm{Pic}^0_X(k) & \longrightarrow & \mathrm{Pic}_X(k) & \longrightarrow & NS(X) & \longrightarrow & 0 \\ & & \downarrow{\scriptstyle m} & & \downarrow{\scriptstyle m} & & \downarrow{\scriptstyle m} & & \\ 0 & \longrightarrow & \mathrm{Pic}^0_X(k) & \longrightarrow & \mathrm{Pic}_X(k) & \longrightarrow & NS(X) & \longrightarrow & 0 \end{array}$$

where the vertical maps are given by multiplication by m, the left one being surjective. We deduce an exact sequence on m-torsion subgroups

$$0 \to {}_m\mathrm{Pic}^0_X(k) \to {}_m\mathrm{Pic}_X(k) \to {}_mNS(X) \to 0.$$

Assuming m prime to the characteristic, we may combine the above with Proposition 5.8.3 and obtain an exact sequence

$$0 \to {}_m\mathrm{Pic}^0_X(k) \to \mathrm{Hom}(\pi_1^{\mathrm{ab}}(X), \mathbf{Z}/m\mathbf{Z}) \to {}_mNS(X) \to 0$$

since in this case $\mathcal{O}_X(X) = k$ is divisible. Dualizing the above sequence of $\mathbf{Z}/m\mathbf{Z}$-modules and bearing in mind the facts recalled in Remark 5.6.11 we obtain:

Corollary 5.8.8 *Let X be a smooth, projective, integral scheme over an algebraically closed field k, and let m be an integer prime to the characteristic of k. We have an exact sequence*

$$0 \to \mathrm{Hom}({}_mNS(X), \mathbf{Z}/m\mathbf{Z}) \to \pi_1^{\mathrm{ab}}(X)/m\pi_1^{\mathrm{ab}}(X) \to \mathrm{Hom}({}_m\mathrm{Pic}^0_X(k), \mathbf{Z}/m\mathbf{Z}) \to 0.$$

In particular, $\pi_1^{\mathrm{ab}}(X)/m\pi_1^{\mathrm{ab}}(X)$ is a finite abelian group that is an extension of $(\mathbf{Z}/m\mathbf{Z})^{2g}$ by a group of order bounded independently of m, where g is the dimension of Pic^0_X.

In the case of a curve or an abelian variety the group $NS(X)$ is torsion-free, and we actually obtain an isomorphism $\pi_1^{\mathrm{ab}}(X)/m\pi_1^{\mathrm{ab}}(X) \cong (\mathbf{Z}/m\mathbf{Z})^{2g}$.

To proceed further, we invoke the theory of Albanese varieties.

Facts 5.8.9 Let X be a geometrically integral separated scheme of finite type over a field k, and assume X has a k-rational point P. There is an abelian variety Alb_X (the *Albanese variety of* X) and a map $\phi_P : X \to \mathrm{Alb}_X$ sending P to 0 such that every k-morphism $X \to A$ into an abelian variety A sending P to 0 factors uniquely through ϕ_P.

The map ϕ_P gives a natural homomorphism $\phi_P^* : \mathrm{Pic}\,(\mathrm{Alb}_X) \to \mathrm{Pic}\,(X)$, induced by pullback of divisors in the smooth case. By a theorem going back to Severi, if we moreover assume k algebraically closed as well as X smooth and projective, the map ϕ_P^* restricts to an isomorphism $\mathrm{Pic}^0_{\mathrm{Alb}_X}(k) \xrightarrow{\sim} \mathrm{Pic}^0_X(k)$ which does not depend on P any more. In fact, this isomorphism comes from an isomorphism of abelian varieties $\mathrm{Pic}^0_{\mathrm{Alb}_X} \xrightarrow{\sim} \mathrm{Pic}^0_X$. In other words, Pic^0_X is the dual abelian variety of Alb_X. When X is a curve, both Alb_X and Pic^0_X are equal to the Jacobian of X, and we recover the classical self-duality of the Jacobian.

For the above facts the best references are still Serre's seminar lectures [86] and [87]. (He assumes k algebraically closed throughout, but the statement of the first paragraph follows via a standard descent argument.)

Finally, recall that by the general theory of abelian varieties ([65], §20) the duality between Alb_X and Pic^0_X induces for every integer m prime to the characteristic of k a perfect pairing

$$ {}_m\mathrm{Alb}_X(k) \times {}_m\mathrm{Pic}^0_X(k) \to \mu_m $$

where μ_m is the group of m-th roots of unity in k. It is called the *Weil pairing*.

Corollary 5.8.10 *Let X be a smooth, projective, integral scheme over an algebraically closed field k. For every prime number ℓ different from the characteristic of k the maximal pro-ℓ-quotient $\pi_1^{\mathrm{ab},(\ell)}(X)$ of $\pi_1^{\mathrm{ab}}(X)$ sits in an exact sequence*

$$ 0 \to \mathrm{Hom}(NS(X)\{\ell\}, \mathbf{Q}_\ell/\mathbf{Z}_\ell) \to \pi_1^{\mathrm{ab},(\ell)}(X) \to T_\ell(\mathrm{Alb}_X) \to 0, $$

where $NS(X)\{\ell\}$ denotes the (finite) ℓ-primary torsion subgroup of $NS(X)$.

In particular, the ℓ-primary torsion subgroup $\pi_1^{\mathrm{ab},(\ell)}(X)\{\ell\}$ of $\pi_1^{\mathrm{ab},(\ell)}(X)$ is finite, and the quotient $\pi_1^{\mathrm{ab},(\ell)}(X)/(\pi_1^{\mathrm{ab},(\ell)}(X)\{\ell\})$ is isomorphic to $\mathbf{Z}_\ell^{2g}$, where g is the dimension of Alb_X.

Proof By choosing a primitive ℓ^r-th root of unity in k for all $r > 0$ we obtain isomorphisms $\mu_{\ell^r} \cong \mathbf{Z}/\ell^r\mathbf{Z}$. Hence we may identify the group ${}_{\ell^r}\mathrm{Alb}_X(k)$ with $\mathrm{Hom}({}_{\ell^r}\mathrm{Pic}^0_X(k), \mathbf{Z}/\ell^r\mathbf{Z})$. The statement then follows from applying Corollary 5.8.8 with $m = \ell^r$ for all $r > 0$, and passing to the inverse limit in the resulting inverse system of exact sequences of finite abelian groups. □

Remark 5.8.11 The exact sequence of the corollary also holds for $\ell = p$. In fact, an argument of Raynaud in flat cohomology (see [59], Proposition III.4.16) implies that there are isomorphisms $\mathrm{Hom}(\pi_1^{\mathrm{ab}}(X), \mathbf{Z}/m\mathbf{Z}) \cong \mathrm{Hom}(\mu_m, \mathrm{Pic}_X)$ for all $m > 0$. The rest of the proof then goes through. Consequently, the whole torsion subgroup of $\pi_1^{\mathrm{ab}}(X)$ is finite, and the quotient is isomorphic to the full Tate module $T(\mathrm{Alb}_X)$.

On the other hand, if we only assume X to be smooth and quasi-projective, Serre's theory of generalized Albanese varieties implies an analogous exact sequence for the abelianized tame fundamental group. See [97], Proposition 4.4.

We now investigate the abelianized fundamental group over more general bases. First assume X is a proper smooth geometrically integral scheme of finite type over an arbitrary perfect base field k. The homotopy exact sequence

$$1 \to \pi_1(\overline{X}, \bar{x}) \to \pi_1(X, \bar{x}) \to \mathrm{Gal}\,(\bar{k}|k) \to 1 \tag{5.5}$$

of Proposition 5.6.1 yields upon abelianization an exact sequence

$$\pi_1^{\mathrm{ab}}(\overline{X}) \to \pi_1^{\mathrm{ab}}(X) \to \mathrm{Gal}\,(\bar{k}|k)^{\mathrm{ab}} \to 0.$$

Recall the canonical outer action of $\mathrm{Gal}\,(\bar{k}|k)$ on $\pi_1(\overline{X}, \bar{x})$: it is induced by the action of $\pi_1(X, \bar{x})$ on itself via conjugation. It thus yields a well-defined action of of $\mathrm{Gal}\,(\bar{k}|k)$ on $\pi_1^{\mathrm{ab}}(\overline{X})$. Moreover, if we equip $\pi_1^{\mathrm{ab}}(X)$ with the trivial $\mathrm{Gal}\,(\bar{k}|k)$-action, the map $\pi_1^{\mathrm{ab}}(\overline{X}) \to \pi_1^{\mathrm{ab}}(X)$ is compatible with the action of $\mathrm{Gal}\,(\bar{k}|k)$. It thus factors through the maximal quotient $\pi_1^{\mathrm{ab}}(\overline{X})_{\mathrm{Gal}\,(\bar{k}|k)}$ of $\pi_1^{\mathrm{ab}}(\overline{X})$ invariant under the action of $\mathrm{Gal}\,(\bar{k}|k)$, the *coinvariants* of the Galois action. We obtain an exact sequence

$$\pi_1^{\mathrm{ab}}(\overline{X})_{\mathrm{Gal}\,(\bar{k}|k)} \to \pi_1^{\mathrm{ab}}(X) \to \mathrm{Gal}\,(\bar{k}|k)^{\mathrm{ab}} \to 0.$$

Remark 5.8.12 In the case when X has a k-rational point the exact sequence (5.5) has a section, so we obtain a semidirect product decomposition

$$\pi_1(X, \bar{x}) \cong \pi_1(\overline{X}, \bar{x}) \rtimes \mathrm{Gal}\,(\bar{k}|k).$$

By a general property of semidirect products, the abelianization then becomes a direct product

$$\pi_1^{\mathrm{ab}}(X) \cong \pi_1^{\mathrm{ab}}(\overline{X})_{\mathrm{Gal}\,(\bar{k}|k)} \times \mathrm{Gal}\,(\bar{k}|k)^{\mathrm{ab}}.$$

Such a decomposition does not exist in general.

When the base field is a finite field, we obtain a finiteness result.

Proposition 5.8.13 *Let X be a smooth projective geometrically integral scheme over a finite field $\mathbf{F}$ of characteristic p. The kernel of the natural map $\pi_1^{\mathrm{ab}}(X) \to \mathrm{Gal}(\overline{\mathbf{F}}|\mathbf{F})$ is the product of a finite group and a pro-p-group.*

In fact, the p-part of the kernel is also finite, but we shall not prove this; see Remark 5.8.15 below.

Proof It suffices to show that the coinvariants of the action of $\mathrm{Gal}(\overline{\mathbf{F}}|\mathbf{F})$ on the Tate module $T_\ell(\mathrm{Alb}_{\bar{X}})$ are finite for all $\ell \neq p$ and trivial for all but finitely many ℓ. As $\mathrm{Gal}(\overline{\mathbf{F}}|\mathbf{F}) \cong \widehat{\mathbf{Z}}$, topologically generated by the Frobenius automorphism $F : x \mapsto x^p$, and the Galois action is continuous, we may identify $T_\ell(\mathrm{Alb}_{\bar{X}})_{\mathrm{Gal}(\overline{\mathbf{F}}|\mathbf{F})}$ with the cokernel of the $\mathbf{Z}_\ell$-linear automorphism $F - \mathrm{id}$ of $T_\ell(\mathrm{Alb}_{\bar{X}})$. Now recall from Remark 5.6.11 that $T_\ell(\mathrm{Alb}_{\bar{X}})$ is a finitely generated free $\mathbf{Z}_\ell$-module. It follows that the map

$$F - \mathrm{id} : T_\ell(\mathrm{Alb}_{\bar{X}}) \to T_\ell(\mathrm{Alb}_{\bar{X}}) \tag{5.6}$$

has finite kernel (resp. cokernel) if and only if the induced map

$$F - \mathrm{id} : T_\ell(\mathrm{Alb}_{\bar{X}}) \otimes_{\mathbf{Z}_\ell} \mathbf{Q}_\ell \to T_\ell(\mathrm{Alb}_{\bar{X}}) \otimes_{\mathbf{Z}_\ell} \mathbf{Q}_\ell$$

is injective (resp. surjective). But the latter is an endomorphism of a finite dimensional $\mathbf{Q}_\ell$-vector space, so it is injective if and only if it is surjective. Therefore the finiteness of the cokernel of (5.6) is equivalent to the finiteness of its kernel. The kernel is indeed finite, being the ℓ-primary torsion subgroup $\mathrm{Alb}_X(\mathbf{F})\{\ell\}$ of the finite group $\mathrm{Alb}_X(\mathbf{F})$. To show that the cokernel is 0 for almost all ℓ we invoke Weil's theorem ([65], §19, Theorem 4) according to which the order of $\mathrm{Alb}_X(\mathbf{F})$ (which is the same as the degree of the endomorphism $F - \mathrm{id}$ of the abelian variety $\mathrm{Alb}_{\overline{X}}$) equals the determinant of the $\mathbf{Z}_\ell$-linear map (5.6). For the ℓ not dividing the order of $\mathrm{Alb}_X(\mathbf{F})$ this determinant is a unit in the ring $\mathbf{Z}_\ell$ and therefore (5.6) is an isomorphism, whence the assertion. □

Using a specialization argument we can prove a much broader statement.

Theorem 5.8.14 (Katz–Lang) *Let S be an integral scheme whose function field K is finitely generated over $\mathbf{Q}$, and let $\phi : X \to S$ be a smooth projective morphism with geometrically integral fibres. The kernel of the natural map $\phi_* : \pi_1^{\mathrm{ab}}(X) \to \pi_1^{\mathrm{ab}}(S)$ is finite.*

Proof First we treat the case $S = \mathrm{Spec}(K)$. As above, we have to show that the coinvariants of the action of $\mathrm{Gal}(\overline{K}|K)$ on $\pi_1^{\mathrm{ab}}(\overline{X})$ are finite. By a similar argument as in the proof of Lemma 5.6.2 we find an integrally closed domain A finitely generated over $\mathbf{Z}$ with fraction field K such that $X \to S$ extends to a proper smooth morphism $\mathcal{X} \to \mathrm{Spec}(A)$ with generic fibre X and all fibres geometrically integral. Pick a point Q of codimension 1 in $\mathrm{Spec}\,(A)$ and a normal closed point P on the closure Y of Q in $\mathrm{Spec}\,(A)$. Let B be the completion of the local ring $\mathcal{O}_{Y,P}$; it is a complete discrete valuation ring with residue field $\kappa(P)$. Denote by p the residue characteristic, and fix geometric points $\bar{\eta}$ (resp. $\bar{s}$) lying above the generic (resp. closed) point of $\mathrm{Spec}\,(B)$. Theorem 5.7.10 implies that the specialization map $sp : \pi_1(\mathcal{X}_{\bar{\eta}})^{\mathrm{ab},(p')} \to \pi_1(\mathcal{X}_{\bar{s}})^{\mathrm{ab},(p')}$ on maximal abelian prime-to-p quotients is an isomorphism. The geometric point η factors as $\eta : \mathrm{Spec}\,(\Omega) \to \mathrm{Spec}\,(\overline{K}) \to \mathrm{Spec}\,(K)$, where $\overline{K}$ is an algebraic closure of K in Ω. Denoting by $\eta_{\overline{K}}$ the geometric point $\mathrm{Spec}\,(\overline{K}) \to \mathrm{Spec}\,(A)$, we obtain from Proposition 5.6.7 an isomorphism $\pi_1(\mathcal{X}_{\bar{\eta}})^{\mathrm{ab},(p')} \xrightarrow{\sim} \pi_1(\mathcal{X}_{\bar{\eta}_{\overline{K}}})^{\mathrm{ab},(p')}$ through which the map sp factors. We conclude that there is an isomorphism $\pi_1(\mathcal{X}_{\bar{\eta}_{\overline{K}}})^{\mathrm{ab},(p')} \xrightarrow{\sim} \pi_1(\mathcal{X}_{\bar{s}})^{\mathrm{ab},(p')}$.

Denote by $K^\wedge$ the fraction field of B. The Galois group $\mathrm{Gal}\,(\overline{K}^\wedge|K^\wedge)$ may be identified with a subgroup $D \subset \mathrm{Gal}\,(\overline{K}|K)$, and its action on $\pi_1(\mathcal{X}_{\bar{\eta}})^{\mathrm{ab},(p')}$ corresponds to the action of D on $\pi_1(\mathcal{X}_{\bar{\eta}_{\overline{K}}})^{\mathrm{ab},(p')}$. It follows that there exists a surjective map

$$\pi_1(\mathcal{X}_{\bar{\eta}})^{\mathrm{ab},(p')}_{\mathrm{Gal}\,(\overline{K}^\wedge|K^\wedge)} \twoheadrightarrow \pi_1(\mathcal{X}_{\bar{\eta}_{\overline{K}}})^{\mathrm{ab},(p')}_{\mathrm{Gal}\,(\overline{K}|K)}. \tag{5.7}$$

But $\mathrm{Gal}\,(\overline{K}^\wedge|K^\wedge)$ acts on $\pi_1(\mathcal{X}_{\bar{\eta}})^{\mathrm{ab},(p')}$ via its quotient $\pi_1(\mathrm{Spec}\,(B),\bar{\eta})$ because of the isomorphism $\pi_1(\mathcal{X}_{\bar{\eta}})^{\mathrm{ab},(p')} \xrightarrow{\sim} \pi_1(\mathcal{X}_{\mathrm{Spec}\,(\overline{B})})^{\mathrm{ab},(p')}$ factoring the specialization isomorphism sp, where $\overline{B}$ is the integral closure of B in $\overline{K}^\wedge$. On the other hand, the choice of a path between $\bar{s}$ and $\bar{\eta}$ yields a map

$$\mathrm{Gal}\,(\overline{\kappa(P)}|\kappa(P)) \cong \pi_1(\mathrm{Spec}\,(\kappa(P),\bar{s}) \to \pi_1(\mathrm{Spec}\,(B),\bar{\eta}).$$

Putting all the above together, we conclude that there is a surjective map

$$\pi_1(\mathcal{X}_{\bar{s}})^{\mathrm{ab},(p')}_{\mathrm{Gal}\,(\overline{\kappa(P)}|\kappa(P))} \twoheadrightarrow \pi_1(\mathcal{X}_{\bar{\eta}})^{\mathrm{ab},(p')}_{\mathrm{Gal}\,(\overline{K}^\wedge|K^\wedge)}. \tag{5.8}$$

But the first group here is finite, as we saw in the proof of the previous proposition. We conclude from the surjections (5.7) and (5.8) that the coinvariants of $\mathrm{Gal}\,(\overline{K}|K)$ on $\pi_1(\overline{X})^{\mathrm{ab},(p')}$ are finite. Working with another closed point of residue characteristic different from p we obtain that the coinvariants on $\pi_1(\overline{X})^{\mathrm{ab},(p)}$ are finite as well. This concludes the proof of the case $S = \mathrm{Spec}\,(K)$.

To treat the general case, fix a geometric point $\bar{x}$ of X and consider the commutative diagram

$$\begin{array}{ccccccc}
\pi_1^{\mathrm{ab}}(X_{\bar{x}}) & \longrightarrow & \pi_1^{\mathrm{ab}}(X \times_S \mathrm{Spec}\,(K)) & \xrightarrow{\phi_{K*}} & \pi_1^{\mathrm{ab}}(\mathrm{Spec}\,(K)) & \longrightarrow & 0 \\
\downarrow{\scriptstyle \mathrm{id}} & & \downarrow & & \downarrow & & \\
\pi_1^{\mathrm{ab}}(X_{\bar{x}}) & \longrightarrow & \pi_1^{\mathrm{ab}}(X) & \xrightarrow{\phi_*} & \pi_1^{\mathrm{ab}}(S) & \longrightarrow & 0
\end{array}$$

whose exact rows result from Proposition 5.6.4 after abelianization. A diagram chase shows that there is a surjective map $\ker(\phi_{K*}) \twoheadrightarrow \ker(\phi_*)$, and we reduce to the special case proven above. □

Remark 5.8.15 If we only assume that the field K is finitely generated over its prime field, the above proof shows that the maximal prime-to-p quotient of $\ker(\phi_*)$ is finite, where p is the characteristic of K. An additional argument sketched in [41] shows that the p-part is finite as well. Katz and Lang also proved the finiteness of the maximal prime-to-p quotient of $\ker(\phi_*)$ under the following assumptions: S is normal, ϕ is smooth and surjective with geometrically integral generic fibre, and K is finitely generated over the prime field, of characteristic $p \geq 0$.

We close this section with a theorem of Lang about the abelianized fundamental group of a separated scheme of finite type over $\mathbf{Z}$. When X is such a scheme, the residue field $\kappa(P)$ of every closed point P is finite by Corollary 4.1.14. The fundamental group of $\mathrm{Spec}\,(\kappa(P))$ is the absolute Galois group of $\kappa(P)$. It is isomorphic to $\widehat{\mathbf{Z}}$, and a topological generator is given by the Frobenius automorphism F_P of a fixed algebraic closure $\overline{\kappa(P)}$.

Given a geometric point $\bar{x}$ of X, the morphism $\mathrm{Spec}\,(\kappa(P)) \to X$ induces a continuous homomorphism $\mathrm{Gal}\,(\overline{\kappa(P)}|\kappa(P)) \to \pi_1(X, \bar{x})$. Its definition involves the choice of a path between the geometric points $\mathrm{Spec}\,(\overline{\kappa(P)})$ and $\bar{x}$ of X, hence it is only defined up to inner automorphism. The element F_P thus gives rise to a conjugacy class of elements in $\pi_1(X, \bar{x})$ called *Frobenius elements*.

When we compose with the projection $\pi_1(X, \bar{x}) \to \pi_1^{\mathrm{ab}}(X)$, the resulting map $\mathrm{Gal}\,(\overline{\kappa(P)}|\kappa(P)) \to \pi_1^{\mathrm{ab}}(X)$ becomes well-defined, and each F_P maps to a unique Frobenius element in $\pi_1^{\mathrm{ab}}(X)$.

Theorem 5.8.16 (Lang) *Let X be an integral separated scheme of finite type over $\mathbf{Z}$. The Frobenius elements generate a dense subgroup in the topological group $\pi_1^{\mathrm{ab}}(X)$.*

Somewhat surprisingly, the proof uses analytic tools.

Facts 5.8.17 Given a separated scheme X of finite type over $\mathbf{Z}$, its *zeta function* ζ_X is defined by the Euler product

$$\zeta_X(s) = \prod_{P \in X_0} \frac{1}{(1 - \mathbf{N}P)^{-s}}$$

where X_0 denotes the set of closed points of X, and $\mathbf{N}P$ is the cardinality of the residue field $\kappa(P)$. The product converges for $\mathrm{Re}\,(s) > \dim(X)$, and can be extended meromorphically to the half-plane $\mathrm{Re}(s) > \dim(X) - 1/2$. At the point $s = \dim(X)$ it has a simple pole.

These facts can be found in [89] without proof. They can, however, be proven in the same way as in the well-known special case of the Dedekind zeta function (where X is the spectrum of the ring of integers in an algebraic number field; see e.g. [50], Chapter VIII, Theorem 5), using estimates on the number of points on varieties over number fields. When X is actually of finite type over a finite field, much more is true: ζ_X is a rational function (Dwork's theorem), there are simple poles at $s = 1$ and $s = d$, and for X smooth and projective the other poles are described by Deligne's theorem (ex Weil conjecture). See e.g. [59], Section VI.12.

Proof of Theorem 5.8.16 Assume the subgroup H generated by the Frobenius elements is not dense in $\pi_1^{ab}(X)$, and denote by $\overline{H}$ its closure. Let U be an open subgroup containing H. It corresponds to a finite étale cover $Y \to X$ (which is moreover Galois with abelian Galois group). Given a closed point P of X, the fibre Y_P above P must be the spectrum of a finite direct product of copies of $\kappa(P)$, because by construction of Y the Frobenius element F_P acts trivially on each geometric point of Y_P. From this we obtain that Y has exactly d closed points lying above each closed point of X, where d is the degree of the cover $Y \to X$. But then by definition $\zeta_Y = \zeta_X^d$, which is impossible because both have a simple pole at $s = \dim(X) = \dim(Y)$. □

Remark 5.8.18 Using a more powerful analytic result one obtains a non-commutative generalization of the theorem. Namely, a consequence of the *generalized Chebotarev density theorem* ([89], Theorem 6) can be stated as follows. Assume given a finite étale Galois cover $Y \to X$ of integral separated schemes of finite type over $\mathbf{Z}$. Then each subset C of the Galois group G that is stable by conjugation contains the image of a conjugacy class of Frobenius elements in $\pi_1(X, \bar{x})$. Consequently, the union D of all conjugacy classes of Frobenius elements in $\pi_1(X, \bar{x})$ meets every single coset of every open normal subgroup, and D is therefore dense in $\pi_1(X, \bar{x})$. Note that this gives a stronger result even for $\pi_1^{ab}(X)$: already the *subset* of the Frobenius elements is dense, not just the subgroup they generate.

Remark 5.8.19 Lang's theorem is the starting point of *unramified class field theory* for schemes of finite type over $\mathbf{Z}$. We review here the main statements. Let X be a regular integral proper scheme over $\mathbf{Z}$. Denote by $Z_0(X)$ the free abelian group with basis the closed points of X; it is called the group of *zero-cycles* on X. By the discussion above, functoriality of the abelianized fundamental group induces a well-defined homomorphism

$$Z_0(X) \to \pi_1^{ab}(X) \tag{5.9}$$

with dense image. A zero-cycle $D \in Z_0(X)$ is called *rationally equivalent to 0* if there exists an integral closed subscheme $C \subset X$ of dimension 1 and a rational function f on the normalization $\widetilde{C}$ of C such that $D = p_*(\mathrm{div}(f))$. Here $p : \widetilde{C} \to X$ is the natural morphism, and the map p_* is induced by sending a closed point $P \in \widetilde{C}$ to the zero-cycle $[\kappa(P) : \kappa(\phi(P))]\phi(P)$. The quotient of $Z_0(X)$ modulo the subgroup of zero-cycles rationally equivalent to 0 is called the *Chow group of zero-cycles* on X, and is denoted by $CH_0(X)$.

In the case when the map $X \to \mathrm{Spec}\,(\mathbf{Z})$ factors through the spectrum of a finite field $\mathbf{F}$, the map (5.9) factors through the quotient $CH_0(X)$, yielding a *reciprocity map* $\rho_X : CH_0(X) \to \pi_1^{\mathrm{ab}}(X)$. Moreover, the natural degree map $CH_0(X) \to \mathbf{Z}$ induced by sending a zero-cycle to the sum of its coefficients sits in a commutative diagram

$$\begin{array}{ccc} CH_0(X) & \xrightarrow{\rho_X} & \pi_1^{\mathrm{ab}}(X) \\ \downarrow & & \downarrow \\ \mathbf{Z} & \longrightarrow & \mathrm{Gal}\,(\overline{\mathbf{F}}|\mathbf{F}). \end{array}$$

Under the assumptions that X is projective and geometrically integral, the reciprocity map ρ_X induces an isomorphism on the kernels of the vertical maps, which are moreover finite abelian groups.

In the case when the map $X \to \mathrm{Spec}\,(\mathbf{Z})$ is surjective, one works with a slight modification of the reciprocity map. Namely, one considers the quotient $\widetilde{\pi}_1^{\mathrm{ab}}(X)$ of $\pi_1^{\mathrm{ab}}(X)$ that classifies finite étale abelian Galois covers $Y \to X$ with the property that for each $\mathbf{R}$-valued point $\mathrm{Spec}\,(\mathbf{R}) \to X$ the base change $Y \times_X \mathrm{Spec}\,(\mathbf{R})$ is a finite disjoint union of copies of $\mathrm{Spec}\,(\mathbf{R})$. From (5.9) one then derives a map $CH_0(X) \to \widetilde{\pi}_1^{\mathrm{ab}}(X)$ that is actually an isomorphism of finite groups.

The above statements are due to Lang, Bloch, Kato and Saito. See [78] for a detailed survey, as well as [83], [84] for generalizations to the tame fundamental group.

Exercises

1. Let A be a commutative ring with unit. Prove that $\mathrm{Spec}\,(A)$ is a connected affine scheme if and only if A contains no idempotent other than 0 and 1.
2. Let G be a profinite group, and let F be the forgetful functor from the category of finite continuous G-sets to the category of sets mapping a G-set to its underlying set. Prove that $\mathrm{Aut}(F) \cong G$. [*Hint:* Begin with the case of finite G.]
3. Show that for a connected scheme S the category of inverse systems $(P_\alpha, \phi_{\alpha\beta})$ indexed by Λ, with Λ and the P_α as in the proof of Proposition 5.4.6, is equivalent to the category of fibre functors $\mathrm{Fib}_{\bar{s}}$ at geometric points of S.

4. Let $\phi : S' \to S$ be a morphism of connected schemes, and let $\bar{s}$ and $\bar{s}'$ be geometric points of S and S', respectively, satisfying $\bar{s} = \phi \circ \bar{s}'$. Show that the induced homomorphism $\phi_* : \pi_1(S', \bar{s}') \to \pi_1(S, \bar{s})$ is surjective if and only if the functor $\mathrm{Fet}|_S \to \mathrm{Fet}_{S'}$ mapping X to $X \times_S S'$ is fully faithful.
5. Let X be a quasi-compact and geometrically integral scheme over a field k. For each finite Galois extension $L|k$ consider the étale Galois cover $X_L \to X$, where $X_L := X \times_{\mathrm{Spec}\,(k)} \mathrm{Spec}\,(L)$. Put $\overline{X} := X \times_{\mathrm{Spec}\,(k)} \mathrm{Spec}\,(k_s)$, and fix a geometric point $\bar{x}$ of $\overline{X}$.
   (a) Establish an isomorphism
   $$\pi_1(\overline{X}, \bar{x}) \cong \varprojlim \pi_1(X_L, \bar{x}),$$
   the inverse limit being taken over all finite Galois extensions of k contained in $\bar{k}$.
   (b) For each $L|k$ as above construct an exact sequence
   $$1 \to \pi_1(X_L, \bar{x}_L) \to \pi_1(X, \bar{x}) \to \mathrm{Aut}(X_L|X)^{op} \to 1.$$
   (c) Give another proof of Proposition 5.6.1 using the above two statements.
6. (Katz–Lang) Let S be a normal integral scheme, and $X \to S$ a smooth surjective morphism. Denote by $\bar{\eta}$ a geometric generic point of S, and by $\bar{x}$ a geometric point of $X_{\bar{\eta}}$.
   Assuming $X_{\bar{\eta}}$ connected, construct an exact sequence
   $$\pi_1(X_{\bar{\eta}}, \bar{x}) \to \pi_1(X, \bar{x}) \to \pi_1(S, \bar{\eta}) \to 1.$$
7. Let A be a complete local ring with maximal ideal M and residue field k (recall that completeness means $A \xrightarrow{\sim} \varprojlim A/M^i$). We say that an A-algebra B is finite étale if the induced morphism of schemes $\mathrm{Spec}\,(B) \to \mathrm{Spec}\,(A)$ is.
   (a) Show that $\mathrm{Spec}\,(B)$ is connected if and only if $\mathrm{Spec}\,(B \otimes_A k)$ is.
   [*Hint*: Observe that the natural maps $\mathrm{Spec}\,(B/M^iB) \to \mathrm{Spec}\,(B/MB)$ are identity maps on the underlying topological spaces, and apply Exercise 1.]
   (b) Show that for every finite separable field extension $L|k$ there is a finite étale A-algebra B with $B \otimes_A k \cong L$.
   (c) Conclude that the natural morphism $\pi_1(\mathrm{Spec}\,(k), \bar{s}) \to \pi_1(\mathrm{Spec}\,(A), \bar{s})$ is an isomorphism for a geometric point $\bar{s}$ lying above the closed point of $\mathrm{Spec}\,(A)$. In particular, $\pi_1(\mathrm{Spec}\,(A), \bar{s}) = \{1\}$ if k is separably closed, and $\pi_1(\mathrm{Spec}\,(A), \bar{s}) \cong \widehat{\mathbf{Z}}$ if k is finite.
   (d) Conclude that the natural functor $\mathrm{Fet}_{\mathrm{Spec}\,(A)} \to \mathrm{Fet}_{\mathrm{Spec}\,(k)}$ induces an equivalence of categories.

   [*Remark:* The statements of this exercise hold more generally for so-called *Henselian* local rings. See [59], §I.4.]
8. Let $k \subset K$ be an extension of algebraically closed fields of characteristic $p > 0$, and fix an element $s \in K \setminus k$. Verify that the map
   $$\mathrm{Spec}\,(K[t, y]/(x^p - x - st)) \to \mathrm{Spec}\,(K[t])$$

defines a finite étale cover of $\mathbf{A}^1_K$ that does not arise by base change from a finite étale cover of $\mathbf{A}^1_k$.

9. Check that if k is an algebraically closed field, then $\pi_1(\mathbf{P}^n_k, \bar{x}) = \{1\}$ for all $n > 0$ and all geometric points $\bar{x}$.
10. Let X be an integral scheme. Recall that a *rational map* $\rho : X \dashrightarrow Y$ is an equivalence class of morphisms from some nonempty open subset of X to Y, two morphisms being equivalent if they coincide over some nonempty open subset where both are defined.
    (a) Let X be a proper regular integral scheme over a field k, and let $\rho : X \dashrightarrow Y$ be a k-rational map to a normal scheme Y of finite type over k. Show that ρ induces a well-defined map $\pi_1(X, \bar{x}) \to \pi_1(Y, \rho \circ \bar{x})$ for every geometric point $\bar{x}$ of X for which $\rho \circ \bar{x}$ is defined.
    [*Hint:* Use Zariski–Nagata purity and the fact that under the above assumptions ρ is defined outside a closed subset of codimension at least 2.]
    (b) Conclude that a birational map between proper regular k-schemes induces an isomorphism on their fundamental groups.
11. Let X be a proper normal integral scheme over an algebraically closed field k for which there exists a rational map $\mathbf{P}^n_k \dashrightarrow X$ with dense image such that the induced extension of function fields $k(\mathbf{P}^n)|k(X)$ is separable. Prove that the fundamental group of S is finite.
    [*Hint:* Use Zariski–Nagata purity to show that the function field of every connected finite étale cover of X can be embedded in the finite extension $k(\mathbf{P}^n)|k(X)$.]

Remark. An X having the property of the exercise is called *separably unirational.* In fact, every separably unirational smooth proper k-scheme has trivial fundamental group. In characteristic 0 this is an old result due to Serre [88] who used Hodge theory. The general case was settled only recently by Kollár using a clever deformation trick and a powerful theorem of de Jong and Starr (see [11], Corollaire 3.6). The argument works more generally for so-called separably rationally connected schemes.

6

Tannakian fundamental groups

The theory of the last chapter established an equivalence between the category of finite étale covers of a connected scheme and the category of finite continuous permutation representations of its algebraic fundamental group. We shall now study a linearization of this concept, also due to Grothendieck and developed in detail by Saavedra [81] and Deligne [14]. The origin is a classical theorem from the theory of topological groups due to Tannaka and Krein: they showed that one may recover a compact topological group from the category of its continuous unitary representations. In Grothendieck's algebraic context the group is a linear algebraic group, or more generally an affine group scheme, and one studies the category of its finite dimensional representations. The key features that enable one to reconstitute the group are the tensor structure on this category and the forgetful functor that sends a representation to its underlying vector space. Having abstracted the conditions imposed on the category of representations, one gets a theorem stating that a category with certain additional structure is equivalent to the category of finite dimensional representations of an affine group scheme. This can be applied in several interesting situations. We shall discuss in some detail the theory of differential Galois groups, and also Nori's fundamental group scheme that creates a link between the algebraic fundamental group and Tannakian theory.

We only treat so-called neutral Tannakian categories, but the reader familiar with Grothendieck's descent theory will have no particular difficulty afterwards in studying the general theory of [14]. Non-commutative generalizations have also been developed in connection with quantum groups; as samples of a vast literature we refer to the books of Chari–Pressley [9] and Majid [55].

6.1 Affine group schemes and Hopf algebras

We have already encountered group schemes in the previous chapter. However, in order to keep the discussion of Tannakian categories at a more elementary level, here we shall work with a more accessible but equivalent definition in the affine case.

Definition 6.1.1 Let k be a field. An *affine group scheme* G over k is a functor from the category of k-algebras to the category of groups that, viewed as a set-valued functor, is representable by some k-algebra A. We call A the *coordinate ring* of G.

Remark 6.1.2 In the last chapter we defined a group scheme over k as a group object in the category of k-schemes, i.e. a k-scheme G together with k-morphisms $m : G \times G \to G$ ('multiplication'), $e : \mathrm{Spec}\,(k) \to G$ ('unit') and $i : G \to G$ ('inverse') subject to the usual group axioms. These morphisms induce a group structure on the set $G(S) := \mathrm{Hom}_k(S, G)$ of k-morphisms into G for each k-scheme S. Therefore the contravariant functor $S \mapsto \mathrm{Hom}_k(S, G)$ on the category of k-schemes represented by G is in fact group-valued. Restricting it to the full subcategory of affine k-schemes we obtain a covariant functor $R \mapsto \mathrm{Hom}_k(\mathrm{Spec}\,(R), G)$. Proposition 5.1.5 shows that when $G = \mathrm{Spec}\,(A)$ is itself affine, this is none but the functor above.

The coordinate ring A of an affine group scheme G carries additional structure coming from the group operations. To see this, note first that the functor $G \times G$ given by $R \mapsto G(R) \times G(R)$ is representable by the tensor product $A \otimes_k A$ in view of the functorial isomorphisms $\mathrm{Hom}(A, R) \times \mathrm{Hom}(A, R) \xrightarrow{\sim} \mathrm{Hom}(A \otimes_k A, R)$ induced by $(\phi, \psi) \mapsto \phi \otimes \psi$ (the inverse map is given by $\lambda \mapsto (a \mapsto \lambda(a \otimes 1), a \mapsto \lambda(1 \otimes a))$. Thus by the Yoneda Lemma (Lemma 1.4.12) the morphism of functors $m : G \times G \to G$ defining the multiplication of G comes from a unique k-algebra homomorphism $\Delta : A \otimes_k A \to A$. The unit and the inverse operation translate similarly to k-algebra maps. We summarize all this by the correspondences

$$\text{multiplication mult} : G \times G \to G \leftrightarrow \text{comultiplication } \Delta : A \to A \otimes_k A$$
$$\text{unit } \{e\} \to G \leftrightarrow \text{counit } \varepsilon : A \to k$$
$$\text{inverse } i : G \to G \leftrightarrow \text{antipode } \iota : A \to A$$

The group axioms imply compatibility conditions for Δ, ε and ι by the uniqueness statement of the Yoneda lemma. Below we indicate diagram translations of the associativity, unit and inverse axioms for groups on the left-hand side, and the corresponding compatibility conditions on A on the right-hand side. They are called the *coassociativity*, *counit* and *antipode (or coinverse) axioms*, respectively.

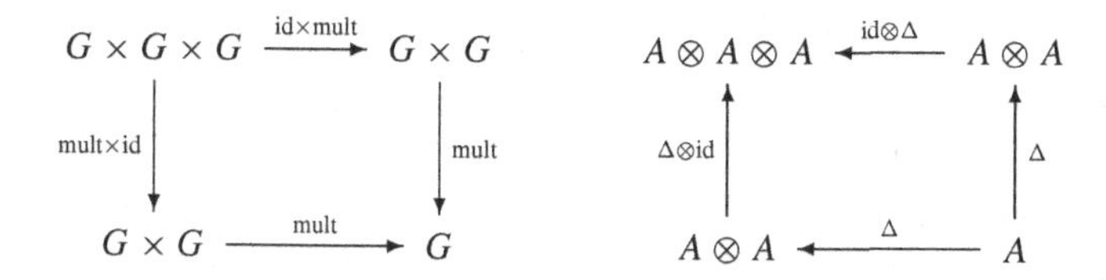

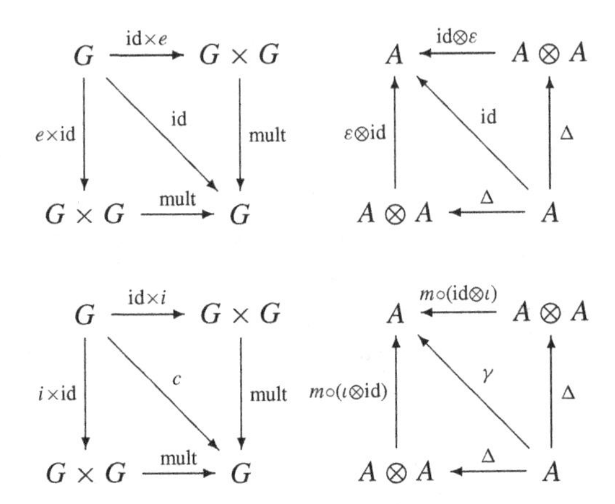

In the two last diagrams c is the constant map $G \to \{e\}$ on the left-hand side, γ the composite $A \to k \to A$ and $m : A \otimes_k A \to A$ the algebra multiplication on the right-hand side. To see that they indeed correspond, observe that m corresponds to the diagonal map $G \to G \times G$ by the Yoneda Lemma.

A not necessarily commutative k-algebra equipped with the above additional structure and satisfying the three axioms is called a *Hopf algebra*. Hopf algebras coming from affine group schemes are always commutative, but interesting noncommutative Hopf algebras arise, for instance, in the theory of quantum groups.

Remark 6.1.3 In calculations it is often useful to write down the Hopf algebra axioms explicitly for concrete elements. For instance, if we write $\Delta(a) = \sum a_i \otimes b_i$ for the comultiplication map, then the counit axiom says $a = \sum \varepsilon(a_i)b_i = \sum a_i\varepsilon(b_i)$, and the antipode axiom says $\varepsilon(a) = \sum \iota(a_i)b_i = \sum a_i\iota(b_i)$. Patient readers will write out the coassociativity axiom.

Tautologically, the category of commutative Hopf algebras over k is anti-equivalent to that of affine group schemes over k. Here are some basic examples of affine group schemes and their Hopf algebras.

Examples 6.1.4

1. The functor $R \mapsto \mathbf{G}_a(R)$ mapping a k-algebra R to its underlying additive group R^+ is an affine group scheme with coordinate ring $k[x]$, in view of the functorial isomorphism $R^+ \cong \mathrm{Hom}_k(k[x], R)$. The comultiplication map on $k[x]$ is given by $\Delta(x) = 1 \otimes x + x \otimes 1$, the counit is the zero map, and the antipode is induced by $x \mapsto -x$.
2. Similarly, the functor $R \mapsto \mathbf{G}_m(R)$ sending a k-algebra R to the subgroup $R^\times$ of invertible elements is an affine group scheme with coordinate ring $k[x, x^{-1}]$, because an invertible element in R corresponds to a k-algebra

homomorphism $k[x, x^{-1}] \to R$. On the coordinate ring the comultiplication map is induced by $\Delta(x) = x \otimes x$, the counit sends x to 1, and the antipode is induced by $x \mapsto x^{-1}$.

3. More generally, sending a k-algebra R to the group $\mathrm{GL}_n(R)$ of invertible matrices with entries in R is an affine group scheme. To find the coordinate ring A, notice that an $n \times n$ matrix M over R is invertible if and only if there exists $r \in R$ with $\det(M)r = 1$. This allows us to recover A as the quotient of the polynomial ring in $n^2 + 1$ variables $k[x_{11}, x_{12}, \ldots, x_{nn}, x]$ by the ideal generated by $\det(x_{ij})x - 1$. The isomorphism $\mathrm{GL}_n(R) \cong \mathrm{Hom}_k(A, R)$ is induced by sending a matrix $M = [m_{ij}]$ to the homomorphism given by $x_{ij} \mapsto m_{ij}$, $x \mapsto \det(m_{ij})^{-1}$. The comultiplication is induced by $x_{ij} \mapsto \sum_l x_{il} \otimes x_{lj}$, the counit sends x_{ij} to δ_{ij} (Kronecker delta), and the antipode comes from the formula for the inverse matrix.

4. Here is a link with the classical theory of linear algebraic groups. Assume k is algebraically closed. Observe that in the previous example we realized $\mathrm{GL}_n(k)$ as a closed subvariety of affine $n^2 + 1$-space; in particular, it inherits a Zariski topology. Let $G(k) \subset \mathrm{GL}_n(k)$ be a closed topological subgroup. By Proposition 4.2.10 this inclusion corresponds to a surjective map $\mathcal{O}(\mathrm{GL}_n) \to \mathcal{O}(G)$ of coordinate rings. Here the k-algebra $\mathcal{O}(\mathrm{GL}_n)$ is just the algebra A of the previous example; in particular, it carries a Hopf algebra structure. But then the quotient map $\mathcal{O}(\mathrm{GL}_n) \to \mathcal{O}(G)$ induces a Hopf algebra structure on $\mathcal{O}(G)$, because the Hopf algebra structure on A corresponds to maps $\mathrm{GL}_n(k) \times \mathrm{GL}_n(k) \to \mathrm{GL}_n(k)$, $k \to \mathrm{GL}_n(k)$ and $\mathrm{GL}_n(k) \to \mathrm{GL}_n(k)$ via Proposition 4.2.10, and $G(k)$ is a subgroup in $\mathrm{GL}_n(k)$. The affine group scheme associated with the Hopf algebra $\mathcal{O}(G)$ is the group scheme determined by the linear algebraic group $G(k)$.

   As concrete examples, one may associate affine group schemes with the other classical groups SL_n, O_n, etc. The construction generalizes to general base fields: an affine group scheme G embeds as a *closed subgroup scheme* in GL_n if there is a morphism of group-valued functors such that the induced map $\mathcal{O}(\mathrm{GL}_n) \to \mathcal{O}(G)$ on Hopf algebras is surjective. Thus Hopf algebra quotients of $\mathcal{O}(\mathrm{GL}_n)$ correspond to closed subgroup schemes in general.

Let us now forget about the k-algebra structure on Hopf algebras for a while. We then obtain the following more general notion.

Definition 6.1.5 A *coalgebra* over k is a k-vector space equipped with a comultiplication $\Delta : A \to A \otimes_k A$ and a counit map $\iota : A \to k$ subject to the coassociativity and counit axioms.

In this definition the maps Δ and ι are only assumed to be maps of k-vector spaces. Coalgebras over k form a category: morphisms are defined as k-linear maps compatible with the k-coalgebra structure.

We now define right comodules over a coalgebra by dualizing the notion of left modules over a k-algebra B. Observe that to give a unitary left B-module is to give a k-vector space V together with a k-linear multiplication $l : B \otimes_k V \to V$ so that the following diagrams commute:

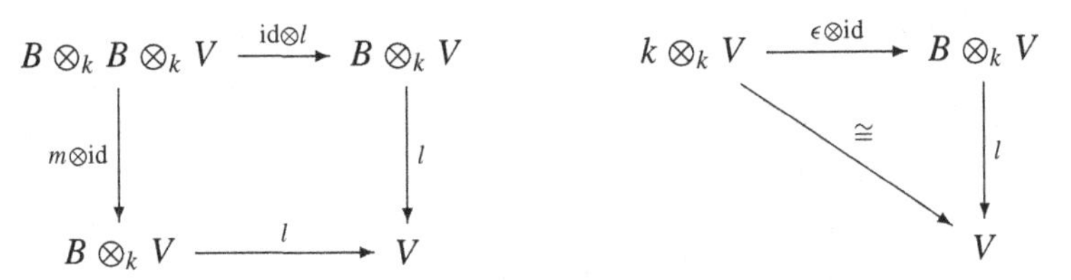

where $\epsilon : k \to B$ is the natural map sending 1 to the unit element of B. The first diagram here corresponds to the axiom $(b_1 b_2)v = b_1(b_2 v)$ for $b_i \in B$ and $v \in V$, and the second to $1 \cdot v = v$. The dual notion for k-coalgebras is the following.

Definition 6.1.6 Let A be a coalgebra over a field k. A right A-*comodule* is a k-vector space M together with a k-linear map $\rho : M \to M \otimes_k A$ so that the diagrams

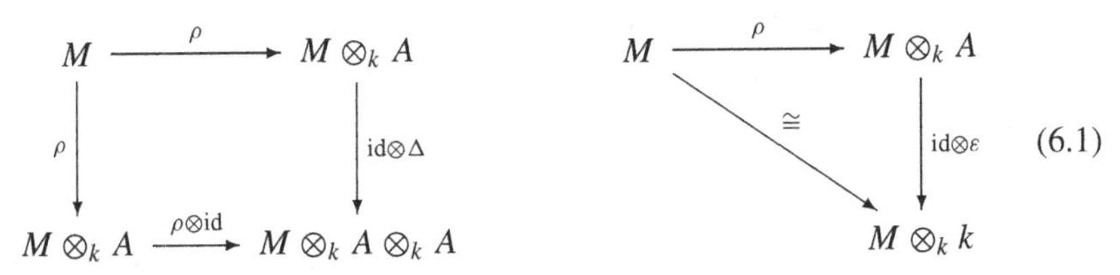

commute.

Remark 6.1.7 We can write out the comodule axioms explicitly on elements as follows. Assume ρ is given by $\rho(m) = \sum m_i \otimes a_i$, and $\rho(m_i) = \sum m_{ij} \otimes c_j$. Use furthermore the notation $\Delta(a_i) = \sum a_{il} \otimes b_l$. Here m, m_i, m_{ij} are in M and the other elements lie in A. Then the commutativity of the first diagram is described by the equality

$$\sum_{i,l} m_i \otimes a_{il} \otimes b_l = \sum_{i,j} m_{ij} \otimes c_j \otimes a_i. \tag{6.2}$$

The second diagram reads

$$\sum_i \varepsilon(a_i) m_i = m. \tag{6.3}$$

Another useful form of the first compatibility is obtained by fixing a k-basis $e_1, \dots, e_n$ of M, and defining $c_{ij} \in A$ via $\rho(e_i) = \sum_j e_j \otimes c_{ij}$. Then $\sum_l \rho(e_l) \otimes c_{il} = \sum_j e_j \otimes \left(\sum_l c_{lj} \otimes c_{il}\right)$ must equal $\sum_j e_j \otimes \Delta(c_{ij})$ by commutativity of the first diagram, which by the linear independence of the e_j holds if and only if

$$\Delta(c_{ij}) = \sum_l c_{lj} \otimes c_{il}. \tag{6.4}$$

A *subcoalgebra* of a coalgebra A is defined as a k-subspace $B \subset A$ with $\Delta(B) \subset B \otimes_k B$. The restrictions of Δ and ε then turn B into a coalgebra over k. One defines a subcomodule of an A-comodule M in a similar way: it is a k-subspace $N \subset M$ with $\rho(N) \subset \rho(N) \otimes_k A$. A subcoalgebra $B \subset A$ is also naturally a subcomodule of A considered as a right comodule over itself.

Subcomodules and subcoalgebras enjoy the following basic finiteness property.

Proposition 6.1.8 *Let A be a coalgebra, M a right A-comodule.*

1. *Each finite set $m_1, \dots, m_n$ of elements of M is contained in a subcomodule $N \subset M$ finite dimensional over k. Consequently, M is a directed union of its finite dimensional subcomodules.*
2. *Each finite set $a_1, \dots, a_n$ of elements of A is contained in a subcoalgebra $B \subset A$ finite dimensional over k. Consequently, A is a directed union of its finite dimensional subcoalgebras.*

Proof For (1), note first that by the k-linearity of $\rho : M \to M \otimes_k A$ the k-linear span of finitely many subcomodules of M is again a subcomodule. Therefore it is enough to prove the case $n = 1$ of the first statement. Fix a (possibly infinite) k-basis $\mathcal{B}$ of A. For $m \in M$ we may write $\rho(m) = \sum m_i \otimes a_i$ with $m_i \in M$ and $a_i \in \mathcal{B}$ (finite sum). Therefore $(\rho \otimes \mathrm{id}_A)(\rho(m)) = \sum_i \rho(m_i) \otimes a_i$. On the other hand, by the first comodule axiom we must have $(\rho \otimes \mathrm{id}_A)(\rho(m)) = \sum_i m_i \otimes \Delta(a_i)$. Writing $\Delta(a_i) = \sum_{j,k} \lambda_{ijk}(a_j \otimes a_k)$, we obtain (after changing running indices)

$$\sum_k \rho(m_k) \otimes a_k = \sum_i m_i \otimes \sum_{j,k} \lambda_{ijk}(a_j \otimes a_k),$$

which by the linear independence of the a_k is equivalent to

$$\rho(m_k) = \sum_i m_i \otimes \sum_j \lambda_{ijk} a_j$$

for all k. This implies that the k-span of m and the m_i is a finite dimensional subcomodule of M.

To prove (2) we again reduce to the case $n = 1$. By part (1), for fixed $a \in A$ we find a finite dimensional k-subspace $N \subset A$ containing a with $\Delta(N) \subset N \otimes_k A$. Fix a k-basis $e_1, \ldots, e_n$ of N, and write $\Delta(e_i) = \sum_j e_j \otimes c_{ij}$ with some $c_{ij} \in A$. By formula (6.4) above we have $\Delta(c_{ij}) = \sum_l c_{lj} \otimes c_{il}$, therefore the k-span of the finitely many elements e_j and c_{ij} is a subcoalgebra containing a. □

In the case when A is moreover a commutative Hopf algebra, an A-comodule M gives rise to a representation of the corresponding affine group scheme G in the following way. Given a k-algebra R, an element of $G(R)$ corresponds to a *k-algebra* homomorphism $\lambda : A \to R$. The composite $\rho : M \to M \otimes_k A \xrightarrow{\mathrm{id}\otimes\lambda} M \otimes_k R$ induces an R-linear map $M \otimes_k R \to M \otimes_k R$ that depends on R in a functorial way. By the comodule axioms, we thus obtain a functorial collection of left group actions $G(R) \times (M \otimes_k R) \to M \otimes_k R$ such that for each $g \in G(R)$ the map $m \mapsto gm$ is R-linear. We call such a collection a *left representation* of the affine group scheme G. If moreover M is finite dimensional over k and we fix a k-basis $m_1 \ldots, m_n$ of M, giving a representation of G becomes equivalent to giving a morphism $G \to \mathrm{GL}_n$ of group-valued functors.

Proposition 6.1.9 *The above construction gives a bijection between right comodules over the commutative Hopf algebra A and left representations of the corresponding affine group scheme G.*

Proof Given a left representation of G on a finite dimensional k-vector space V, the element in $G(A)$ corresponding to the identity morphism of A gives rise to an A-linear map $V \otimes_k A \to V \otimes_k A$. By composition with the natural map $V \to V \otimes A$ sending $v \in V$ to $v \otimes 1$ we obtain a k-linear map $V \to V \otimes A$. The reader will check that this is an A-comodule structure on V, and the two constructions are inverse to each other. □

We conclude this section by discussing dualities between algebras and coalgebras. Given a k-coalgebra A, the k-linear dual $A^* := \mathrm{Hom}_k(A, k)$ of the underlying vector space of A carries additional structure. Namely, the comultiplication $\Delta : A \to A \otimes_k A$ induces a k-bilinear map $m : A^* \otimes_k A^* \to A^*$ sending a pair (ϕ, ψ) of k-linear maps $A \to k$ to the k-linear map $(\phi \otimes \phi) \circ \Delta$. We may view m as a multiplication map on A^*; the coassociativity axiom for A implies that it is associative. Furthermore, the k-linear dual $k \to A^*$ of the counit map $e : A \to k$ is determined by the image of $1 \in k$ in A^*; by the counit axiom on A it is a unit element for the multiplication on A^*. We thus obtain a k-algebra with unit that is not necessarily commutative. The rule $A \mapsto A^*$ is a contravariant functor: dualizing a morphism $A_1 \to A_2$ of k-coalgebras gives a k-algebra homomorphism $A_2^* \to A_1^*$.

Conversely, if we start with a k-algebra B with unit, the multiplication map $m : B \otimes_k B \to B$ induces a map $B^* \to (B \otimes_k B)^*$ on k-linear duals. But there is a caveat here: it does not necessarily induce a comultiplication map on B^* as the natural map $B^* \otimes_k B^* \to (B \otimes_k B)^*$ sending $\phi \otimes \psi \in B^* \otimes_k B^*$ of to the k-linear map given by $a \otimes b \mapsto \phi(a)\psi(b)$ is not necessarily an isomorphism (Exercise 1). However, for B finite dimensional over k it is, being an injective map between vector spaces of the same dimension. So in this case we can equip the k-linear dual B^* with a k-coalgebra structure by reversing the procedure above, and obtain:

Proposition 6.1.10 *The contravariant functor $A \mapsto A^*$ induces an anti-isomorphism between the category of k-coalgebras finite dimensional over k and that of not necessarily commutative finite dimensional k-algebras with unit.*

Under the duality of the proposition each right A-comodule structure on a finite dimensional vector space V gives rise to a natural left A^*-module structure on V^*, and we have:

Corollary 6.1.11 *Given a finite dimensional k-coalgebra A, the contravariant functor $V \mapsto V^*$ induces an anti-isomorphism between the category of finitely generated right A-comodules and that of finitely generated left A^*-modules.*

Combining the above corollary with Proposition 6.1.8 is often useful in calculations. Here is such an application that will be needed later.

Proposition 6.1.12 *Let A be a coalgebra, M a right A-comodule. The sequence*

$$0 \longrightarrow M \xrightarrow{\ \rho\ } M \otimes_k A \xrightarrow{\rho \otimes \mathrm{id}_A - \mathrm{id}_M \otimes \Delta} M \otimes_k A \otimes_k A \qquad (6.5)$$

is exact.

Proof The first comodule axiom implies that the sequence is a complex, and the second axiom implies the injectivity of ρ. It remains to see that each element $\alpha = \sum m_i \otimes a_i$ in the kernel of $\rho \otimes \mathrm{id}_A - \mathrm{id}_M \otimes \Delta$ is in the image of ρ. Using Proposition 6.1.8 we find a finite dimensional subcomodule $M' \subset M$ and a finite dimensional subcoalgebra $A' \subset A$ with $\alpha \in M' \otimes_k A'$. Thus we reduce to the case when A and M are finite dimensional over k. Taking k-linear duals we then obtain a finite dimensional k-algebra $B = A^*$, a left B-module $N = M^*$, and the sequence becomes

$$B \otimes_k B \otimes_k N \to B \otimes_k N \xrightarrow{\rho^*} N \to 0.$$

Here ρ^* is the map giving the left B-module structure on N, whereas the unnamed map is the difference of the maps $b \otimes b' \otimes n \mapsto b(b' \otimes n)$ and $b \otimes b' \otimes n \mapsto (bb') \otimes n$. Exactness of this sequence is a tautology, and the dual exact sequence is (6.5) by Corollary 6.1.11. □

6.2 Categories of comodules

A basic fact about a finite dimensional algebra B over a field k is that it is determined up to isomorphism by the category Modf_B of finitely generated left B-modules; in fact, it can be recovered as the endomorphism algebra of the forgetful functor from Modf_B to the category of finite dimensional k-vector spaces (see Exercise 2). A similar statement holds for arbitrary k-algebras if we allow B-modules that are infinite dimensional over k. By dualizing the finite dimensional statement we obtain a result about comodules over finite dimensional k-coalgebras, but as we have seen, in infinite dimension the dualizing procedure breaks down.

Still, it is possible to recover an arbitrary k-coalgebra A from the category Comodf_A of finite dimensional k-vector spaces carrying a right A-coalgebra structure. To do so, denote by ω the forgetful functor from Comodf_A to the category Vecf_k of finite dimensional k-vector spaces. For an arbitrary k-vector space V denote by $\omega \otimes V$ the functor $M \mapsto \omega(M) \otimes_k V$ from Comodf_A to the category Vec_k of k-vector spaces. Write $\mathrm{Hom}(\omega, \omega \otimes V)$ for the set of functor morphisms $\omega \to \omega \otimes V$, where ω is considered as a Vec_k-valued functor under the natural embedding.

Proposition 6.2.1 *The underlying k-vector space of A represents the functor $V \mapsto \mathrm{Hom}(\omega, \omega \otimes V)$ on the category* Vec_k.

In other words, for each k-vector space V there are functorial isomorphisms

$$\mathrm{Hom}(A, V) \xrightarrow{\sim} \mathrm{Hom}(\omega, \omega \otimes V). \tag{6.6}$$

The proof below is taken from [82].

Proof of Proposition 6.2.1 To begin with, there is a canonical morphism of functors $\Pi : \omega \to \omega \otimes A$ given for each object M of Comodf_A by the comodule structure map $\omega(M) \to \omega(M) \otimes_k A$. A morphism $\phi : A \to V$ induces maps $(\mathrm{id} \otimes \phi) : \omega(M) \otimes A \to \omega(M) \otimes V$ for each M, whence a morphism of functors $(\mathrm{id} \otimes \phi) : \omega \otimes A \to \omega \otimes V$. Sending ϕ to $(\mathrm{id} \otimes \phi) \circ \Pi$ therefore defines a map $\Psi_V : \mathrm{Hom}(A, V) \to \mathrm{Hom}(\omega, \omega \otimes V)$ that is functorial in V, whence a morphism of functors $\Psi : \mathrm{Hom}(A, \_) \to \mathrm{Hom}(\omega, \omega \otimes \_)$. We now construct a morphism Ξ in the reverse direction. Consider A as a right comodule over itself, and for each $a \in A$ fix a finite dimensional subcomodule $N \subset A$

containing A; such an N exists by Proposition 6.1.8 (1). Then for a morphism $\Phi \in \mathrm{Hom}(\omega, \omega \otimes V)$ we define $\Xi(\Phi) \in \mathrm{Hom}(A, V)$ to be the morphism $a \mapsto (\epsilon|_N \otimes \mathrm{id}_V)(\Phi_N(a))$. This definition does not depend on the choice of N, as we can always embed finite dimensional subcomodules N, N' into a larger finite dimensional subcomodule N'', and use the fact that Φ is a morphism of functors.

To check that $\Xi \circ \Psi$ is the identity, take $\phi \in \mathrm{Hom}(A, V)$, $a \in A$ and a finite dimensional subcomodule $N \subset A$ containing a. As N is a subcomodule, $\Delta(a) = \sum n_i \otimes b_i$ with some $n_i \in N$ and $b_i \in A$, and $\Psi_N(\phi)(a) = \sum n_i \otimes \phi(b_i)$. Then $(\Xi \circ \Psi)(\phi)(a) = \sum \varepsilon(n_i)\phi(b_i) = \phi(a)$ by the counit axiom.

We finally show that the composite $\Psi \circ \Xi$ is the identity. Fix a finite dimensional A-comodule N and a morphism of functors $\Phi \in \mathrm{Hom}(\omega, \omega \otimes V)$. Using Proposition 6.1.8 (2) we find a finite dimensional subcoalgebra $B \subset A$ with $\rho_N(N) \subset N \otimes_k B$, where $\rho_N : N \to N \otimes_k A$ is the comodule structure map. In particular, N is a right comodule over B. The statement to be proven is that Φ_N equals the composite map

$$N \stackrel{\rho_N}{\to} N \otimes B \stackrel{\mathrm{id} \otimes \Phi_B}{\longrightarrow} N \otimes_k B \otimes_k V \stackrel{\mathrm{id} \otimes \varepsilon \otimes \mathrm{id}}{\longrightarrow} N \otimes_k k \otimes_k V \stackrel{\sim}{\to} N \otimes_k V,$$

where we have omitted the ω's to ease notation. To prove it, notice first that the map $\mathrm{id}_N \otimes \Delta : N \otimes_k B \to N \otimes_k B \otimes_k B$ defines a right B-comodule structure on the k-vector space $N \otimes_k B$. The fact that N is a B-comodule means precisely that $\rho_N : N \to N \otimes_k B$ is a morphism of B-comodules. As Φ is a morphism of functors, we have a commutative diagram

$$\begin{array}{ccccccc}
N & \stackrel{\rho_N}{\longrightarrow} & N \otimes_k B & \stackrel{\mathrm{id} \otimes \varepsilon}{\longrightarrow} & N \otimes_k k & \stackrel{\cong}{\longrightarrow} & N \\
{\scriptstyle \Phi_N} \downarrow & & {\scriptstyle \Phi_{N \otimes_k B}} \downarrow & & & & {\scriptstyle \Phi_N} \downarrow \\
N \otimes_k V & \stackrel{\rho_N \otimes \mathrm{id}}{\longrightarrow} & N \otimes_k B \otimes_k V & \stackrel{\mathrm{id} \otimes \varepsilon \otimes \mathrm{id}}{\longrightarrow} & N \otimes_k k \otimes_k V & \stackrel{\cong}{\longrightarrow} & N \otimes_k V
\end{array}$$

where the composites of the maps in the horizontal lines are identity maps by the second comodule axiom. It then suffices to see that the second vertical map equals $\mathrm{id}_N \otimes \Phi_B$. This holds because choosing a k-basis of N identifies $N \otimes_k B$ as a B-comodule with a finite direct sum of copies of B, and Φ commutes with direct sums. □

One has the corollary:

Corollary 6.2.2 *The k-coalgebra A is determined up to unique isomorphism by the category* Comodf_A *and the functor ω.*

Proof By the proposition, the underlying vector space of A is determined up to unique isomorphism. To recover the comultiplication $\Delta : A \to A \otimes_k A$, we apply (6.6) with $V = A \otimes_k A$ to see that it corresponds to a canonical morphism of functors $\omega \to \omega \otimes (A \otimes_k A)$. In order to exhibit this morphism we use the morphism of functors $\Pi : \omega \to \omega \otimes A$ defined at the beginning of the above proof. Iterating Π we obtain a morphism $(\Pi \otimes \mathrm{id}) \circ \Pi : \omega \to (\omega \otimes A) \otimes A \cong \omega \otimes (A \otimes_k A)$, which is the one we were looking for. Finally, the counit map $A \to k$ corresponds under (6.6) to the natural isomorphism $\omega \xrightarrow{\sim} \omega \otimes k$. □

Assume now that A is moreover a Hopf algebra. We shall investigate how the additional structure on A is reflected by the category Comod_A. We treat the multiplication, unit and antipode maps one by one. Let us first concentrate on the multiplication map $m : A \otimes_k A \to A$. Observe that the coalgebra structure maps Δ and ε are algebra homomorphisms with respect to m if and only if m is a k-coalgebra morphism. So let a k-coalgebra morphism $m : A \otimes_k A \to A$ be given. For a pair (M, N) of A-comodules, it enables us to define an A-comodule structure on the tensor product $M \otimes_k N$ of vector spaces by

$$\begin{aligned} M \otimes_k N &\xrightarrow{\rho_M \otimes \rho_N} M \otimes_k A \otimes_k N \otimes_k A \xrightarrow{\sim} M \otimes_k N \otimes_k A \otimes_k A \\ &\xrightarrow{\mathrm{id} \otimes m} M \otimes_k N \otimes_k A. \end{aligned} \tag{6.7}$$

Denote the tensor product comodule obtained in this way by $M \otimes_m N$.

To be more precise, we have actually defined an A-comodule structure on $\omega(M) \otimes_k \omega(N)$, where ω is the above forgetful functor. In the same way we see that each k-linear map $A \otimes_k A \to V$ to a k-vector space V induces a k-linear map $\omega(M) \otimes_k \omega(N) \to \omega(M) \otimes_k \omega(N) \otimes_k V$. Denote by $\omega \otimes \omega$ the functor $(M, N) \mapsto \omega(M) \otimes_k \omega(N)$ on $\mathrm{Comodf}_A \times \mathrm{Comodf}_A$. We then have the following analogue of Proposition 6.2.1.

Proposition 6.2.3 *The underlying k-vector space of $A \otimes_k A$ represents the functor $V \mapsto \mathrm{Hom}(\omega \otimes \omega, \omega \otimes \omega \otimes V)$ on the category Vec_k. In particular, we have a bijection*

$$\mathrm{Hom}(A \otimes_k A, A) \xrightarrow{\sim} \mathrm{Hom}(\omega \otimes \omega, \omega \otimes \omega \otimes A).$$

Proof We have just seen that each k-linear map $A \otimes_k A \to V$ induces a k-linear map $\omega(M) \otimes_k \omega(N) \to \omega(M) \otimes_k \omega(N) \otimes_k V$ functorial in V for each pair (M, N) of objects of Comodf_A. This gives a morphism of functors $\mathrm{Hom}(A \otimes_k A, \_) \to \mathrm{Hom}(\omega \otimes \omega, \omega \otimes \omega \otimes \_)$. The proof that it is an isomorphism is analogous to the proof of the previous proposition. □

By the proposition the multiplication map $m : A \otimes_k A \to A$ can be recovered as the map corresponding to the morphism of functors given on an object

(M, N) of $\mathrm{Comodf}_A \times \mathrm{Comodf}_A$ by the composite

$$\omega(M) \otimes_k \omega(N) \xrightarrow{\sim} \omega(M \otimes_m N) \to \omega(M \otimes_m N) \otimes_k A$$
$$\xrightarrow{\sim} \omega(M) \otimes_k \omega(N) \otimes_k A,$$

where the map in the middle is the one defining the comodule structure of $M \otimes_m N$. As we have seen in the proof of Proposition 6.2.1, it is induced by the morphism of functors $\omega \to \omega \otimes A$ corresponding to the identity map of A.

Corollary 6.2.4

1. *The multiplication map m is commutative if and only if for all M, N in Comodf_A the isomorphism $\omega(M) \otimes_k \omega(N) \xrightarrow{\sim} \omega(N) \otimes_k \omega(M)$ of k-vector spaces comes from an isomorphism $M \otimes_m N \xrightarrow{\sim} N \otimes_m M$ of A-comodules via ω.*
2. *The map m is associative if and only if for all M, N, P in Comodf_A the isomorphism $(\omega(M) \otimes_k \omega(N)) \otimes_k \omega(P) \xrightarrow{\sim} \omega(M) \otimes_k (\omega(N) \otimes_k \omega(P))$ of k-vector spaces comes from an isomorphism*

$$(M \otimes_m N) \otimes_m P \xrightarrow{\sim} M \otimes_m (N \otimes_m P)$$

of A-comodules via ω.

Proof For commutativity, note that the multiplication m is commutative if and only if $m = m \circ \sigma$, where $\sigma : A \otimes_k A \to A \otimes_k A$ is the map $a \otimes b \mapsto b \otimes a$. By the above discussion this holds if and only if the k-linear maps $\omega(M \otimes_m N) \to \omega(M \otimes_m N) \otimes_k A$ and $\omega(M \otimes_{m\circ\sigma} N) \to \omega(M \otimes_{m\circ\sigma} N) \otimes_k A$ are the same. The construction of the comodule structure on $M \otimes_m N$ in (6.7) shows that this is exactly the case when the isomorphism $M \otimes_k N \cong N \otimes_k M$ of k-vector spaces is compatible with the comodule structure of both sides. The proof for associativity is similar. □

We now turn to the unit element e for the multiplication in Hopf algebras. It is determined by the morphism $k \to A$ sending 1 to e that we also denote by e. As in the definition of a Hopf algebra we imposed that the counit map is a k-algebra homorphism, we now have to require dually that $e : k \to A$ be compatible with the coalgebra structures on k and A, where k is equipped with the comultiplication sending 1 to $1 \otimes 1$. This holds if and only if $\Delta(e) = e \otimes e$, which is also precisely the condition for e to equip k with a right comodule structure. Now we have:

Proposition 6.2.5 *An element $e \in A$ is a unit for the multiplication defined by m compatible with the coalgebra structure on A if and only if the map $e : k \to A$ defines an A-comodule structure on k, and moreover the k-linear isomorphisms $k \otimes_k \omega(M) \cong \omega(M) \otimes_k k \cong \omega(M)$ come from A-comodule isomorphisms $k \otimes_m M \cong M \otimes_m k \cong M$ for each A-comodule M.*

Proof For a finite dimensional k-comodule M consider the composite map

$$M \xrightarrow{\sim} M \otimes_k k \xrightarrow{\rho_M \otimes e} M \otimes A \otimes_k A \xrightarrow{\mathrm{id} \otimes m} M \otimes_k A.$$

It corresponds by Proposition 6.2.1 to the map $A \to A$ given by $a \mapsto m(a \otimes e)$. Therefore the right unit property holds if and only if the above composite equals the comodule structure map $\rho_M : M \to M \otimes_k A$ for all M. This means precisely that the isomorphism $M \otimes_k k \cong M$ is compatible with the A-comodule structures on $M \otimes_m k$ and M. Similar arguments apply to the left unit property, and the proposition follows. □

It remains to discuss the antipode $\iota : A \to A$. It turns out that ι induces an A-coalgebra structure on the dual k-vector space M^* of each A-comodule M. Before going into more detail about this, let us recall some easy facts about duals of vector spaces.

Given k-vector spaces V and W, there is a natural map

$$\tau_{V,W} : V^* \otimes_k W \to \mathrm{Hom}(V, W)$$

given by $\phi \otimes w \mapsto \phi \otimes \phi_w$, where $\phi_w \in \mathrm{Hom}(k, W)$ is the map $\lambda \mapsto \lambda w$. For V finite dimensional the map $\tau_{V,W}$ is an isomorphism, as it commutes with direct sums, and is trivially an isomorphism for $\dim V = 1$. In particular, for $V = W$ finite dimensional we have an isomorphism $\tau_{V,V} : V^* \otimes V \xrightarrow{\sim} \mathrm{End}(V)$. Sending 1 to $\tau_{V,V}^{-1}(\mathrm{id}_V)$ defines a canonical k-linear map $\delta : k \to V^* \otimes_k V$ called the *coevaluation map*. The name comes from the fact that dually there is an *evaluation map* $\epsilon : V \otimes_k V^* \to k$ (for V of arbitrary dimension) given by $v \otimes \phi \mapsto \phi(v)$. They are related by the commutative diagrams

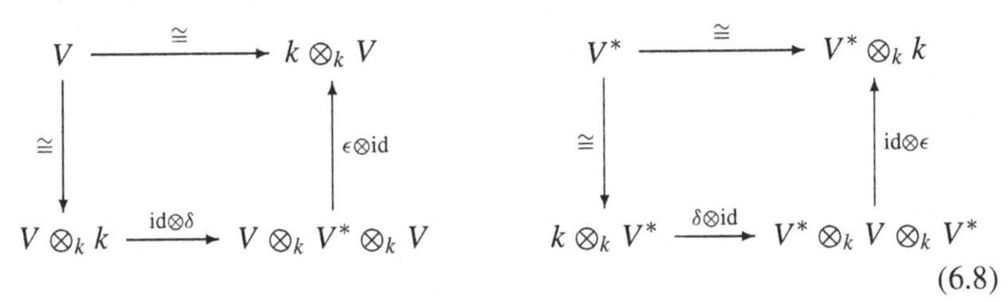

(6.8)

where the marked isomorphisms are the usual ones sending v to $1 \otimes v$ or $v \otimes 1$. Commutativity of the diagrams may be easily checked by choosing bases $v_1, \ldots, v_n$ (resp. $\phi_1, \ldots, \phi_n$) in V (resp. V^*) so that $\phi_i(v_j) = \delta_{ij}$ (Kronecker delta), and noticing that δ sends 1 to $\sum v_i \otimes \phi_i$.

Now assume A is a Hopf algebra and M is an object of Comodf_A. Define a map $\rho^* : M^* \to M^* \otimes_k A \cong \mathrm{Hom}(M, A)$ by sending $\phi \in M^*$ to the composite map

$$M \to M \otimes_k A \xrightarrow{\phi \otimes \iota} k \otimes_k A \xrightarrow{\sim} A,$$

where the first map is given by the comodule structure on M and ι is the antipode.

Lemma 6.2.6 *The map ρ^* defines an A-comodule structure on M^* so that the evaluation (resp. coevaluation) maps $\epsilon : M \otimes_k M^* \to k$ (resp. $\delta : k \to M^* \otimes_k M$) are A-comodule homomorphisms.*

Here the tensor products are equipped with the A-comodule structure coming from the multiplication of A, and k with the one coming from the unit element $1 \in A$.

Proof This is just calculation with the axioms. We first check that ρ^* defines an A-comodule structure on M^*. The first diagram in (6.1) for M^* can be rewritten as

$$\begin{array}{ccc} M^* & \xrightarrow{\rho^*} & \mathrm{Hom}(M, A) \\ {\scriptstyle \rho^*}\downarrow & & \downarrow \\ \mathrm{Hom}(M, A) & \longrightarrow & \mathrm{Hom}(M, A \otimes_k A) \end{array}$$

where the right vertical map is induced by Δ and the bottom map is defined similarly as ρ^*, with A in place of k. With the notation of Remark 6.1.7 the composite of the upper and right maps in the diagram sends $\phi \in M^*$ to the map $m \mapsto \sum_{i,l} \phi(m_i)\iota(a_{il}) \otimes \iota(b_l)$, and the composite of the left and bottom maps sends it to $m \mapsto \sum_{i,j} \phi(m_{ij})\iota(c_j) \otimes \iota(a_i)$. Equality of the two follows from Equation (6.2) of Remark 6.1.7. Finally the counit axiom $\phi(m) = \sum_i \phi(m)\varepsilon(\iota(a_i))$ follows from (6.3) and the identity $\varepsilon \circ \iota = \varepsilon$, which is an easy consequence of the antipode axiom.

Checking the compatibility of the evaluation map $\epsilon : M \otimes_k M^* \to k$ with the comodule structures amounts to checking the commutativity of the diagram

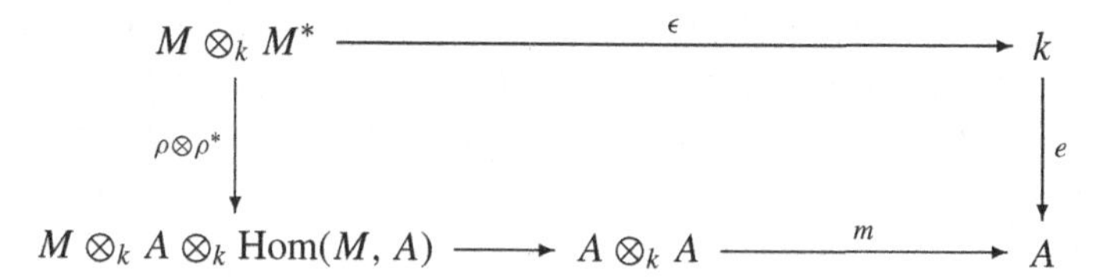

where the unnamed map is induced by the evaluation $M \otimes_k \mathrm{Hom}(M, A) \to A$ after permuting the first two factors. In the above notation, the composite of the left and bottom maps sends $m \otimes \phi \in M \otimes_k M^*$ to $\sum_{i,j} \phi(m_{ij})\iota(c_j)a_i$ which, by the same argument as above, equals $\sum_{i,l} \phi(m_i)\iota(a_{il})b_l$. By the coalgebra axioms (Remark 6.1.3) we have

$$\sum_i \phi(m_i) \sum_l \iota(a_{il})b_l = \sum_i \phi(m_i)\varepsilon(a_i) = \phi(m),$$

as required. The proof of compatibility with the coevaluation map is similar. $\square$

The next proposition provides a converse to this lemma.

Proposition 6.2.7 *Let A be a coalgebra over k equipped with a multiplication map $m : A \otimes_k A \to A$ and a unit map $k \to A$ that are compatible with the coalgebra structure.*

Assume moreover that for each object M of Comodf_A *the dual k-vector space M^* has an A-comodule structure so that the evaluation (resp. coevaluation) maps $\epsilon : M \otimes_k M^* \to k$ (resp. $\delta : k \to M^* \otimes_k M$) are A-comodule homomorphisms. Then A has an antipode map $\iota : A \to A$ making it into a Hopf algebra.*

The following proof is due to Ulbrich [105].

Proof We begin by constructing a morphism of functors $\omega \to \omega \otimes A$. Given M in Comodf_A, we construct $\omega(M) \to \omega(M) \otimes A$ as the composite

$$\begin{aligned}\omega(M) &\xrightarrow{\mathrm{id}\otimes\delta} \omega(M) \otimes_k \omega(M)^* \otimes_k \omega(M) \xrightarrow{=} \\ &\to \omega(M) \otimes_k \omega(M^*) \otimes_k \omega(M) \xrightarrow{\mathrm{id}\otimes\rho_{M^*}\otimes\mathrm{id}} \\ &\to \omega(M) \otimes_k \omega(M^*) \otimes_k A \otimes_k \omega(M) \xrightarrow{\cong} \\ &\to \omega(M) \otimes_k \omega(M^*) \otimes_k \omega(M) \otimes_k A \xrightarrow{=} \\ &\to \omega(M) \otimes_k \omega(M)^* \otimes_k \omega(M) \otimes_k A \xrightarrow{\epsilon\otimes\mathrm{id}} \omega(M) \otimes_k A.\end{aligned}$$

Here the equality $\omega(M^*) = \omega(M)^*$ that we used twice is of course a tautology, but we wrote it out explicitly, because for later use it is important to note that ω transforms duals in Comodf_A (which exist by assumption) to duals in Vecf_k.

By Proposition 6.2.1 the resulting morphism of functors $\omega \to \omega \otimes A$ yields a map of comodules $\iota : A \to A$. It remains to show that it satisfies the antipode axiom, i.e. that the composite map

$$A \xrightarrow{\Delta} A \otimes_k A \xrightarrow{\mathrm{id}\otimes\iota} A \otimes_k A \xrightarrow{m} A$$

is the identity. Again by Proposition 6.2.1 this is equivalent to saying that for all M in Comodf_A with comodule structure map $\rho_M : \omega(M) \to \omega(M) \otimes_k A$ the composite map

$$\begin{aligned}\rho_1 : \omega(M) &\xrightarrow{\rho_M} \omega(M) \otimes_k A \xrightarrow{\mathrm{id}\otimes\Delta} \omega(M) \otimes_k A \otimes_k A \xrightarrow{\mathrm{id}\otimes\mathrm{id}\otimes\iota} \\ &\longrightarrow \omega(M) \otimes_k A \otimes_k A \xrightarrow{\mathrm{id}\otimes m} \omega(M) \otimes_k A\end{aligned}$$

equals ρ_M. This we shall check in two steps. First we shall show that ρ_1 equals the composite map

$$\begin{aligned}\rho_2 : \omega(M) &\xrightarrow{\mathrm{id}\otimes\delta} \omega(M)\otimes_k \omega(M)^*\otimes_k \omega(M) \xrightarrow{\mathrm{id}\otimes\rho_{M^*}\otimes\mathrm{id}}\\ &\to \omega(M)\otimes_k \omega(M)^*\otimes_k \omega(M)\otimes_k A \xrightarrow{\mathrm{id}\otimes\rho_M\otimes\mathrm{id}}\\ &\to \omega(M)\otimes_k \omega(M)^*\otimes_k \omega(M)\otimes_k A\otimes_k A \xrightarrow{\mathrm{id}\otimes m}\\ &\to \omega(M)\otimes_k \omega(M)^*\otimes_k \omega(M)\otimes_k A \xrightarrow{\epsilon\otimes\mathrm{id}} \omega(M)\otimes_k A\end{aligned}$$

(where we have not written out the equality $\omega(M^*) = \omega(M)^*$ and permutation of components any more), and then we check that $\rho_2 = \rho_M$.

To check $\rho_1 = \rho_2$, note first that the diagram

$$\begin{array}{ccccc} \omega(M) & \xrightarrow{\rho_M} & \omega(M)\otimes_k A & \xrightarrow{\mathrm{id}\otimes\iota} & \omega(M)\otimes_k A \\ {\scriptstyle\rho_M}\downarrow & & {\scriptstyle\rho_M\otimes\mathrm{id}}\downarrow & & {\scriptstyle\rho_M\otimes\mathrm{id}}\downarrow \\ \omega(M)\otimes_k A & \xrightarrow{\mathrm{id}\otimes\Delta} & \omega(M)\otimes_k A\otimes_k A & \xrightarrow{\mathrm{id}\otimes\mathrm{id}\otimes\iota} & \omega(M)\otimes_k A\otimes_k A \end{array}$$

commutes: the first square by the comodule axiom, and the second by construction. Therefore ρ_1 equals the composite map

$$\begin{aligned}\rho_1' : \omega(M) &\xrightarrow{\rho_M} \omega(M)\otimes_k A \xrightarrow{\mathrm{id}\otimes\iota} \omega(M)\otimes_k A \xrightarrow{\rho_M\otimes\mathrm{id}} \omega(M)\otimes_k A\otimes_k A\\ &\xrightarrow{\mathrm{id}\otimes m} \omega(M)\otimes_k A.\end{aligned}$$

Here the composite of the first two maps equals

$$\begin{aligned}\omega(M) &\xrightarrow{\mathrm{id}\otimes\delta} \omega(M)\otimes_k \omega(M)^*\otimes_k \omega(M) \xrightarrow{\mathrm{id}\otimes\rho_{M^*}\otimes\mathrm{id}}\\ &\to \omega(M)\otimes_k \omega(M^*)\otimes_k \omega(M)\otimes_k A \xrightarrow{\epsilon\otimes\mathrm{id}} \omega(M)\otimes_k A,\end{aligned}$$

by the very definition of ι. Thus to show $\rho_1' = \rho_2$ it remains to note the commutativity of the diagram

$$\begin{array}{ccc} \omega(M)\otimes_k \omega(M)^*\otimes_k \omega(M)\otimes_k A & \xrightarrow{\epsilon\otimes\mathrm{id}} & \omega(M)\otimes_k A \\ {\scriptstyle\mathrm{id}\otimes\rho_M\otimes\mathrm{id}}\downarrow & & \downarrow{\scriptstyle\rho_M\otimes\mathrm{id}} \\ \omega(M)\otimes_k \omega(M)^*\otimes_k \omega(M)\otimes_k A\otimes_k A & \xrightarrow{\epsilon\otimes\mathrm{id}} & \omega(M)\otimes_k A\otimes_k A \\ {\scriptstyle\mathrm{id}\otimes m}\downarrow & & \downarrow{\scriptstyle\mathrm{id}\otimes m} \\ \omega(M)\otimes_k \omega(M)^*\otimes_k \omega(M)\otimes_k A & \xrightarrow{\epsilon\otimes\mathrm{id}} & \omega(M)\otimes_k A \end{array}$$

which follows from the compatibility of the evaluation map ϵ with the comodule structure, and the fact that the composite of the right vertical maps is the identity, the map m being a morphism of comodules.

Finally, to show that $\rho_2 = \rho_M$, note that by definition of the comodule structure on $M^* \otimes_m M$ and the compatibility of ω with tensor products the map ρ_2 equals the composite

$$\omega(M) \xrightarrow{\mathrm{id}\otimes\delta} \omega(M) \otimes_k \omega(M^* \otimes_m M) \xrightarrow{\sim} (\omega(M \otimes_m M^*) \otimes \omega(M) \xrightarrow{\rho_{M\otimes_m M^*}\otimes\mathrm{id}}$$
$$\to \omega(M \otimes_m M^*) \otimes_k \omega(M) \otimes_k A \xrightarrow{\epsilon\otimes\mathrm{id}} \omega(M) \otimes_k A.$$

Using the compatibility of ϵ with the comodule structure we may rewrite this map as the composite

$$\omega(M) \xrightarrow{\mathrm{id}\otimes\delta} \omega(M \otimes_m M^* \otimes_m M) \xrightarrow{\epsilon\otimes\mathrm{id}} \omega(M) \xrightarrow{\rho_M} \omega(M) \otimes_k A,$$

but the composite of the first two maps is the identity by the first diagram in (6.8) (we have dropped tensorizations with k throughout to ease notation). □

6.3 Tensor categories and the Tannaka–Krein theorem

In order to elucidate the relation between Hopf algebra structures and comodule categories completely, it is convenient to axiomatize the properties of comodule categories that were used in the previous section. They lead us to abstract category-theoretical notions that we now define formally.

A *tensor category* is a category $\mathcal{C}$ together with a functor $\mathcal{C} \times \mathcal{C} \to \mathcal{C}$ and an isomorphism Φ of functors from $\mathcal{C} \times \mathcal{C} \times \mathcal{C}$ to $\mathcal{C}$ given on a triple (X, Y, Z) of objects by

$$\Phi_{X,Y,Z} : (X \otimes Y) \otimes Z \xrightarrow{\sim} X \otimes (Y \otimes Z)$$

so that the diagram

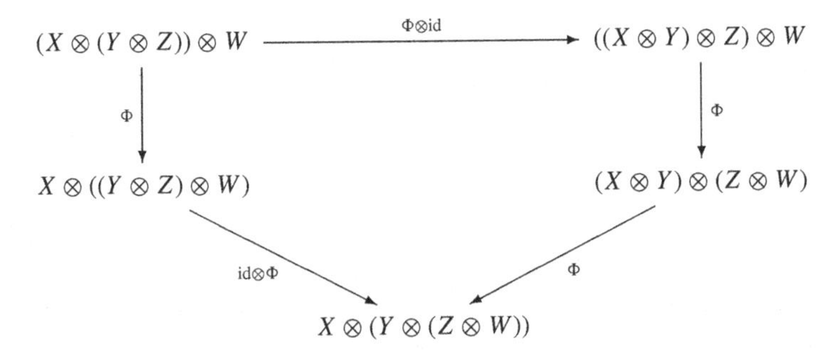

commutes for each four-tuple (X, Y, Z, W) of objects in $\mathcal{C}$. It is customary to call the isomorphism Φ the *associativity constraint*. Not surprisingly, the commutativity of the above diagram is usually referred to as the *pentagon axiom*.

A *unit object* in a tensor category $\mathcal{C}$ is an object 1 together with an isomorphism $\nu : 1 \to 1 \otimes 1$ so that moreover the functors $X \mapsto 1 \otimes X$ and $X \mapsto X \otimes 1$ are fully faithful. In what follows we shall often be sloppy and forget about the isomorphism ν. In the tensor category of k-vector spaces a unit object is given by k itself, together with one of the canonical isomorphisms $k \xrightarrow{\sim} k \otimes_k k$.

Remark 6.3.1 For each object X in $\mathcal{C}$ there exist canonical functorial isomorphisms $\alpha_X^1 : 1 \otimes X \xrightarrow{\sim} X$ and $\beta_X^1 : X \otimes 1 \xrightarrow{\sim} X$; in particular, the functors $X \mapsto 1 \otimes X$ and $X \mapsto X \otimes 1$ induce category equivalences of $\mathcal{C}$ with itself. To construct α_X^1, start with the isomorphism $\nu \otimes \mathrm{id}_X : 1 \otimes 1 \otimes X \xrightarrow{\sim} 1 \otimes X$, and then define α_X as the morphism $1 \otimes X \to X$ that induces $\nu \otimes \mathrm{id}_X$ via tensoring by 1 on the left. Such an α_X^1 exists and is unique as the functor $X \mapsto 1 \otimes X$ is fully faithful, and it must be an isomorphism because so is $\nu \otimes \mathrm{id}_X$. The construction of β_X^1 is similar.

Given two unit objects 1 and $1'$, the composite $\alpha_{1'}^1 \circ (\beta_1^{1'})^{-1} : 1 \xrightarrow{\sim} 1 \otimes 1' \xrightarrow{\sim} 1'$ defines a canonical isomorphism between 1 and $1'$. It is the unique isomorphism ϕ making the diagram

$$\begin{array}{ccc} 1 \otimes 1 & \xrightarrow{\phi \otimes \phi} & 1' \otimes 1' \\ {\scriptstyle \nu}\downarrow & & \downarrow{\scriptstyle \nu'} \\ 1 & \xrightarrow{\phi} & 1' \end{array}$$

commute. Consequently, a unit object is unique up to unique isomorphism.

In what follows we shall assume that *all tensor categories under consideration have a unit object.* Such tensor categories are often called *monoidal categories*; this terminology goes back to MacLane. (There is no universally accepted terminology concerning tensor categories in the literature; here we are adopting that of [42].)

A *commutativity constraint* on a tensor category is an isomorphism Ψ of functors from $\mathcal{C} \times \mathcal{C}$ to $\mathcal{C}$ given on a pair (X, Y) of objects by

$$\Psi_{X,Y} : X \otimes Y \xrightarrow{\sim} Y \otimes X$$

such that $\Psi_{Y,X} \circ \Psi_{X,Y} = \mathrm{id}_{X \otimes Y}$ for all X, Y. A tensor category $\mathcal{C}$ is *commutative* if there is a commutativity constraint on $\mathcal{C}$ so that the diagram

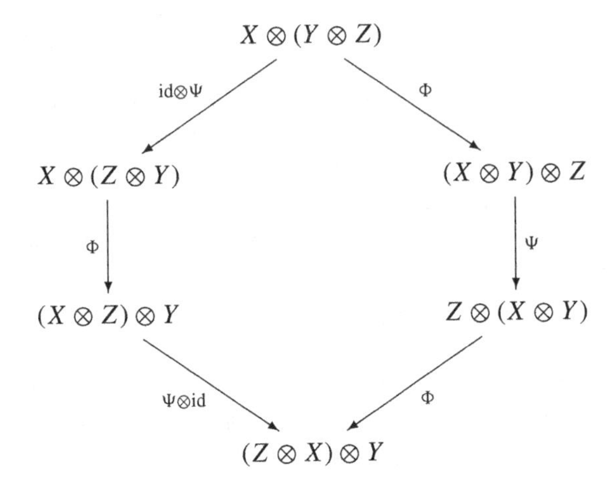

commutes for each triple (X, Y, Z) of objects in $\mathcal{C}$. This compatibility is called the *hexagon axiom*.

A *tensor functor* between two tensor categories $\mathcal{C}$ and $\mathcal{C}'$ is a functor $F : \mathcal{C} \to \mathcal{C}'$ together with an isomorphism Λ of functors from $\mathcal{C} \times \mathcal{C}$ to $\mathcal{C}'$ given on a pair (X, Y) of objects of $\mathcal{C}$ by

$$\Lambda_{X,Y} : F(X \otimes Y) \xrightarrow{\sim} F(X) \otimes F(Y)$$

such that moreover for a triple (X, Y, Z) of objects of $\mathcal{C}$ the diagram

$$\begin{array}{ccccc}
F((X \otimes Y) \otimes Z) & \xrightarrow{\Lambda_{X\otimes Y,Z}} & F(X \otimes Y) \otimes F(Z) & \xrightarrow{\Lambda_{X,Y}\otimes \mathrm{id}} & (F(X) \otimes F(Y)) \otimes F(Z) \\
{\scriptstyle F(\Phi_{X,Y,Z})}\downarrow & & & & \downarrow{\scriptstyle \Phi_{F(X),F(Y),F(Z)}} \\
F(X \otimes (Y \otimes Z)) & \xrightarrow{\Lambda_{X,Y\otimes Z}} & F(X) \otimes F(Y \otimes Z) & \xrightarrow{\mathrm{id}\otimes \Lambda_{Y,Z}} & F(X) \otimes (F(Y) \otimes F(Z))
\end{array}$$

commutes. Moreover, if 1 and $1'$ denote unit objects of $\mathcal{C}$ and $\mathcal{C}'$, respectively, we require that $F(1) = 1'$, the isomorphism $1' \to 1' \otimes 1'$ being given by $\Lambda_{1,1} \circ F(\nu)$.

We define a tensor category to be *rigid* if each object X has a dual X^*. The precise definition is as follows: there exist morphisms $\epsilon : X \otimes X^* \to 1$ and $\delta : 1 \to X^* \otimes X$ so that the diagrams

$$\begin{array}{ccc}
X & \xrightarrow{\cong} & 1 \otimes X \\
{\scriptstyle \cong}\downarrow & & \uparrow{\scriptstyle \epsilon\otimes\mathrm{id}} \\
X \otimes 1 & \xrightarrow{\mathrm{id}\otimes\delta} & X \otimes X^* \otimes X
\end{array}
\qquad
\begin{array}{ccc}
X^* & \xrightarrow{\cong} & X^* \otimes 1 \\
{\scriptstyle \cong}\downarrow & & \uparrow{\scriptstyle \mathrm{id}\otimes\epsilon} \\
1 \otimes X^* & \xrightarrow{\delta\otimes\mathrm{id}} & X^* \otimes X \otimes X^*
\end{array}
\qquad (6.9)$$

commute. Here the isomorphisms are the inverses of the canonical isomorphisms α^1 and β^1 constructed in Remark 6.3.1. The reader will recognize the formalism of dual vector spaces discussed in the previous section. Note that while the preceding axioms are satisfied by the usual tensor product of vector spaces, here we have to restrict to finite dimensional spaces in order to get examples. The following general lemma implies that up to isomorphism the k-linear dual V^* of a finite dimensional vector space V is the only dual object of V in the tensor category of finite dimensional vector spaces.

Lemma 6.3.2 *A dual X^* of an object X that satisfies the above properties is uniquely determined up to isomorphism. If we fix one of the maps ϵ or δ, then this isomorphism is unique.*

Proof We fix X and ϵ, and show that X^* represents the contravariant functor $Z \mapsto \mathrm{Hom}(X \otimes Z, 1)$, from which the uniqueness statements will follow. For representability we show that for each map $\phi : X \otimes Z \to 1$ we can find a map $\nu : Z \to X^*$ making the diagram

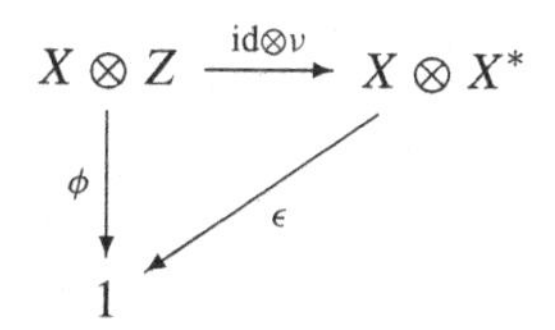

commute. Define ν as the composite

$$Z \xrightarrow{\sim} 1 \otimes Z \xrightarrow{\delta \otimes \mathrm{id}_Z} X^* \otimes X \otimes Z \xrightarrow{\mathrm{id}_{X^*} \otimes \phi} X^* \otimes 1 \xrightarrow{\sim} X^*$$

(with some δ as in the definition of a dual). Commutativity of the diagram then holds because the first diagram in (6.9) commutes. □

In a rigid tensor category every morphism $\phi : X \to Y$ has a *transpose*: it is the morphism $\phi^t : Y^* \to X^*$ defined as the composite

$$Y^* \xrightarrow{\sim} 1 \otimes Y^* \xrightarrow{\delta \otimes \mathrm{id}} X^* \otimes X \otimes Y^* \xrightarrow{\mathrm{id} \otimes \phi \otimes \mathrm{id}} X^* \otimes Y \otimes Y^* \xrightarrow{\mathrm{id} \otimes \epsilon} X^* \otimes 1 \xrightarrow{\sim} X^*.$$

The map $\phi \mapsto \phi^t$ induces a bijection $\mathrm{Hom}(X, Y) \xrightarrow{\sim} \mathrm{Hom}(Y^*, X^*)$.

A *morphism of tensor functors* between two tensor categories is a morphism of functors $F \to G$ compatible with the isomorphisms $\Lambda_{X,Y}$ for F and G occurring in the definition of tensor functors, and for which the composite isomorphism $1' \cong F(1) \xrightarrow{\sim} G(1) \cong 1'$ is the identity of $1'$. An isomorphism of tensor functors is a morphism as above that has a two-sided inverse that is again a morphism of tensor functors. We shall need the following lemma later.

Lemma 6.3.3 *A morphism of tensor functors between rigid tensor categories is always an isomorphism.*

Proof Given a morphism of tensor functors $\Phi : F \to G$ and an object X, we have isomorphisms $F(X^*) \cong F(X)^*$ and $G(X^*) \cong G(X)^*$ by the previous lemma. The morphism Φ_{X^*} therefore induces a morphism $F(X)^* \to G(X)^*$ with transpose $G(X) \to F(X)$. Applying this to all X we obtain a morphism of functors $\Phi^* : G \to F$. Verifying that Φ^* is a morphism of tensor functors inverse to Φ on both sides is a formal exercise left to the readers. □

Having introduced all this terminology, we can now summarize the discussion of the previous section in the following theorem.

Theorem 6.3.4 *Let A be a coalgebra over a field k, and ω the forgetful functor from the category* Comodf_A *of right A-comodules finite dimensional over k to the category* Vecf_k *of finite dimensional k-vector spaces.*

Assume that there is a tensor category structure on Comodf_A *for which ω becomes a tensor functor when* Vecf_k *carries its usual tensor structure.*

1. *There is a canonical k-algebra structure with unit on A defined by means of coalgebra morphisms.*
2. *If moreover the tensor category structure on* Comodf_A *is rigid, then A has the structure of a Hopf algebra.*
3. *Assume moreover the tensor category structure on* Comodf_A *is commutative, and ω transforms the commutativity constraint on* Comodf_A *to that on* Vecf_k. *Then A is a commutative Hopf algebra, and* Comodf_A *becomes equivalent to the category* Rep_G *of finite dimensional representations of the associated affine group scheme G.*

The last statement of course uses Proposition 6.1.9.

We close this section by two key properties of the category Rep_G of finite dimensional representations of an affine group scheme G. Before stating them, observe that given a commutative k-algebra R, the forgetful functor ω on Rep_G induces a tensor functor $\omega \otimes R : V \mapsto V \otimes_k R$ with values in the tensor category of R-modules. We can thus define set-valued functors $\mathbf{End}(\omega)$ (resp. $\mathbf{End}^{\otimes}(\omega)$, $\mathbf{Aut}^{\otimes}(\omega)$) on the category of k-algebras by sending R to the set of R-linear functor morphisms (resp. tensor functor morphisms, tensor functor isomorphisms) $\omega \otimes R \to \omega \otimes R$. The functor $\mathbf{Aut}^{\otimes}(\omega)$ is actually group-valued.

Proposition 6.3.5 *There is a canonical isomorphism of group-valued functors $G \xrightarrow{\sim} \mathbf{Aut}^{\otimes}(\omega)$. Consequently, $\mathbf{Aut}^{\otimes}(\omega)$ is an affine group scheme.*

Proof The proof is in three steps. First we show that we have functorial isomorphisms $\mathbf{End}(\omega)(R) \cong \mathrm{Hom}_{\mathrm{Mod}_R}(A \otimes_k R, R)$, where A is the Hopf algebra of G. Note that here we are considering R-module homomorphisms. In fact, we have $\mathrm{Hom}_{\mathrm{Mod}_R}(A \otimes_k R, R) \cong \mathrm{Hom}_{\mathrm{Vec}_k}(A, R)$, the isomorphism being induced by composition with the map $A \to A \otimes_k R$ given by $a \mapsto a \otimes 1$. Similarly, we obtain $\mathrm{Hom}_R(\omega \otimes R, \omega \otimes R) \cong \mathrm{Hom}_k(\omega, \omega \otimes R)$. Now we can conclude by Proposition 6.2.1, with the additional remark that it also holds with Vec_k replaced by the category of underlying vector spaces of k-algebras, because all we needed in the proof is that A is an object of the category.

Next we check that elements of $\mathbf{End}(\omega)(R)$ lying in $\mathbf{End}^{\otimes}(\omega)(R)$ correspond to k-algebra homomorphisms $A \to R$. By definition, a morphism of functors $\Phi_R : \omega \otimes R \to \omega \otimes R$ is a morphism of tensor functors if the diagram

$$\begin{array}{ccc} \omega(\_ \otimes \_) \otimes R & \xrightarrow{\Phi_R} & \omega(\_ \otimes \_) \otimes R \\ \downarrow & & \downarrow \\ \omega(\_) \otimes \omega(\_) \otimes R & \xrightarrow{\Phi_R \otimes \Phi_R} & \omega(\_) \otimes \omega(\_) \otimes R \end{array}$$

commutes, where the vertical maps are induced by the isomorphisms $\Lambda_{X,Y}$ in the definition of tensor functors. By Propositions 6.2.1 and 6.2.3 the maps Φ_R and $\Phi_R \otimes \Phi_R$ correspond to k-linear maps $\lambda : A \to R$ and $\lambda \circ m : A \otimes_k A \to R$, respectively, where $m : A \otimes_k A \to R$ is the multiplication map of A. The diagram thus expresses the fact that λ is a k-algebra homomorphism.

Finally, Lemma 6.3.3 implies that the natural morphism $\mathbf{Aut}^{\otimes}(\omega) \to \mathbf{End}^{\otimes}(\omega)$ of functors is an isomorphism. □

The proposition shows that one may recover an affine group scheme from the tensor category of its representations. This is the algebraic analogue of the classical Tannaka–Krein theorem on topological groups ([35], §30).

Now let G and G' be affine group schemes. Given a group scheme homomorphism $\phi : G \to G'$, every finite dimensional representation of G' yields a representation of G via composition with ϕ. In this way we obtain a tensor functor $\phi^* : \mathrm{Rep}_{G'} \to \mathrm{Rep}_G$ satisfying $\omega \circ \phi^* = \omega'$, where ω' is the forgetful functor on $\mathrm{Rep}_{G'}$.

Corollary 6.3.6 *The rule $\phi \mapsto \phi^*$ induces a bijection between group scheme homomorphisms $G \to G'$ and tensor functors $F : \mathrm{Rep}_{G'} \to \mathrm{Rep}_G$ satisfying $\omega \circ F = \omega'$.*

Proof Every tensor automorphism of ω yields a tensor automorphism of ω' via composition with a tensor functor F as above. The same holds for $\omega \otimes R$ for a k-algebra R, and therefore we obtain a morphism of group-valued functors $\mathbf{Aut}^{\otimes}(\omega) \to \mathbf{Aut}^{\otimes}(\omega')$. By the previous proposition it may be identified with a group scheme homomorphism $G \to G'$. Readers will check that this construction yields an inverse to the map $\phi \mapsto \phi^*$. □

6.4 Second interlude on category theory

To proceed further, we need to recall some basic notions about abelian categories. These are obtained by axiomatizing some properties of the category of abelian groups.

To begin with, if $\mathcal{C}$ is a category and A_1, A_2 two objects of $\mathcal{C}$, a *product* of A_1 and A_2 (if it exists) is by definition an object together with morphisms $p_i : A_1 \times A_2 \to A_i$, such that each pair $\phi_i : C \to A_i$ $(i = 1, 2)$ of morphisms from an object C factors uniquely as $\phi_i = p_i \circ \phi$ with a morphism $\phi : C \to A_1 \times A_2$. In other words, $A_1 \times A_2$ represents the functor $C \mapsto \mathrm{Hom}(C, A_1) \times \mathrm{Hom}(C, A_2)$. As such, it is determined up to unique isomorphism if it exists. Dually, a *coproduct* $A_1 \coprod A_2$ is an object representing the functor $C \mapsto \mathrm{Hom}(A_1, C) \times \mathrm{Hom}(A_2, C)$. One defines similarly arbitrary finite products and coproducts, and even (co)products over an infinite index set; we shall not need the latter.

If moreover A_1 and A_2 are equipped with morphisms $\psi_i : A_i \to A$ into a fixed object A, a *fibre product* $A_1 \times_A A_2$, if it exists, is an object representing the set-valued functor

$$C \mapsto \{\phi_1, \phi_2) \in \mathrm{Hom}(C, A_1) \times \mathrm{Hom}(C, A_2) : \psi_1 \circ \phi_1 = \psi_2 \circ \phi_2\}.$$

There is also a dual notion of an amalgamated sum that the reader will formulate.

This being said, an *additive category* is a category $\mathcal{A}$ in which each pair of objects has a product, and moreover the sets $\mathrm{Hom}(A, B)$ carry the structure of an abelian group so that the composition map $(\phi, \psi) \mapsto \phi \circ \psi$ is $\mathbf{Z}$-bilinear. If moreover the $\mathrm{Hom}(A, B)$ are k-vector spaces over a fixed field k and composition of maps is k-bilinear, we speak of a *k-linear additive category*.

Note that in an additive category each set $\mathrm{Hom}(A, B)$ has a zero element, i.e. there is a canonical morphism $0 : A \to B$ between A and B whose composite with other morphisms is again 0. This allows us to define the *kernel* $\ker(\phi)$ of a morphism $\phi : A \to B$ (if it exists) as the fibre product of the morphisms ϕ and $0 : A \to B$ over B. Dually, the *cokernel* $\mathrm{coker}(\phi)$ of ϕ is the amalgamated sum of ϕ and 0 over A, or equivalently, the kernel of ϕ in the opposite category of $\mathcal{A}$. Assume that $\ker(\phi)$ exists. We then define the *coimage* $\mathrm{coim}(\phi)$ of ϕ as the cokernel of the natural map $\ker(\phi) \to A$ (if it exists). Similarly, assuming the existence of $\mathrm{coker}(\phi)$, we define $\mathrm{im}(\phi)$ as the kernel of the natural map $B \to \mathrm{coker}(\phi)$ (if it exists). Note that if ϕ has an image and a coimage, it induces a natural map $\mathrm{coim}(\phi) \to \mathrm{im}(\phi)$.

Definition 6.4.1 An *abelian category* is an additive category in which every morphism ϕ has a kernel and a cokernel (hence also an image and a coimage), and moreover the natural map $\mathrm{coim}(\phi) \to \mathrm{im}(\phi)$ is an isomorphism. An abelian category is k-linear for some field k if it is k-linear as an additive category.

Plainly, the categories of abelian groups, modules over a fixed ring or abelian sheaves over a topological space are abelian, and the category of vector spaces over a field k is a k-linear abelian category.

In an abelian category it is customary to speak of direct products and direct sums instead of products and coproducts. Also, one says that $\phi : A \to B$ is a *monomorphism* (resp. *epimorphism*) if the morphism $\ker(\phi) \to A$ (resp. $B \to \operatorname{coker}(\phi)$) is the zero morphism. We say abusively that A' is a *subobject* (resp. A'' is a *quotient*) of A if there is a monomorphism $A' \to A$ (resp. an epimorphism $A \to A''$). We shall often use the notations $A' \subset A$ and A/A' for subobjects and quotients. An object A is *simple* if for each subobject $\phi : A' \to A$ the monomorphism ϕ is either 0 or an isomorphism. A *composition series* of an object A, if it exists, is a descending series $A = F^0 \supset F^1 \supset F^2 \supset \cdots$ of subobjects such that the quotients F^i/F^{i+1} are simple.

We say that A has *finite length* if it has a finite composition series. One proves as in the case of abelian groups that in this case every chain of subobjects in A can be refined to a composition series. Thus A is both Noetherian and Artinian, i.e. all ascending and descending series of subobjects in A stabilize. Moreover, all composition series of A are finite of the same length, and the finite set of the isomorphism classes of the F^i/F^{i+1} is the same up to permutation.

The usual notion of an exact sequence carries over without change to abelian categories. A functor between abelian categories is said to be exact if it takes short exact sequences to short exact sequences; there are also the usual weaker properties of left and right exactness that the appropriate Hom-functors enjoy. An object P of an abelian category $\mathcal{A}$ is *projective* if the functor $\operatorname{Hom}(P, \_)$ is exact. As the Hom-functor is always left exact, this is equivalent to requiring that given an epimorphism $A \twoheadrightarrow B$, each map $P \to B$ can be lifted to a map $P \to A$. The reader will define the dual notion of injective objects.

An object G of $\mathcal{A}$ is a *generator* if the functor $\operatorname{Hom}(G, \_)$ is faithful. This amounts to saying that for each nonzero morphism $\phi : A \to B$ in $\mathcal{A}$ there is a morphism $G \to A$ such that the composite $G \to A \to B$ is again nonzero. In the case when G is projective, this is moreover equivalent to the condition $\operatorname{Hom}(G, A) \neq 0$ for all $A \neq 0$ (for the nontrivial implication, use the projectivity of G to lift a nonzero morphism $G \to \operatorname{im}(\phi)$ to a morphism $G \to A$).

In the next section we shall need the following variant of a theorem of Mitchell and Freyd (for the original result, see [24], Exercise 4F).

Proposition 6.4.2 *Let $\mathcal{A}$ be an abelian category such that every object of $\mathcal{A}$ has finite length. Assume that $\mathcal{A}$ has a projective generator P. Then the functor* $\operatorname{Hom}(P, \_)$ *induces an equivalence of $\mathcal{A}$ with the category* $\operatorname{Modf}_{\operatorname{End}(P)}$ *of finitely generated right* $\operatorname{End}(P)$*-modules.*

We first prove a lemma.

Lemma 6.4.3 *Under the assumptions of the proposition for each A in $\mathcal{A}$ there is an epimorphism $P^{\oplus r} \to A$ from a finite direct power of P.*

Proof Start with a nonzero morphism $\phi_1 : P \to A$. If it is an epimorphism, we are done. Otherwise there is a nonzero morphism $P \to A/\mathrm{im}\,(\phi_1)$ that lifts to a morphism $\phi_2 : P \to A$. The image of $(\phi_1, \phi_2) : P \oplus P \to A$ is then strictly larger than $\mathrm{im}\,(\phi)$. As A has finite length, by continuing the procedure we obtain after finitely many steps an epimorphism $(\phi_1, \ldots, \phi_r) : P^{\oplus r} \to A$. □

Proof The left action of $\mathrm{End}(P)$ on P induces a right $\mathrm{End}(P)$-module structure on $\mathrm{Hom}(P, A)$ for each A via composition of maps. To see that we obtain a finitely generated module, consider an epimorphism $P^{\oplus r} \to A$ as in the lemma. Applying the functor $\mathrm{Hom}(P, \_)$ and noting the isomorphism $\mathrm{End}(P)^{\oplus r} \cong \mathrm{Hom}(P, P^{\oplus r})$ we obtain a surjection $\mathrm{End}(P)^{\oplus r} \twoheadrightarrow \mathrm{Hom}(P, A)$.

Next we show that $\mathrm{Hom}(P, \_)$ is fully faithful. As P is a generator by assumption, this boils down to showing that every morphism $\phi : \mathrm{Hom}(P, A) \to \mathrm{Hom}(P, B)$ comes from a morphism $A \to B$. Consider epimorphisms $P^{\oplus r} \to A$, $P^{\oplus s} \to B$ given by the lemma. Applying the functor $\mathrm{Hom}(P, \_)$ we obtain a diagram

$$\begin{array}{ccccc} \mathrm{End}(P)^{\oplus r} & \longrightarrow & \mathrm{Hom}(P, A) & \longrightarrow & 0 \\ & & \downarrow{\scriptstyle \phi} & & \\ \mathrm{End}(P)^{\oplus s} & \longrightarrow & \mathrm{Hom}(P, B) & \longrightarrow & 0 \end{array}$$

with exact rows. As $\mathrm{End}(P)^{\oplus r}$ is a free, hence projective $\mathrm{End}(P)$-module, there is a map $\psi : \mathrm{End}(P)^{\oplus r} \to \mathrm{End}(P)^{\oplus s}$ of free $\mathrm{End}(P)$-modules making the diagram commute. Here ψ is defined by multiplication with an $r \times s$ matrix of elements in $\mathrm{End}(P)$. But such a matrix defines a morphism $\bar{\psi} : P^{\oplus r} \to P^{\oplus s}$ that gives rise to ψ after applying the functor $\mathrm{Hom}(P, \_)$. By construction, the composite map $\mathrm{End}(P)^{\oplus r} \xrightarrow{\psi} \mathrm{End}(P)^{\oplus s} \to \mathrm{Hom}(P, B)$ annihilates the kernel of the map $\mathrm{End}(P)^{\oplus r} \to \mathrm{Hom}(P, A)$. Therefore by exactness of the functor $\mathrm{Hom}(P, \_)$ the composite $P^{\oplus r} \xrightarrow{\bar{\psi}} P^{\oplus s} \to B$ annihilates the kernel of $P^{\oplus r} \to A$, i.e. induces a map $A \to B$. This is the map we were looking for.

Finally, for essential surjectivity write a finitely generated $\mathrm{End}(P)$-module M as the cokernel of a map $\alpha : \mathrm{End}(P)^{\oplus r} \to \mathrm{End}(P)^{\oplus s}$ of free modules (this is possible as $\mathrm{End}(P)$ is right Noetherian by the finiteness assumption). By what we have just proven α comes from a map $P^{\oplus r} \to P^{\oplus s}$. Let C denote its cokernel; we then have $\mathrm{Hom}(P, C) \cong M$ by the exactness of $\mathrm{Hom}(P, \_)$, i.e. the projectivity of P. □

We shall also need a description of certain subcategories of a category as in the above proposition. Observe that given a homomorphism $\phi : \mathrm{End}(P) \to R$ to some ring R, there is an induced functor $\mathrm{Modf}_R \to \mathrm{Modf}_{\mathrm{End}(P)}$ sending a right R-module to its underlying abelian group equipped with the $\mathrm{End}(P)$-module structure coming from ϕ. If moreover ϕ is surjective, i.e. R is of the form $\mathrm{End}(P)/I$ for some two-sided ideal I, this functor is fully faithful, so that Modf_R identifies with a full subcategory of $\mathrm{Modf}_{\mathrm{End}(P)}$. Besides, the subcategory thus obtained is closed under subobjects, quotients and finite direct sums. The next proposition gives a converse.

Proposition 6.4.4 *Keep the assumptions of the previous proposition, and assume moreover that $\mathcal{B}$ is a full subcategory of $\mathcal{A}$ closed under subobjects, quotients and finite direct sums. There exist a uniquely determined ideal $I \subset \mathrm{End}(P)$ and an equivalence of categories between $\mathcal{B}$ and the category $\mathrm{Modf}_{\mathrm{End}(P)/I}$ under which the inclusion functor $\mathcal{B} \to \mathcal{A}$ becomes identified with the functor $\mathrm{Modf}_{\mathrm{End}(P)/I} \to \mathrm{Modf}_{\mathrm{End}(P)}$ described above.*

The assumption on $\mathcal{B}$ is to be understood in the sense that every object of $\mathcal{A}$ isomorphic to a subquotient of a finite direct sum of objects in $\mathcal{B}$ lies in $\mathcal{B}$. For the proof we need a lemma.

Lemma 6.4.5 *For each object A of $\mathcal{A}$ there is a maximal quotient $q_{\mathcal{B}}(A)$ of A lying in $\mathcal{B}$. More precisely, there is an object $q_{\mathcal{B}}(A)$ in $\mathcal{B}$ and an epimorphism $\rho_A : A \twoheadrightarrow q_{\mathcal{B}}(A)$ such that every epimorphism $A \twoheadrightarrow B$ with an object B of $\mathcal{B}$ factors as a composite $\lambda \circ \rho_A$ for some $\lambda : q_{\mathcal{B}}(A) \to B$.*

By its very definition, the rule $A \to q_{\mathcal{B}}(A)$ induces a functor $\mathcal{A} \to \mathcal{B}$. Experts in category theory will recognize that it is a left adjoint to the inclusion functor $\mathcal{B} \to \mathcal{A}$.

Proof Let $K_{\mathcal{B}}(A)$ be the intersection of the kernels of all epimorphisms $A \twoheadrightarrow B$ with B in $\mathcal{B}$, and set $q_{\mathcal{B}}(A) := A/K_{\mathcal{B}}(A)$. By construction $q_{\mathcal{B}}(A)$ satisfies the required universal property, but we still have to show that it lies in $\mathcal{B}$. As A has finite length, there exist finitely many objects $B_1, \dots, B_n$ in $\mathcal{B}$ together with epimorphisms $\rho_i : A \twoheadrightarrow B_i$ so that $K_{\mathcal{B}}(A) = \bigcap \ker(\phi_i)$. Here $A/\bigcap \ker(\phi_i)$ is isomorphic to a subobject of $\bigoplus (A/\ker(\phi_i)) \cong \bigoplus B_i$, so it lies in $\mathcal{B}$ by the assumption on $\mathcal{B}$. □

Proof of Proposition 6.4.4 Given a projective generator P of $\mathcal{A}$, the object $q_{\mathcal{B}}(P)$ is a projective generator in $\mathcal{B}$, as exactness and faithfulness of the functor $\mathrm{Hom}_{\mathcal{B}}(q_{\mathcal{B}}(P), \_)$ immediately follow from the corresponding properties of $\mathrm{Hom}_{\mathcal{A}}(P, \_)$. By Proposition 6.4.2 the functor $\mathrm{Hom}_{\mathcal{B}}(q_{\mathcal{B}}(P), \_)$ establishes an equivalence of categories between $\mathcal{B}$ and $\mathrm{Modf}_{\mathrm{End}(q_{\mathcal{B}}(P))}$. As $q_{\mathcal{B}}$ is a functor, there is a natural map $q : \mathrm{End}(P) \to \mathrm{End}(q_{\mathcal{B}}(P))$. This map is surjective, because given an endomorphism $\psi \in \mathrm{End}(q_{\mathcal{B}}(P))$, we may lift the composite $P \to q_{\mathcal{B}}(P) \xrightarrow{\psi} q_{\mathcal{B}}(P)$ to an endomorphism of P by the projectivity of P. The ideal $I := \ker(q)$ and the equivalence between $\mathcal{B}$ and $\mathrm{Modf}_{\mathrm{End}(q_{\mathcal{B}}(P))} \cong \mathrm{Modf}_{\mathrm{End}(P)/I}$ then satisfy the requirements of the proposition. □

6.5 Neutral Tannakian categories

We can now state the main results of this chapter. First the long-awaited definition:

Definition 6.5.1 A *neutral Tannakian category* over a field k is a rigid k-linear abelian tensor category $\mathcal{C}$ whose unit 1 satisfies $\mathrm{End}(1) \cong k$, and is moreover equipped with an exact faithful tensor functor $\omega : \mathcal{C} \to \mathrm{Vecf}_k$ into the category of finite dimensional k-vector spaces. The functor ω is called a (neutral) *fibre functor*.

Remark 6.5.2 The term 'k-linear abelian tensor category' involves a compatibility condition relating the tensor and abelian category structures on $\mathcal{C}$. It means that the tensor operation

$$\mathrm{Hom}(X, Y) \times \mathrm{Hom}(Z, W) \to \mathrm{Hom}(X \otimes Z, Y \otimes W), \quad (\phi, \psi) \mapsto \phi \otimes \psi$$

should be k-bilinear with respect to the k-vector space structures on the Hom-sets involved. This also explains the presence of the condition $\mathrm{End}(1) \cong k$. Namely, the isomorphism $1 \otimes X \xrightarrow{\sim} X$ induces a map $\mathrm{End}(1) \to \mathrm{End}(X)$ that endows the group $\mathrm{Hom}(X, Y)$ with the structure of an $\mathrm{End}(1)$-module via composition, and similarly for endomorphisms of Y. We require that the isomorphism $\mathrm{End}(1) \cong k$ transforms these to the k-linear structure on $\mathrm{Hom}(X, Y)$. These requirements are of course satisfied in tensor categories of k-vector spaces.

Given an affine group scheme G over k, the category Rep_G of finite dimensional representations of G with its usual tensor structure and the forgetful functor as fibre functor is a neutral Tannakian category. Conversely, we have:

Theorem 6.5.3 *Every neutral Tannakian category $(\mathcal{C}, \omega)$ over k is equivalent to the category Rep_G of finite dimensional representations of an affine group scheme G over k.*

By Proposition 6.3.5 the group scheme G as a functor is isomorphic to the functor $\mathbf{Aut}^{\otimes}(\omega)$ of tensor automorphisms of the fibre functor ω. We call it the *Tannakian fundamental group* of $(\mathcal{C}, \omega)$.

Theorem 6.5.3 immediately follows from Theorem 6.3.4 and the following statement that does not involve the tensor structure.

Theorem 6.5.4 *Let $\mathcal{C}$ be a k-linear abelian category equipped with an exact faithful k-linear functor $\omega : \mathcal{C} \to \mathrm{Vecf}_k$. There exists a k-coalgebra $A_{\mathcal{C}}$ so that each $\omega(X)$ carries a natural right $A_{\mathcal{C}}$-comodule structure for X in $\mathcal{C}$. Moreover, ω induces an equivalence of categories between $\mathcal{C}$ and the category $\mathrm{Comodf}_{A_{\mathcal{C}}}$ of finite dimensional right $A_{\mathcal{C}}$-comodules.*

We shall give a proof of Theorem 6.5.4 due to Deligne and Gabber which is extracted from Deligne's fundamental paper [14]; see also [38], §7 and [82] for other approaches. For the readers' convenience we begin with a brief overview of the main steps of the argument. First some general notation: for an object X in an abelian category $\mathcal{A}$ we denote by $\langle X \rangle$ the full subcategory of $\mathcal{A}$ spanned by the objects of $\mathcal{A}$ isomorphic to a subquotient of a finite direct sum of copies of X.

Step 1. Let $\mathcal{C}$ be as in the theorem, and let X be an object of $\mathcal{C}$. There exist a finite dimensional k-algebra R and an equivalence of categories between $\langle X \rangle$ and the category Modf_R of finitely generated right R-modules.

Step 2. Moreover, there exists a finitely generated left R-module M so that under the equivalence of Step 1 the functor ω becomes identified with the functor $N \to N \otimes_R M$ from Modf_R to Vecf_k.

Step 3. There is a canonical k-coalgebra structure on $A := M^* \otimes_R M$ and a right A-comodule structure on $N \otimes_R M$ for each right R-module N so that the functor $N \to N \otimes_R M$ induces an equivalence of the previous categories with the category Comodf_A of finite dimensional right A-comodules.

Step 4. The theorem follows by writing $\mathcal{C}$ as a directed union of subcategories of the form $\langle X \rangle$ and passing to the direct limit.

We now give details on the first step. The crucial statement is:

Proposition 6.5.5 (Gabber) *Let $\mathcal{A}$ be a k-linear abelian category in which every object has a finite composition series and the k-vector spaces $\mathrm{Hom}(A, B)$ are finite dimensional for all A, B in $\mathcal{A}$. Then for each object X the full subcategory $\langle X \rangle$ has a projective generator P.*

The following lemma summarizes the basic strategy of the proof of the proposition.

Lemma 6.5.6 *Denote by $\mathcal{S}$ the finite set of isomorphism classes of simple objects occurring in a composition series of X. Assume given for each representative S of a class $[S] \in \mathcal{S}$ an epimorphism $\phi_S : P_S \to S$ in $\langle X \rangle$ with P_S projective. Then $P := \bigoplus_{[S]\in\mathcal{S}} P_S$ is a projective generator in $\langle X \rangle$.*

Proof Finite direct sums of projective objects are again projective, so we only have to exhibit a nonzero morphism $P \to X'$ for each nonzero object X' of $\langle X \rangle$.

By definition the composition factors of X' lie in $\mathcal{S}$, so in particular there is an epimorphism $X' \to S$ with $[S] \in \mathcal{S}$. We may define an epimorphism $P \to S$ by taking it to be ϕ_S on P_S and extending by 0 on the other components. By projectivity of P it lifts to a nonzero morphism $P \to X'$, as required. □

To construct the objects $P_S \to S$ we need the notion of *essential extension* of an object Y in an abelian category. By definition, this is an object E together with an epimorphism $\alpha : E \to Y$ such that there is no subobject $E' \subset E$ distinct from E so that the composite $E' \to E \to Y$ is still an epimorphism. When Y is simple, this is the same as saying that all $E' \subset E$ distinct from E are contained in the kernel of α, because for Y simple each nonzero morphism $E' \to Y$ is an epimorphism.

Lemma 6.5.7 *Let S be a simple object in an abelian category, and $\alpha : E \to S$ an essential extension. For each simple object T the natural map*

$$\alpha^* : \mathrm{Hom}(S, T) \to \mathrm{Hom}(E, T)$$

induced by α is an isomorphism. In particular, $\mathrm{Hom}(E, T) = 0$ *for $T \not\cong S$.*

Proof Given a nonzero map $\phi : E \to T$, we have $\ker(\phi) \subset \ker(\alpha)$ by the above. The induced map $A/\ker(\phi) \to A/\ker(\alpha)$ must be an isomorphism because both objects are simple, whence $\ker(\phi) = \ker(\alpha)$ and $\phi = \phi' \circ \alpha$ for some $\phi' : S \to T$. The map $\phi \mapsto \phi'$ is then an inverse to α^*. The second statement follows because $\mathrm{Hom}(S, T) = 0$ for nonisomorphic simple objects. □

Let us now return to the k-linear category $\langle X \rangle$ of Proposition 6.5.5. Let S be a simple object in $\langle X \rangle$, and $E \to S$ an essential extension. The following lemma gives a criterion for E to be projective.

Lemma 6.5.8 *Let $E \to S$ be an essential extension as above. For all Y in $\langle X \rangle$ we have an inequality*

$$\dim_k \mathrm{Hom}_k(E, Y) \leq \ell_S(Y) \dim_k \mathrm{End}(S), \qquad (6.10)$$

where $\ell_S(Y)$ denotes the number of composition factors of Y isomorphic to S. Moreover, the following statements are equivalent.

1. *The essential extension E of S is projective.*
2. *There is equality in (6.10) for all Y in $\langle X \rangle$.*
3. *There is equality in (6.10) for $Y = X$.*

Proof The right-hand side of (6.10) is additive for short exact sequences of the form

$$0 \to Y' \to Y \to Y'' \to 0. \qquad (6.11)$$

Concerning the left-hand side, we have the inequality

$$\dim_k \mathrm{Hom}_k(E, Y) \le \dim_k \mathrm{Hom}_k(E, Y') + \dim_k \mathrm{Hom}_k(E, Y'') \tag{6.12}$$

by left exactness of the Hom-functor; it is an equality for all short exact sequences (6.11) if and only if the Hom-functor is exact, i.e. E is projective. By the previous lemma (6.10) holds with equality for all simple objects Y in $\langle X \rangle$. For arbitrary Y we may consider a composition series and conclude from the previous arguments that (6.10) holds for Y; moreover it holds with equality for all Y if and only if E is projective. This shows the first statement of the lemma as well as the equivalence of (1) and (2). The implication (2) $\Rightarrow$ (3) is obvious. To prove the converse, note that given a short exact sequence (6.11), we infer from (6.12) and (6.10) that there is equality in (6.10) for Y if and only if there is equality for Y' and Y'', and moreover equality holds in (6.12). This shows that equality in (6.10) for $Y = X$ implies equality for all finite direct powers of X, as well as for their subquotients. □

Proof of Proposition 6.5.5 In view of Lemma 6.5.6 we have to construct a projective essential extension $P_S \to S$ for each simple object S in $\langle X \rangle$. We do so by considering a composition series $X = F^0 \supset F^1 \supset \cdots \supset F^r = \{0\}$, and constructing by induction on i essential extensions $P_i \to S$ satisfying

$$\dim_k \mathrm{Hom}_k(P_i, X/F^i) = \ell_S(X/F^i) \dim_k \mathrm{End}(S). \tag{6.13}$$

By the previous lemma $P_S = P_r$ will be a good choice. We start by setting $P_1 := S$. Assuming that P_{i-1} has been constructed, consider a k-basis $\phi_1, \dots, \phi_n$ of $\mathrm{Hom}_k(P_i, X/F^i)$. For each $1 \le j \le n$, take the fibre product Q_j defined by the square

$$\begin{array}{ccc} Q_j & \longrightarrow & X/F^{i+1} \\ \downarrow & & \downarrow \\ P_i & \xrightarrow{\phi_j} & X/F^i, \end{array}$$

where the right vertical map is the natural projection; note here that the vertical maps are epimorphisms. Let Q be the fibre product of the Q_j over P_i for $1 \le j \le n$, and let $P_{i+1} \subset Q$ be a minimal subobject with the property that the composite $P_{i+1} \to Q \to P_i$ is an epimorphism. It is then an essential extension of P_i, hence of S.

To show that P_{i+1} satisfies (6.13) for $i+1$, we begin by constructing a retraction for the natural map $\mathrm{Hom}(P_{i+1}, X/F^{i+1}) \to \mathrm{Hom}(P_{i+1}, X/F^i)$. To do so, observe first that since $P_{i+1} \to P_i$ is an epimorphism, the induced map $\mathrm{Hom}(P_i, X/F^i) \to \mathrm{Hom}(P_{i+1}, X/F^i)$ is injective. But the dimension of the first k-vector space here is $\ell_S(X/F^i) \dim_k \mathrm{End}(S)$ by the inductive hypothesis, and that of the second is at most $\ell_S(X/F^i)\dim_k \mathrm{End}(S)$ by (6.10) applied with $E = P_{i+1}$ and $Y = X/F^i$. This is only possible if

$$\mathrm{Hom}(P_i, X/F_i) \xrightarrow{\sim} \mathrm{Hom}(P_{i+1}, X/F_i). \tag{6.14}$$

From the diagram above we infer that each morphism ϕ_j gives rise to a map $\psi_j : Q_j \to X/F^{i+1}$ by base change. As the ϕ_j form a k-basis of $\mathrm{Hom}_k(P_i, X/F^i)$, composing the ψ_j with the epimorphisms $P_{i+1} \to Q_j$ induces a homomorphism $\mathrm{Hom}(P_i, X/F^i) \to \mathrm{Hom}(P_{i+1}, X/F^{i+1})$ whose composite with the inverse isomorphism of (6.14) yields the required retraction.

All in all, we obtain that the last map in the exact sequence

$$0 \to \mathrm{Hom}(P_{i+1}, F^i/F^{i+1}) \to \mathrm{Hom}(P_{i+1}, X/F^{i+1}) \to \mathrm{Hom}(P_{i+1}, X/F^i)$$

is surjective. The dimension of the last term is $\ell_S(X/F^i)\dim_k \mathrm{End}(S)$ by the inductive hypothesis and (6.14), and that of the first is $\ell_S(F^i/F^{i+1})\dim_k \mathrm{End}(S)$ by Lemma 6.5.7. The required formula for $i + 1$ follows. □

Combining the proposition with Proposition 6.4.2 immediately yields:

Corollary 6.5.9 *In the above situation the functor* $A \mapsto \mathrm{Hom}(P, A)$ *induces an equivalence of the subcategory* $\langle X \rangle$ *with the category* $\mathrm{Modf}_{\mathrm{End}(P)}$ *of finitely generated right* $\mathrm{End}(P)$*-modules.*

This completes Step 1 of the proof of Theorem 6.5.4 outlined above, with $R = \mathrm{End}(P)$. We now turn to Step 2, and show that $M = \omega(P)$ is a good choice for the R-module M required there. For one thing, there is indeed a natural left $\mathrm{End}(P)$-module structure on $\omega(P)$, the multiplication $\mathrm{End}(P) \times \omega(P) \to \omega(P)$ being defined for a pair (ϕ, a) by $\omega(\phi)(a)$. The statement of Step 2 is then:

Proposition 6.5.10 *Via the category equivalence of the previous corollary the functor* ω *becomes isomorphic to the functor mapping a right* $\mathrm{End}(P)$*-module* N *to the underlying* k*-vector space of* $N \otimes_{\mathrm{End}(P)} \omega(P)$.

Proof For each object A in $\langle X \rangle$ the rule $\phi \otimes x \mapsto \omega(\phi)(x)$ defines a natural map $\mathrm{Hom}(P, A) \otimes_{\mathrm{End}(P)} \omega(P) \to \omega(A)$ that is moreover functorial in A. It is tautologically an isomorphism for $A = P$ and, being compatible with finite direct sums, for $A = P^{\oplus r}$ for all $r > 0$. Given an arbitrary object A, we choose an epimorphism $\lambda : P^{\oplus r} \to A$ with the benediction of Lemma 6.4.3, and consider the commutative diagram

$$\begin{array}{ccccccc}
\mathrm{Hom}(P, K) \otimes \omega(P) & \longrightarrow & \mathrm{Hom}(P, P^{\oplus r}) \otimes \omega(P) & \longrightarrow & \mathrm{Hom}(P, A) \otimes \omega(P) & \longrightarrow & 0 \\
\downarrow & & \downarrow \cong & & \downarrow & & \\
\omega(K) & \longrightarrow & \omega(P^{\oplus r}) & \longrightarrow & \omega(A) & \longrightarrow & 0
\end{array}$$

where $K := \ker(\lambda)$ and tensor products are taken over $\mathrm{End}(P)$. The lower row is exact by exactness of ω (it is even exact on the left), and the upper row by projectivity of P and right exactness of the tensor product. We have just seen that the middle vertical map is surjective, which implies the surjectivity of the map on the right. As this holds for all A, we get surjectivity of the left vertical map as well. But then the injectivity of the middle vertical map implies the injectivity of the one on the right. □

Our assumption on ω then yields:

Corollary 6.5.11 *The functor $N \mapsto N \otimes_{\mathrm{End}(P)} \omega(P)$ is exact and faithful on* $\mathrm{Modf}_{\mathrm{End}(P)}$.

We now turn to Step 3 of the proof of Theorem 6.5.4, and use the notation $R := \mathrm{End}(P)$ and $M := \omega(P)$ from now on. Note first that the tensor product $A := M^* \otimes_R M$ makes sense as a k-vector space because the dual k-vector space M^* carries a right R-module structure induced from that of M. Our next task is to define a k-coalgebra structure on A.

Quite generally, given a right R-module N, we have a natural map

$$\mathrm{id}_{N\otimes_R M} \otimes \delta_M : N \otimes_R M \to N \otimes_R M \otimes_k M^* \otimes_R M, \qquad (6.15)$$

where $\delta_M : k \to M^* \otimes_k M$ is the coevaluation map. For $N = M^*$ this defines a comultiplication $\Delta : A \to A \otimes_k A$ whose coassociativity the reader will verify. Together with the counit $A \to k$ given by the map $M^* \otimes_R M \to k$ sending $\phi \otimes m$ to $\phi(m)$ we obtain a k-coalgebra structure on A. Moreover, for each right R-module N the map (6.15) equips $N \otimes_R M$ with a right A-comodule structure. In this way we obtain a functor $N \mapsto N \otimes_R M$ from the category of right R-modules to that of right A-comodules.

Proposition 6.5.12 *The above functor induces an equivalence of categories, where finitely generated R-modules correspond to A-comodules finite dimensional as a k-vector space.*

Proof Start with a k-vector space V equipped with an A-comodule structure $\rho : V \to V \otimes_k A$. Recalling that $A = M^* \otimes_R M$, we have two natural maps of right R-modules $V \otimes_k M^* \to V \otimes_k M^* \otimes_R M \otimes_k M^*$: one is $\rho \otimes \mathrm{id}_{M^*}$, the other is $\mathrm{id}_V \otimes \delta_M \otimes \mathrm{id}_{M^*}$. Write λ for the difference of these two maps, and set $N := \ker(\lambda)$. Tensoring λ with id_M over R we obtain a map $V \otimes_k A \to V \otimes_k A \otimes_k A$ that is none but the map $\rho \otimes \mathrm{id}_A - \mathrm{id}_V \otimes \Delta$. Proposition 6.1.12 together with the exactness of the functor $N \mapsto N \otimes_R M$ (Corollary 6.5.11) then yields an isomorphism $N \otimes_R M \cong V$, which shows that the said functor is essentially surjective. By the same corollary it is also faithful, so for the equivalence of categories it remains to be shown that each A-coalgebra map

$\phi : N \otimes_R M \to N' \otimes_R M$ comes from a map of R-modules $N \to N'$. This is shown by applying the above construction simultaneously for the source and the target of ϕ. Finally, the second statement of the proposition follows from the finite dimensionality of R over k. □

This completes Step 3 in the proof of Theorem 6.5.4, so it remains to give some details on the limit procedure of Step 4, which will complete the proof.

Proof of Theorem 6.5.4 Write the category $\mathcal{C}$ as a union of the full subcategories $\langle X \rangle$ for each object X. The system of these subcategories is partially ordered by inclusion; the partial order is directed, because given two objects X and Y, both $\langle X \rangle$ and $\langle Y \rangle$ are full subcategories of $\langle X \oplus Y \rangle$. By Proposition 6.5.5 each $\langle X \rangle$ is equivalent to a module category Modf_R, and by Proposition 6.4.4 a full subcategory $\langle Y \rangle \subset \langle X \rangle$ corresponds to the category $\text{Modf}_{R/I}$ for some ideal $I \subset R$. From Proposition 6.5.12 we obtain a further equivalence of Modf_R and Comodf_A, where $A = M^* \otimes_R M$. In particular, by Proposition 6.5.10 we obtain an A-comodule structure on $\omega(X)$. In what follows we write A_X in place of A in order to emphasize the dependence of A on X. The coalgebra A_Y corresponding to the full subcategory $\langle Y \rangle \cong \text{Modf}_{R/I}$ is

$$(M \otimes_R R/I)^* \otimes_{R/I} (M \otimes_R R/I) \cong (M \otimes_R R/I)^* \otimes_R M,$$

whose natural map in $A_X = M^* \otimes_R M$ is injective, because tensoring with M is an exact functor by Corollary 6.5.11. The direct limit of the coalgebras A_X with respect to these maps is a coalgebra $A_{\mathcal{C}}$ in which each A_X is in fact a subcoalgebra. We thus obtain an $A_{\mathcal{C}}$-comodule structure on $\omega(X)$ by extending the A_X-comodule structure to $A_{\mathcal{C}}$. To show that ω induces an equivalence between $\mathcal{C}$ and $\text{Comodf}_{A_{\mathcal{C}}}$, we check fully faithfulness and essential surjectivity as usual. The bijection between $\text{Hom}_{\mathcal{C}}(X, Y)$ and $\text{Hom}_{A_{\mathcal{C}}}(\omega(X), \omega(Y))$ already follows from Steps 2 and 3, as the morphisms in $\text{Hom}_{\mathcal{C}}(X, Y)$ all lie in $\langle X \oplus Y \rangle$. Finally, as $A_{\mathcal{C}}$ is a directed union of the subcoalgebras A_X, each object of $\text{Comodf}_{A_{\mathcal{C}}}$ comes from some A_X-comodule by base extension, and hence is of the form $\omega(Z)$ for a $Z \in \langle X \rangle$, again by Steps 2 and 3. □

Remark 6.5.13 We now briefly discuss the theory of general Tannakian categories, for which we have to use the language of schemes. Assume that $\mathcal{C}$ is, as before, a rigid k-linear abelian tensor category whose unit $\mathbf{1}$ satisfies $\text{End}(\mathbf{1}) \cong k$. A general (non-neutral) fibre functor is an exact faithful tensor functor ω on $\mathcal{C}$ with values in the k-linear abelian tensor category LF_S of locally free sheaves on a k-scheme S. The pair $(\mathcal{C}, \omega)$ defines a *Tannakian category.*

Given a morphism $\phi : T \to S$, the functor ω induces a fibre functor $\phi^*\omega$ with values in LF_T via $X \mapsto \omega(X) \otimes_{\mathcal{O}_S} \mathcal{O}_T$. Define the functor $\text{Aut}_k^{\otimes}(\omega)$ on the category of schemes over $S \times_k S$ by sending the scheme $T \to S \times_k S$ to the set of $\mathcal{O}_T$-linear isomorphisms of tensor functors $\pi_2^*\omega \xrightarrow{\sim} \pi_1^*\omega$, where $\pi_1, \pi_2 : T \to S$ are the composite maps of $T \to S \times_k S$ with the two projections $S \times_k S \to S$. It can be shown that this functor is representable by an affine, flat and surjective morphism $\Pi_\omega \to S \times_k S$. Moreover, the scheme Π_ω carries the structure of an *affine groupoid scheme* over k acting on S, a notion we now explain. The reader will notice the similarity with the definition of groupoid covers in Exercise 7 of Chapter 2.

Let Π be a k-scheme equipped with two k-morphisms $s, t : \Pi \to S$; these maps enable us to view Π as an $S \times_k S$-scheme. To turn Π into a k-groupoid scheme acting on S, three additional morphisms of $S \times_k S$-schemes are required: a multiplication map $m : \Pi \times_S \Pi \to \Pi$ (where the fibre product over S is with respect to the maps s and t), a unit map $e : S \to \Pi$ (where S is considered as an $S \times_k S$-scheme by means of the diagonal morphism), and an inverse map $\iota : \Pi \to \Pi$ which satisfies $t \circ \iota = s$, $s \circ \iota = t$. These are subject to the following conditions:

$$m(m \times \mathrm{id}) = m(\mathrm{id} \times m), \quad m(e \times \mathrm{id}) = m(\mathrm{id} \times e) = \mathrm{id},$$
$$m(\mathrm{id} \times \iota) = e \circ t, \quad m(\iota \times \mathrm{id}) = e \circ s.$$

Note that the multiplication m induces a product operation $(\pi, \pi') \mapsto \pi\pi'$ for morphisms $\pi, \pi' : T \to \Pi$ satisfying $s \circ \pi = t \circ \pi'$.

A *representation* of Π is a locally free sheaf $\mathcal{F}$ on S together with morphisms of $\mathcal{O}_T$-modules $\rho_\pi : (s \circ \pi)^*\mathcal{F} \to (t \circ \pi)^*\mathcal{F}$ for each k-scheme T and k-morphism $\pi : T \to \Pi$, such that the following hold: we have $\rho_{\pi\pi'} = \rho_\pi \circ \rho_{\pi'}$ whenever $s \circ \pi = t \circ \pi'$; the map ρ_π is the identity whenever $\pi = e \circ \psi$ for some $\psi : T \to S$; and finally all ρ_π are compatible with respect to morphisms $T' \to T$.

The main theorem is then the following: the affine $S \times_k S$-scheme Π_ω representing the functor $\mathbf{Aut}_k^{\otimes}(\omega)$ has the additional structure of a k-groupoid scheme, and the Tannakian category $(\mathcal{C}, \omega)$ is equivalent to the category of representations of Π_ω. The proof of this theorem occupies most of Deligne's paper [14]. When $S = \mathrm{Spec}\,(k)$, identifying $k \otimes_k k$ with k via the multiplication map $a \otimes b \mapsto ab$ turns Π_ω into an affine k-group scheme, and we recover Theorem 6.5.3.

We return to neutral Tannakian categories, and close this section by discussing an almost tautological example which will nevertheless have interesting consequences.

Example 6.5.14 Let Γ be an arbitrary group, and let k be a field. Denote by $\mathrm{Repf}_k(\Gamma)$ the category of finite dimensional representations of Γ over k. The usual tensor product and dual operations for representations equip $\mathrm{Repf}_k(\Gamma)$ with the structure of a k-linear rigid abelian tensor category. With the forgetful functor $\mathrm{Repf}_k(\Gamma) \to \mathrm{Vecf}_k$ as fibre functor we get a neutral Tannakian category. Its Tannakian fundamental group is called the *algebraic hull* (or *algebraic envelope*) of Γ over k; we denote it by Γ^{alg}.

A similar construction can be applied to the category of finite dimensional continuous real or complex linear representations of a topological group.

The algebraic hull Γ^{alg} is usually huge, even for $\Gamma = \mathbf{Z}$, so it is mainly of theoretical interest. However, we can understand small quotients of it rather well, as the following construction will show.

Quite generally, given a Tannakian category $(\mathcal{C}, \omega)$ and an object X of $\mathcal{C}$, we shall denote by $\langle X \rangle_\otimes$ the smallest full Tannakian subcategory of $\mathcal{C}$ containing X. Its objects are subquotients of finite direct sums of objects of the form $X^{\otimes r} \otimes (X^*)^{\otimes s}$, with some $r, s \geq 0$.

This being said, let us return to the category $\mathrm{Repf}_k(\Gamma)$ and fix an object V. After the choice of a basis it corresponds to a homomorphism $\rho_V : \Gamma \to \mathrm{GL}_n(k)$. Assume for simplicity that k is algebraically closed.

Proposition 6.5.15 *The Tannakian fundamental group of the full Tannakian subcategory $\langle V \rangle_\otimes$ of $\mathrm{Repf}_k(\Gamma)$ is canonically isomorphic to the Zariski closure of the image of ρ_V in $\mathrm{GL}_n(k)$.*

Here we view the closure $\overline{\mathrm{im}\,(\rho_V)}$ as an affine group scheme as in Example 6.1.4 (4). For the proof we need a standard lemma from the theory of linear algebraic groups.

Lemma 6.5.16 *Let K be a field, and G a closed subgroup scheme of GL_n over K. Denote by W the n-dimensional representation of G corresponding to its action on K^n via the inclusion $G \to \mathrm{GL}_n$. The full Tannakian subcategory $\langle W \rangle_\otimes \subset \mathrm{Rep}_G$ is the whole of Rep_G.*

We include a proof for the readers' convenience, taken from [111].

Proof Let M be an arbitrary object of Rep_G. We first reduce to the case where M is a finite dimensional K-subcomodule of the Hopf algebra A of G. It will be enough to embed M in a finite direct sum A^m of copies of A, because each subcomodule in A^m is the direct sum of its projections to the components. Now let $m = \dim_K(M)$, and consider $M \otimes_K A \cong A^m$. It is an A-comodule via the map $\mathrm{id}_M \otimes \Delta : M \otimes_K A \to M \otimes_K A \otimes_K A$. Now look at the two diagrams in Definition 6.1.6: the first one says that $\rho : M \to M \otimes_K A$ is a morphism of A-comodules, and the second one implies that ρ is injective.

The closed embedding $G \to \mathrm{GL}_n$ corresponds to a surjection of Hopf algebras $B \twoheadrightarrow A$, where $B \cong K[x_{11}, x_{12}, \ldots, x_{nn}, \det(x_{ij}^{-1})]$ by Example 6.1.4 (3). As every finite dimensional A-subcomodule of A is contained in the image of a finite dimensional B-subcomodule of B (for instance as a consequence of Proposition 6.1.8 (1)), we reduce to the case $B = A$, or in other words $G = \mathrm{GL}_n$.

Given integers $r, s \geq 0$, define a K-subspace

$$V_{r,s} := \{\det(x_{ij})^{-r} f \,|\, f \in K[x_{11}, x_{12}, \ldots, x_{nn}],\ \deg(f) \leq s\}$$

of B. Each finite dimensional subcomodule of B is contained in some $V_{r,s}$ for r, s large enough. Actually, $V_{r,s}$ is a finite dimensional B-subcomodule of B, because the comultiplication on B is given by $x_{ij} \mapsto x_{ik} \otimes x_{kj}$. Now W is just the standard representation of GL_n on K^n, so its comodule structure is given by $v_j \mapsto \sum v_k \otimes x_{kj}$, where the v_j are the standard basis elements of K^n. Thus for each i the map $v_k \mapsto x_{ik}$ induces a morphism of B-comodules $W \to B$; it is actually an embedding. Taking the direct sum over all i we obtain an isomorphism of W^n with the K-subspace $V_1 \subset B$ spanned by the x_{ij}; in particular $V_1 \subset \langle W \rangle_\otimes$. As the subspace V_s of homogeneous polynomials of degree s is the symmetric product of s copies of V_1 and hence a quotient of $V_1^{\otimes s}$, we conclude $V_s \subset \langle W \rangle_\otimes$. The one-dimensional subspace $\langle \det(x_{ij}) \rangle$ lies in V_n, so the tensor powers of its dual $\langle \det(x_{ij})^{-1} \rangle$ also lie in $\langle W \rangle_\otimes$. As by definition $V_{r,s}$ is the tensor product of $\langle \det(x_{ij}^{-1}) \rangle^{\otimes r}$ with the direct sum of the V_i for $i \leq s$, we conclude that $V_{r,s} \subset \langle W \rangle_\otimes$, which suffices to conclude. □

Proof of Proposition 6.5.15 Denote by G the affine group scheme over k defined by $\overline{\mathrm{im}\,(\rho_V)}$. The representation ρ_V gives rise to a representation of the affine group scheme G in a natural way; we denote it by $\overline{V}$. Similarly, each object of $\langle V \rangle_\otimes$ gives rise to an object of Rep_G, and we obtain an equivalence between $\langle V \rangle_\otimes$ and the full Tannakian subcategory $\langle \overline{V} \rangle_\otimes$ of Rep_G. In particular, the two categories have isomorphic Tannakian fundamental groups. But $\langle \overline{V} \rangle_\otimes$ is the whole of Rep_G by the lemma, whence the proposition. □

Remark 6.5.17 As $\mathrm{Repf}_k(\Gamma)$ is the direct limit of the full subcategories $\langle V \rangle_\otimes$, the proposition implies the following description of Γ^{alg}. Consider pairs (G, ϕ), where G is a closed subgroup of $\mathrm{GL}_n(k)$ for some n, and $\phi : \Gamma \to G$ is a morphism with dense image. The system of the (G, ϕ) carries a natural partial order where $(G, \phi) \geq (G', \phi')$ if there is a morphism $\lambda : G \to G'$ of algebraic groups with $\lambda \circ \phi = \phi'$. Notice that if λ exists, it is unique by our assumption that ϕ has dense image. Thus the (G, ϕ) together with the λ form a natural inverse system whose inverse limit is precisely Γ^{alg}.

6.6 Differential Galois groups

We now turn to more interesting examples of Tannakian categories. We begin by revisiting some constructions of Chapter 2 in the Tannakian context.

Example 6.6.1 Consider a connected and locally simply connected topological space X, and fix a base point $x \in X$. The category LS_X of complex local systems on X is a $\mathbf{C}$-linear abelian category, and the customary tensor product and dual space constructions of linear algebra induce a commutative and rigid tensor structure on it. Taking the stalk of a local system at x yields a fibre functor with values in $\mathrm{Vecf}_{\mathbf{C}}$. Hence LS_X is a neutral Tannakian category.

In Corollary 2.6.2 we established an equivalence between LS_X and the category $\mathrm{Repf}_{\pi_1(X,x)}$ of finite dimensional left representations of the fundamental group $\pi_1(X, x)$. This is moreover an equivalence of neutral Tannakian categories if we consider $\mathrm{Repf}_{\pi_1(X,x)}$ with its Tannakian structure described in Example 6.5.14. Consequently, the Tannakian fundamental group of LS_X is isomorphic to the algebraic hull of $\pi_1(X, x)$ over $\mathbf{C}$. Moreover, by Proposition 6.5.15 the Tannakian fundamental group of each full Tannakian subcategory $\langle \mathcal{L} \rangle_\otimes$ of LS_X is isomorphic to the Zariski closure of the image of $\rho_{\mathcal{L},x} : \pi_1(X, x) \to \mathrm{GL}_n(\mathbf{C})$, where $\rho_{\mathcal{L},x}$ is the monodromy representation of $\mathcal{L}$ at the point x.

In the next example we exploit the Riemann–Hilbert correspondence of Proposition 2.7.5.

Example 6.6.2 Consider now a connected open subset $D \subset \mathbf{C}$, and let x be a point of D. We introduce a neutral Tannakian structure on the category Conn_D of holomorphic connections on D as follows. The tensor product of two connections $(\mathcal{E}_1, \nabla_1)$ and $(\mathcal{E}_2, \nabla_2)$ is to be $(\mathcal{E}_1 \otimes_{\mathcal{O}} \mathcal{E}_2, \nabla_1 \otimes \nabla_2)$, with the connection $\nabla_1 \otimes \nabla_2 : \mathcal{E}_1 \otimes_{\mathcal{O}} \mathcal{E}_2 \to \mathcal{E}_1 \otimes_{\mathcal{O}} \mathcal{E}_2 \otimes_{\mathcal{O}} \Omega^1_D$ given by

$$(\nabla_1 \otimes \nabla_2)(s_1 \otimes s_2) = \nabla_1(s_1) \otimes s_2 + s_1 \otimes \nabla_2(s_2).$$

The dual connection to $(\mathcal{E}, \nabla)$ is $(\mathcal{E}^*, \nabla^*)$, the locally free sheaf $\mathcal{E}^*$ being given by $U \mapsto \mathrm{Hom}(\mathcal{E}|_U, \mathcal{O}|_U)$ and ∇^* over $\phi \in \mathrm{Hom}(\mathcal{E}|_U, \mathcal{O}|_U)$ by

$$\nabla^*(\phi)(s) = 1 \otimes d\phi(s) - \left(\phi \otimes \mathrm{id}_{\Omega^1_D}\right)(\nabla(s)).$$

Finally, the fibre functor is given by $(\mathcal{E}, \nabla) \mapsto \mathcal{E}^\nabla_x$. One checks that the category equivalence $(\mathcal{E}, \nabla) \mapsto \mathcal{E}^\nabla$ of Proposition 2.7.5 between Conn_D and LS_D is compatible with the tensor structures and fibre functors, so the Tannakian fundamental group of Conn_D with respect to the above fibre functor is again the algebraic hull of $\pi_1(D, x)$ over $\mathbf{C}$.

The above construction generalizes to arbitrary complex manifolds if one considers flat connections as in Example 2.7.6.

Notice a subtlety in the above argument: it is a priori not obvious at all that Conn_D is a $\mathbf{C}$-linear abelian category, which is part of the definition of Tannakian categories. We deduced this fact from its equivalence with LS_D.

The next example is, in some sense, a differential analogue of Theorem 3.3.7. We relate holomorphic connections on D to algebraically defined objects over $\mathbf{C}(z)$, the field of meromorphic functions D.

Example 6.6.3 Consider the field $\mathbf{C}(t)$ equipped with its usual derivation $f \mapsto f'$. A *differential module* over $\mathbf{C}(t)$ is a finite dimensional $\mathbf{C}(t)$-vector space V together with a $\mathbf{C}$-linear map $\nabla : V \to V$ satisfying

$$\nabla(fv) = f'v + f\nabla(v) \tag{6.16}$$

for all $f \in \mathbf{C}(t)$ and $v \in V$. Morphisms of differential modules are to be morphisms of $\mathbf{C}(t)$-vector spaces compatible with ∇. This defines a category which is in fact abelian: a submodule of (V, ∇) is a $\mathbf{C}(t)$-subspace $W \subset V$ stable by ∇, and for such a W we obtain an induced differential module structure on the quotient space V/W.

Next we introduce tensor products and duals for differential modules in a similar way as above. Given (V_1, ∇_1) and (V_2, ∇_2), we define $\nabla_1 \otimes \nabla_2$ on $V_1 \otimes V_2$ by

$$(\nabla_1 \otimes \nabla_2)(v_1 \otimes v_2) = \nabla_1(v_1) \otimes v_2 + v_1 \otimes \nabla_2(v_2), \tag{6.17}$$

and ∇^* on $V^* = \mathrm{Hom}_{\mathbf{C}(t)}(V, \mathbf{C}(t))$ by

$$\nabla^*(\phi)(v) = (\phi(v))' - \phi(\nabla(v)). \tag{6.18}$$

In this way we obtain a $\mathbf{C}$-linear rigid tensor abelian category $\mathrm{Diffmod}_{\mathbf{C}(t)}$.

If we identify an n-dimensional $\mathbf{C}(t)$-vector space V with $\mathbf{C}(t)^n$, we can define a differential module (V, d) by $d(f_1, \dots, f_n) = (f_1', \dots, f_n')$. Given another differential module structure (V, ∇) on V, the difference $\nabla - d$ is $\mathbf{C}(t)$-linear by (6.16), so it is defined by an $n \times n$ matrix $[f_{ij}]$ of functions in $\mathbf{C}(t)$ called the *connection matrix* of ∇. Setting $A = -[f_{ij}]$ we see that ∇ corresponds to the system of linear differential equations $y' = Ay$. As the reader has noticed, all this is completely parallel to the discussion in Example 2.7.3.

Given an open subset $D \subset \mathbf{C}$ whose complement is finite, we know from Proposition 2.7.7 that every holomorphic connection $(\mathcal{E}, \nabla)$ on D extends to a connection on $\mathbf{P}^1(\mathbf{C})$ with simple poles outside D. In particular, the entries of its connection matrix are meromorphic functions on $\mathbf{P}^1(\mathbf{C})$, i.e. elements of $\mathbf{C}(t)$. We can thus associate to $(\mathcal{E}, \nabla)$ the differential module (V, ∇) over $\mathbf{C}(t)$ with the same connection matrix. This induces an equivalence of the full subcategory $\langle(\mathcal{E}, \nabla)\rangle_\otimes$ of Conn_D with the full subcategory $\langle(V, \nabla)\rangle_\otimes$ of

$\text{Diffmod}_{\mathbf{C}(t)}$. A fibre functor on the former category induces one on the latter, so $\langle (V, \nabla)\rangle_{\otimes}$ is a neutral Tannakian category. By Proposition 6.5.15 and the two previous examples, its Tannakian fundamental group is isomorphic to the Zariski closure of the image of the monodromy representation of the local system $\mathcal{E}^{\nabla}$.

A shortcoming of the above example was that we introduced the fibre functor on $\langle (V, \nabla)\rangle_{\otimes}$ in a transcendental way, by comparison with the holomorphic theory. It is natural to ask whether this is possible using a direct construction. After all, the fibre functor is given by local horizontal sections of the connection, i.e. solutions of the corresponding differential equation. These are not elements of $\mathbf{C}(z)$ any more, but typically exponential functions. By adjoining the solution functions to $\mathbf{C}(z)$ we obtain a large transcendental extension which, however, can be studied in a purely algebraic way. This is the purpose of differential Galois theory, a vast subject of which we now explain the rudiments.

Let K be a field of characteristic 0. A *derivation* on K is an additive function $\partial : f \mapsto f'$ satisfying the Leibniz rule $(fg)' = f'g + fg'$. The pair (K, ∂) is called a *differential field*. The kernel $\ker(\partial)$ is a subfield of K; we shall call it the *constant field* of (K, ∂) and denote it by k. We define a *differential module* over (K, ∂) by direct generalization of the above special case: it is a pair (V, ∇) consisting of a finite dimensional K-vector space V and a k-linear map $\nabla : V \to V$ satisfying (6.16). The subset $V^{\nabla} := \ker(\nabla)$ is a k-subspace of V; its elements are called *horizontal vectors.* As above, the category Diffmod_K of differential modules over K is a rigid k-linear abelian tensor category, the tensor structure being defined by (6.17) and (6.18). Given an extension of differential fields $(L, \partial) \supset (K, \partial)$ (i.e. a field extension $L|K$ together with a derivation extending ∂) and a differential module (V, ∇) over K, there is a natural notion of base change $(V \otimes_K L, \nabla_L)$, where ∇_L has the same connection matrix as ∇ in a basis of V.

We can now give the formal definition of the extension generated by the local solutions.

Definition 6.6.4 *Let (V, ∇) be a differential module over K. A Picard–Vessiot extension for (V, ∇) is a differential field extension $(L, \partial) \supset (K, \partial)$ satisfying the following properties:*

1. *(L, ∂) has the same constant field k as K.*
2. *$V \otimes_K L$ is generated as an L-vector space by the horizontal vectors of ∇_L.*
3. *The coordinates of the horizontal vectors of ∇_L in any L-basis of $V \otimes_K L$ coming from a K-basis of V generate the field extension $L|K$.*

In more down-to-earth terms, property (3) says that L is generated over K by a system of local solutions of the system of differential equations corresponding to (V, ∇), and (2) ensures that the functions in the system are linearly independent over k.

Now observe that if L satisfies property (2) of the above definition for (V, ∇), then so does each object of the full subcategory $\langle(V, \nabla)\rangle_\otimes$ by our definition of the abelian and tensor structure on the category $\mathrm{Diffmod}_K$. Therefore such an L defines a fibre functor ω_L on $\langle(V, \nabla)\rangle_\otimes$ with values in Vecf_k via

$$\omega_L : (W, \nabla) \mapsto (W \otimes_K L)^{\nabla_L}.$$

Remark 6.6.5 In the last section of [14] Deligne and Bertrand show that Picard–Vessiot extensions for (V, ∇) correspond *bijectively* to neutral fibre functors on $\langle(V, \nabla)\rangle_\otimes$. The proof is not very hard, and uses a construction akin to Lemma 6.7.17 below.

On the other hand, Corollary 6.20 of the same paper implies the much deeper fact that over an algebraically closed field k a neutral fibre functor always exists on a rigid abelian k-linear tensor category of the form $\langle X\rangle_\otimes$ that has some (non-neutral) fibre functor into a category Vecf_K for an extension $K \supset k$. As the forgetful functor on $\langle(V, \nabla)\rangle_\otimes$ is such a fibre functor, we obtain that Picard–Vessiot extensions always exist for differential modules over differential fields with algebraically closed constant field (they may not exist in general). In [107], Propositions 1.20 and 1.22 a simple direct proof is given for this fundamental fact. There it is also proven that if a Picard–Vessiot extension exists, it is unique up to an isomorphism of differential fields.

Definition 6.6.6 Assume that there exists a Picard–Vessiot extension (L, ∂) for (V, ∇). The *differential Galois group scheme* $\mathrm{Gal}\,(V, \nabla)$ of (V, ∇) is defined as the Tannakian fundamental group of the category $\langle(V, \nabla)\rangle_\otimes$ equipped with the above fibre functor ω_L.

By definition, $\mathrm{Gal}\,(V, \nabla)$ is an affine group scheme over k. A priori it depends on the Picard–Vessiot extension L, but the last sentence of the above remark shows that its isomorphism class does not.

In the classical literature the differential Galois group is defined as the group of automorphisms of the field extension $L|K$ preserving ∂. We now make the link with the above definition. The first step is to turn the automorphism group of $(L, \partial) \supset (K, \partial)$ into an affine k-group scheme.

Given a k-algebra R, extend the derivation ∂ on L to $L \otimes_k R$ by the rule $f \otimes r \mapsto f' \otimes r$. Define a group-valued functor $\mathrm{Gal}_\partial(L|K)$ on the category of k-algebras by sending R to the group of $K \otimes_k R$-algebra automorphisms of $L \otimes_k R$ commuting with ∂. Given a morphism $\phi : R_1 \to R_2$, we define

$\mathrm{Gal}_{\partial}(L|K)(\phi)$ by sending an automorphism of $L \otimes_k R_1$ to the automorphism of $L \otimes_k R_2$ induced by base change via ϕ.

Lemma 6.6.7 *The functor* $\mathrm{Gal}_{\partial}(L|K)$ *defines an affine group scheme over* k.

Proof We have to show that $\mathrm{Gal}_{\partial}(L|K)$ is representable by a finitely generated k-algebra A. The proof will actually realize A as a quotient of the coordinate ring $B = k[x_{11}, \ldots, x_{nn}, \det(x_{ij})^{-1}]$ of GL_n, where $n = [L : K]$. Fixing a K-basis of L as in defining property (3) of Picard–Vessiot extensions, we can write the coordinates of a k-basis of horizontal vectors in a nonsingular $n \times n$ matrix $[f_{ij}]$. An automorphism $\sigma \in \mathrm{Gal}_{\partial}(L|K)(k)$ multiplies this matrix by a matrix $M_\sigma \in \mathrm{GL}_n(k)$; as the f_{ij} generate L over K, this determines σ. Similarly, elements $\sigma_R \in \mathrm{Gal}_{\partial}(L|K)(R)$ for a k-algebra R correspond to matrices $M_{\sigma_R} \in \mathrm{GL}_n(R)$.

There is a natural K-algebra surjection $B \otimes_k K \to L$ induced by $x_{ij} \mapsto f_{ij}$, compatible with the natural derivations on B and L. Its kernel is a maximal ideal $P \subset B \otimes_k K$. Each $\sigma_R : L \otimes_k R \to L \otimes_k R$ lifts to a K-algebra map $\widetilde{\sigma}_R : B \otimes_k K \otimes_k R \to L \otimes_k R$ with $\widetilde{\sigma}_R(P_R) = 0$, where P_R is the ideal generated by P in $B \otimes_k K \otimes_k R$. Choosing an (infinite) k-basis $\{e_\lambda : \lambda \in \Lambda\}$ of L, we may write $\widetilde{\sigma}_R(b) = \sum \alpha_\lambda^{\widetilde{\sigma}_R}(b) e_\lambda$ for all $b \in B \otimes_k K \otimes_k R$, where the sum is finite and the coefficients are in R. The condition is therefore that the $\alpha_\lambda^{\widetilde{\sigma}_R}(b)$ must vanish for $b \in P_R$; it suffices to verify this for a system $p_1, \ldots, p_m$ of generators of P.

Now choose $R = B$, and let σ_B be the 'universal' automorphism corresponding to $M_{\sigma_B} = [x_{ij}]$. Write A for the quotient of B by the ideal generated by the $\alpha_\lambda^{\widetilde{\sigma}_B}(p_l)$ for $1 \leq l \leq m$ and $\lambda \in \Lambda$. Given a k-algebra R and an element $\sigma_R \in \mathrm{Gal}_{\partial}(L|K)(R)$ corresponding to a matrix $M_{\sigma_R} \in \mathrm{GL}_n(R)$, define a map $B \to R$ by sending x_{ij} to the (i, j)-term of M_{σ_R}. As σ_R comes from a $\widetilde{\sigma}_R$ annihilating P_R, our map factors through A. This shows that A represents $\mathrm{Gal}_{\partial}(L|K)$. □

Now assume given a differential module (V, ∇) over K, and let L be a Picard–Vessiot extension for (V, ∇). Observe that if $\sigma : L \to L$ is a K-automorphism of L commuting with ∂, then the induced automorphism of $V \otimes_K L$ preserves the space $(V \otimes_K L)^{\nabla_L}$ of horizontal vectors. This holds for each object of $\langle (V, \nabla) \rangle_\otimes$, so the fibre functor ω_L takes its values not only in the category of finite dimensional k-vector spaces, but actually in the category of finite dimensional representations of $\mathrm{Gal}_{\partial}(L|K)(k)$. For each object (W, ∇) of $\langle (V, \nabla) \rangle_\otimes$ we have a functor on the category of k-algebras given by $R \mapsto (W \otimes_K L \otimes_K R)^{\nabla_{L \otimes_K R}}$, where the base change $\nabla_{L \otimes_K R}$ is defined in a straightforward way. This shows that ω_L induces a functor from $\langle (V, \nabla) \rangle_\otimes$

to the category $\mathrm{Rep}_{\mathrm{Gal}_\partial(L|K)}$ of representations of the affine group scheme $\mathrm{Gal}_\partial(L|K)$. It is moreover a tensor functor as ω_L is.

Proposition 6.6.8 *If k is algebraically closed, the above tensor functor induces an equivalence of neutral Tannakian categories between $\langle(V,\nabla)\rangle_\otimes$ and $\mathrm{Rep}_{\mathrm{Gal}_\partial(L|K)}$. Consequently, the affine group schemes $\mathrm{Gal}(V,\nabla)$ and $\mathrm{Gal}_\partial(L|K)$ are isomorphic.*

We give a proof following [107]. One additional algebraic fact is needed which does not result from the Tannakian theory.

Fact 6.6.9 Let (K,∂) be a differential field with algebraically closed constant field k, and (L,∂) a Picard–Vessiot extension for the differential module (V,∇) over K. Then the fixed field for the action of $\mathrm{Gal}_\partial(L|K)(k)$ on L is exactly K. Thus Picard–Vessiot extensions play the role of Galois extensions in the differential theory.

For proofs, see [107], Theorem 1.27 (3) or [39], Theorem 5.7.

Proof of Proposition 6.6.8 For fully faithfulness, assume given two objects (V_1,∇) and (V_2,∇) of $\langle(V,\nabla)\rangle_\otimes$. By the definition of duals in a rigid tensor category we have $\mathrm{Hom}((V_1,\nabla),(V_2,\nabla))\cong\mathrm{Hom}(K,(V_1^*\otimes_K V_2,\nabla))$, and similarly for the corresponding representations of $\mathrm{Gal}_\partial(L|K)$, so we may assume that (V_1,∇) is K equipped with the trivial connection. Elements of $\mathrm{Hom}(K,(V_2,\nabla))$ correspond bijectively to horizontal elements of V_2. As k is algebraically closed, elements in the image of $\mathrm{Hom}(K,(V_2,\nabla))$ by ω_L correspond to elements of $(V_2\otimes_K L)^{\nabla_L}$ invariant by $\mathrm{Gal}_\partial(L|K)(k)$. But by the above fact the $\mathrm{Gal}_\partial(L|K)(k)$-invariant elements in $V_2\otimes_k L$ are exactly the elements of V_2, so we again obtain the horizontal elements of V_2.

For essential surjectivity, note that by property (3) of Picard–Vessiot extensions the functor ω_L maps (V,∇) to a faithful representation of $\mathrm{Gal}_\partial(L|K)$, or in other words the induced map $\mathrm{Gal}_\partial(L|K)\to\mathrm{GL}_n(k)$ is injective. We then conclude by Lemma 6.5.16. □

When k is algebraically closed, the group $\mathrm{Gal}_\partial(L|K)(k)$ of k-points of the affine group scheme $\mathrm{Gal}(M,\nabla)\cong\mathrm{Gal}_\partial(L|K)$ is called the *differential Galois group* of (M,∇). In this case the above considerations imply an analogue of the main theorem of Galois theory (Theorem 1.2.5) in the differential context.

Theorem 6.6.10 *Let (K,∂) be a differential field with algebraically closed constant field k, and (L,∂) a Picard–Vessiot extension for the differential module (V,∇) over K. The rule $H(k)\mapsto L^{H(k)}$ induces a bijection between closed subgroups of $\mathrm{Gal}_\partial(L|K)(k)$ and differential subfields of (L,∂) containing (K,∂). Here closed normal subgroups correspond to Picard–Vessiot extensions of (K,∂), and the associated differential Galois group is $(\mathrm{Gal}_\partial(L|K)/H)(k)$.*

Proof If (M, ∂) is an intermediate differential field between (K, ∂) and (L, ∂), then (L, ∂) is a Picard–Vessiot extension for the base change (V_M, ∇_M). This identifies $\mathrm{Gal}_\partial(L|M)(k)$ with a closed subgroup $H(k)$ of $\mathrm{Gal}_\partial(L|K)(k)$ (as it is closed in $\mathrm{GL}_n(k)$ with the same GL_n as for $\mathrm{Gal}_\partial(L|K)(k)$), and by Fact 6.6.9 we have $L^{H(k)} = M$. Similarly, a closed subgroup $H(k) \subset \mathrm{Gal}_\partial(L|K)(k)$ fixes some M with $\mathrm{Gal}(V_M, \nabla_M)(k) \cong H(k)$.

Given a closed normal subgroup scheme $H \subset \mathrm{Gal}_\partial(L|K)$, fix a representation ρ in $\mathrm{Rep}_{\mathrm{Gal}_\partial(L|K)}$ with kernel H. (Such a representation exists; see e.g. [111], §16.1.) The full subcategory $\langle\rho\rangle_\otimes$ is equivalent to $\mathrm{Rep}_{\mathrm{Gal}_\partial(L|K)/H}$ by Lemma 6.5.16 on the one hand, and to $\langle(W, \nabla)\rangle_\otimes$ for some (W, ∇) in $\langle(V, \nabla)\rangle_\otimes$ on the other hand. As we have already noted, (L, ∂) satisfies defining properties (1) and (2) of Picard–Vessiot extensions for (W, ∇), therefore by defining property (3) it contains a Picard–Vessiot extension (M, ∂) for (W, ∇). By construction we have $\mathrm{Gal}(W, \nabla) \cong \mathrm{Gal}(V, \nabla)/H$, and $\mathrm{Gal}_\partial(L|M)(k) \cong H(k)$ by the first part of the proof. □

6.7 Nori's fundamental group scheme

In this final section we draw the ties closer between the Tannakian theory and Grothendieck's algebraic fundamental group discussed in the previous chapter. Recall that the fundamental group of a scheme S is a profinite group whose finite quotients correspond to finite étale Galois covers $X \to S$. We observed in Proposition 5.3.16 (1) that the Galois cover $X \to S$ can be viewed as a torsor under a finite constant group scheme. In his influential paper [72], Nori observed that by means of the theory of Tannakian categories one may construct a fundamental group classifying torsors under arbitrary finite group schemes in the case when S is a proper integral scheme over a field k and has a k-rational point. Nori's fundamental group is actually an affine group scheme which is a projective limit of finite group schemes. When k is algebraically closed of characteristic 0, there is a comparison theorem with Grothendieck's fundamental group, but in general Nori's group is larger.

From now on we again consider group schemes as schemes and not as functors. The starting observation is the following. Assume k is a field, and $S \to \mathrm{Spec}(k)$ a scheme over k. Given a finite group scheme G over k, by a (left) G-torsor over S we mean a (left) torsor over S with structure group $G_S = G \times_{\mathrm{Spec}(k)} S$ (note that G_S is a finite flat group scheme over S by construction).

Lemma 6.7.1 *Let $\phi : X \to S$ be a G-torsor over S. The locally free sheaf $\mathcal{E}_X := \phi_*\mathcal{O}_X$ satisfies $\mathcal{E}_X^{\otimes 2} \cong \mathcal{E}_X^{\oplus n}$, where n is the order of the group $G(\bar{k})$ for an algebraic closure $\bar{k}$ of k.*

Proof Consider the isomorphism $G_S \times_S X \cong X \times_S X$ given by the definition of torsors. The $\mathcal{O}_S$-algebra corresponding via Remark 5.1.24 to the left-hand side is isomorphic *as an* $\mathcal{O}_S$*-module* to $\mathcal{E}_X^{\oplus n}$ because $G_S \times_S X \cong G \times_k X$ and G is a finite k-group scheme. The right-hand side corresponds to $\mathcal{E}_X^{\otimes 2}$, whence an isomorphism $\mathcal{E}_X^{\otimes 2} \cong \mathcal{E}_X^{\oplus n}$ as required. □

The property of $\mathcal{E}_X$ in the lemma is a special case of a general definition introduced by Weil in [112]. Given a scheme S, an $\mathcal{O}_S$-module $\mathcal{F}$ and a polynomial $f = a_n x^n + a_{n-1}x^{n-1} + \cdots + a_0$ with non-negative integer coefficients, we define

$$f(\mathcal{F}) := \bigoplus_{i=0}^{n} (\mathcal{F}^{\otimes i})^{\oplus a_i}.$$

By convention, here $\mathcal{F}^{\otimes 0} := \mathcal{O}_S$ and $\mathcal{F}^{\oplus 0} = 0$.

Definition 6.7.2 A locally free sheaf $\mathcal{E}$ is called *finite* if there exist polynomials $f \neq g$ with non-negative integer coefficients such that $f(\mathcal{E}) \cong g(\mathcal{E})$.

Here and below the term 'locally free sheaf' will always mean 'locally free sheaf of finite rank'.

To obtain another characterization of finite locally free sheaves, call a locally free sheaf indecomposable if it is not isomorphic to a direct sum of nonzero locally free sheaves. For a general locally free sheaf $\mathcal{E}$ write $I(\mathcal{E})$ for the set of isomorphism classes of indecomposable locally free sheaves $\mathcal{E}'$ for which there exists a locally free $\mathcal{E}''$ with $\mathcal{E} \cong \mathcal{E}' \oplus \mathcal{E}''$.

Lemma 6.7.3 *If S is proper over k, the set $I(\mathcal{E})$ is finite.*

Proof The k-algebra End $(\mathcal{E}) =$ **End** $(\mathcal{E})(S)$ is finite dimensional because S is proper, so $\mathcal{E}$ as a coherent sheaf has only finitely many direct summands up to isomorphism. This is a consequence of the Krull–Remak–Schmidt theorem for coherent sheaves which is proven in the same way as the classical theorem for modules (see [48], Chapter X, Theorem 7.5). Now if $\mathcal{E}$ has only finitely many nonisomorphic indecomposable direct summands as a coherent sheaf, then it has the corresponding property as a locally free sheaf. □

Denote by $S(\mathcal{E})$ the union of the finite sets $I(\mathcal{E}^{\otimes i})$ for all $i > 0$.

Proposition 6.7.4 *Assume S is proper over k. A locally free sheaf $\mathcal{E}$ is finite if and only if $S(\mathcal{E})$ is a finite set.*

Proof The 'if' part is proven using the well-known 'determinant trick'. Write L for the free abelian group generated by isomorphism classes of indecomposable locally free sheaves on S. Those elements in L that have non-negative coefficients correspond to isomorphism classes of locally free sheaves, and L carries a structure of a commutative ring induced by tensor product. Consider the free additive subgroup $L_{\mathcal{E}} \subset L$ with basis $S(\mathcal{E})$. It is preserved by the $\mathbf{Z}$-linear map $m_{\mathcal{E}}$ given by multiplication with the class of $\mathcal{E}$. The characteristic polynomial χ is a monic polynomial in $\mathbf{Z}[x]$ with $\chi(m_{\mathcal{E}}) = 0$. Writing $\chi = f - g$, where f and g have non-negative coefficients, we obtain polynomials with $f(\mathcal{E}) = g(\mathcal{E})$.

Conversely, if there exist such f and g, then plugging the class of $\mathcal{E}$ in L into $f - g \in \mathbf{Z}[x]$ we obtain 0. Writing d for the degree of $f - g$, this implies that each indecomposable direct summand of $\mathcal{E}^{\otimes d}$ must be a direct summand of some $\mathcal{E}^{\otimes j}$ for $j < d$. Applying the above to the polynomials $(f - g)x^i$ for $i > 0$ and using induction on i, we obtain that each element of $S(\mathcal{E})$ may be represented by a direct summand of some $\mathcal{E}^{\otimes j}$ for $j < d$. We then conclude by the previous lemma. □

Corollary 6.7.5 *An invertible sheaf $\mathcal{L}$ is finite if and only if $\mathcal{L}^{\otimes m} \cong \mathcal{O}_S$ for some $m > 0$.*

Proof This follows from the proposition because the tensor powers $\mathcal{L}^{\otimes i}$ are invertible for all $i > 0$, and invertible sheaves are indecomposable. □

Corollary 6.7.6 *The category of finite sheaves is stable under taking direct sums, direct summands, tensor products of two sheaves and duals.*

Here recall that the dual $\mathcal{E}^\vee$ of a locally free sheaf $\mathcal{E}$ on S is the sheaf $U \mapsto \mathrm{Hom}_{\mathcal{O}_U}(\mathcal{E}|_U, \mathcal{O}_S|_U)$, where the Hom-group means homomorphisms of $\mathcal{O}_U$-modules. It is again a locally free sheaf of the same rank. The easy proof of the corollary is left to the readers.

We now see that the category of finite locally free sheaves on a proper S is a rigid commutative tensor category with unit $\mathcal{O}_S$. If moreover S is equipped with a morphism $S \to \mathrm{Spec}\,(k)$ for a field k and has a k-rational point s : $\mathrm{Spec}\,(k) \to S$, the rule $\mathcal{E} \mapsto s^*\mathcal{E}$ yields a faithful tensor functor to the category Vecf_k. (Recall that if $\mathrm{Spec}\,(A) \subset S$ is an affine open subset containing the image of s and $\mathcal{E} = \widetilde{E}$ for an A-module E, the pullback $s^*\mathcal{E}$ is just given by the k-vector space $E \otimes_A k$ on the one-point space $\mathrm{Spec}\,(k)$, where the map $A \to k$ corresponds to s.) Our category thus satisfies all requirements for a neutral Tannakian category except that we don't know yet whether it is abelian. It will turn out to be abelian when k is of characteristic 0 and S is proper and integral, but this is not obvious to prove. Nori's idea was to use the theory of semi-stable sheaves, so we need first to recall some facts about these.

Facts 6.7.7 Let S be a scheme, and $\mathcal{E}$ a locally free sheaf on S. Consider the dual sheaf $\mathcal{E}^\vee$ and the $\mathcal{O}_S$-algebra $S(\mathcal{E}^\vee) := \bigoplus_{i\geq 0} (\mathcal{E}^\vee)^{\otimes i}$. Let $\mathbf{V}(\mathcal{E}) \to S$ be the affine morphism corresponding to $S(\mathcal{E}^\vee)$ via Remark 5.1.24; it is called the *vector bundle* associated with $\mathcal{E}$. For an open subset $U \subset S$ where $\mathcal{E}|_U$ is a free sheaf of rank n we have $\mathbf{V}(\mathcal{E}) \times_S U \cong U \times \mathbf{A}^n$; this is because the tensor algebra of a free module is a polynomial ring. We have worked with the dual sheaf $\mathcal{E}^\vee$ in order to obtain a covariant functor $\mathcal{E} \mapsto \mathbf{V}(\mathcal{E})$. It is also possible to define a vector bundle as an affine morphism satisfying such a local triviality condition; see e.g. [34], Exercise II.5.18. In the literature locally free sheaves are often called vector bundles.

A locally free subsheaf $\mathcal{E}' \subset \mathcal{E}$ is called a *subbundle* if the corresponding morphism $\mathbf{V}(\mathcal{E}') \to \mathbf{V}(\mathcal{E})$ is a closed immersion. In this case $\mathcal{E}'$ is a direct summand in $\mathcal{E}$ with locally free complement ([95], Chapter VI, §1.3, Proposition), so the quotient $\mathcal{E}/\mathcal{E}'$ is locally free; it is called a *quotient bundle*. A sufficient condition for the kernel and the image of a morphism $\phi : \mathcal{E} \to \mathcal{F}$ of locally free sheaves to be subbundles is that for all geometric points $\bar{s} : \mathrm{Spec}\,(\Omega) \to S$ the Ω-linear maps $\bar{s}^*\mathcal{E} \to \bar{s}^*\mathcal{F}$ have the same rank ([53], Proposition 1.7.2).

Assume now that C is an integral proper normal curve over a field, and $\mathcal{E}$ is a locally free sheaf of rank r on C. The determinant $\det(\mathcal{E})$ (which is by definition the r-th exterior power of $\mathcal{E}$) is an invertible sheaf, so by [34], Propositions II.6.11 and II.6.13 it corresponds to a divisor D on C. The degree d of D is called the *degree* of $\mathcal{E}$, and the quotient $\mu(\mathcal{E}) := d/r$ the *slope* of $\mathcal{E}$. For locally free sheaves $\mathcal{E}$ and $\mathcal{F}$ on C one has

$$\mu(\mathcal{E} \otimes_{\mathcal{O}_C} \mathcal{F}) = \mu(\mathcal{E}) + \mu(\mathcal{F}). \tag{6.19}$$

A quick way to prove this formula is to identify the degree of a locally free sheaf with the degree of its first Chern class ([27], Theorem 3.2 (f) and Remark 3.2.3 (c)), and then to use the multiplicativity of the Chern character ([27], Example 3.2.3). See also Exercise 9 for a direct argument.

Also, using the Riemann–Roch formula for vector bundles ([53], Theorems 2.6.3 and 2.6.9) one can calculate the degree of a locally free sheaf $\mathcal{E}$ of rank r as

$$\deg(\mathcal{E}) = \chi(\mathcal{E}) - r\chi(\mathcal{O}_C), \tag{6.20}$$

where $\chi(\mathcal{G})$ denotes the Euler–Poincaré characteristic of a coherent sheaf $\mathcal{G}$ ([53], Section 2.3).

One says that $\mathcal{E}$ is *semi-stable* if $\mu(\mathcal{E}') \leq \mu(\mathcal{E})$ for all nonzero subbundles $\mathcal{E}' \subset \mathcal{E}$. Equivalently (see [53], p. 74), the sheaf $\mathcal{E}$ is semi-stable if $\mu(\mathcal{E}'') \geq \mu(\mathcal{E})$ for all nonzero quotient bundles $\mathcal{E}''$ of $\mathcal{E}$. The kernel and the image of a morphism $\mathcal{E} \to \mathcal{F}$ of semi-stable sheaves of slope μ are subbundles, and the cokernel is a quotient bundle that is semi-stable of rank μ.

Consequently, the full subcategory $\mathcal{C}(\mu)$ of the category of locally free sheaves on C spanned by semi-stable sheaves of slope μ is an *abelian category*. For a proof of these facts, see [53], Proposition 5.3.6.

The relevance of semi-stability to our topic is shown by the following proposition.

Proposition 6.7.8 *A finite locally free sheaf $\mathcal{E}$ on an integral proper normal curve is semi-stable of slope 0.*

We first prove a lemma.

Lemma 6.7.9 *Let C be an integral proper normal curve.*

1: *For a locally free sheaf $\mathcal{F}$ on C the slopes of subbundles $\mathcal{F}' \subset \mathcal{F}$ are bounded from above.*
2. *If we denote by $\mu_{\max}(\mathcal{F})$ the maximum of the slopes of subbundles of $\mathcal{F}$, and $\mathcal{F}_1, \mathcal{F}_2$ are locally free sheaves on C, then*

$$\mu_{\max}(\mathcal{F}_1 \oplus \mathcal{F}_2) \leq \max(\mu_{\max}(\mathcal{F}_1), \mu_{\max}(\mathcal{F}_2)).$$

Proof Denoting by r the rank of $\mathcal{F}$, for (1) it suffices to show that the degrees of rank r' subbundles $\mathcal{F}' \subset \mathcal{F}$ are bounded for all $1 \leq r' \leq r$. But by the Riemann–Roch formula (6.20) the degree of such an $\mathcal{F}'$ equals $\chi(\mathcal{F}') - r'\chi(\mathcal{O}_C)$. As the Euler–Poincaré characteristic χ is additive for short exact sequences, we see that $\chi(\mathcal{F}') \leq \chi(\mathcal{F})$, whence statement (1).

For statement (2) consider a subbundle $\mathcal{G} \subset \mathcal{F}_1 \oplus \mathcal{F}_2$, and denote by $j : \operatorname{Spec}(K(C)) \to C$ the inclusion of the generic point of C. The pullback $j^*\mathcal{G}$ decomposes as a direct sum $j^*\mathcal{G} \cong \mathcal{G}_1' \oplus \mathcal{G}_2'$, where the $\mathcal{G}_i'$ are locally free sheaves on $\operatorname{Spec}(K(C))$ corresponding to subspaces of the vector spaces $j^*\mathcal{F}_i$. The $j_*(\mathcal{G}_i')$ are locally free subsheaves of the $\mathcal{F}_i$, hence there exist subbundles $j_*(\mathcal{G}_i') \subset \mathcal{G}_i \subset \mathcal{F}_i$ that have the same rank as the $j_*(\mathcal{G}_i')$ (see [53], p. 73). It then suffices to prove $\mu(\mathcal{G}_1 \oplus \mathcal{G}_2) \leq \max(\mu(\mathcal{G}_1), \mu(\mathcal{G}_2))$, which is an easy calculation. □

Proof of Proposition 6.7.8 We use the characterization of Proposition 6.7.4. Denote by μ the maximum of the numbers $\mu_{\max}(\mathcal{E}_i)$, where $\mathcal{E}_i$ runs over a set of representatives of $S(\mathcal{E})$. By part (2) of the above lemma and the definition of $S(\mathcal{E})$ we have $\mu(\mathcal{F}) \leq \mu$ for every locally free sheaf $\mathcal{F}$ that is a subbundle of some $\mathcal{E}^{\otimes i}$. For every $j > 0$ the tensor power $\mathcal{F}^{\otimes j}$ satisfies $\mu(\mathcal{F}^{\otimes j}) = j\mu(\mathcal{F})$ by (6.19). On the other hand, it is a subbundle of $\mathcal{E}^{\otimes ij}$, so $\mu(\mathcal{F}^{\otimes j}) \leq \mu$ must hold by what we have just proven. This is only possible if $\mu(\mathcal{F}) \leq 0$ for all $\mathcal{F}$; in particular, we have $\mu(\mathcal{E}) \leq 0$. But the dual sheaf $\mathcal{E}^\vee$ is again finite by Corollary 6.7.6, so $\mu(\mathcal{E}^\vee) \leq 0$ must hold as well. Since $\mu(\mathcal{E}^\vee) = -\mu(\mathcal{E})$, this shows $\mu(\mathcal{E}) = 0$, and therefore also $\mu(\mathcal{E}') \leq \mu(\mathcal{E})$ for all subbundles $\mathcal{E}' \subset \mathcal{E}$. □

The above considerations are sufficient for constructing a Tannakian category out of finite locally free sheaves when the base scheme is a proper normal curve. However, we first discuss how to generalize the theory to higher dimensional base schemes.

Let S be a proper integral scheme over a field k. Following Nori, we say that a locally free sheaf $\mathcal{E}$ on S is *semi-stable of slope 0* if for all integral closed subschemes $C \subset X$ of dimension 1 with normalization $\widetilde{C} \to C$ the pullback of $\mathcal{E}$ via the composite morphism $\widetilde{C} \to C \to X$ is a semi-stable sheaf of slope 0 on the proper normal curve $\widetilde{C}$.

Proposition 6.7.10 *Let S be a proper integral scheme over a field k.*

1. *The full subcategory of the category of locally free sheaves on S spanned by semi-stable sheaves of slope 0 is an abelian category.*
2. *Every finite locally free sheaf on S is semi-stable of slope 0.*

Proof To prove (1), consider a morphism $\phi : \mathcal{E} \to \mathcal{F}$ of semi-stable locally free sheaves of slope 0 on S. Given an integral closed subscheme $C \subset X$ of dimension 1 with normalization $\widetilde{C}$, the pullback $\phi_{\widetilde{C}}$ of ϕ to $\widetilde{C}$ is a morphism of semi-stable sheaves of slope 0, so by the theory of semi-stable sheaves on curves (last paragraph of Facts 6.7.7) the kernel and the image are subbundles, and both the kernel and the cokernel are semi-stable of slope 0. In particular, for every geometric point $\bar{s} : \mathrm{Spec}\,(\Omega) \to \widetilde{C}$ the Ω-linear maps $\bar{s}^*\mathcal{E} \to \bar{s}^*\mathcal{F}$ have the same rank. Now observe that after base change to an algebraic closure of k any two points of S lie on a closed connected subscheme of dimension 1. When S is projective, this can be proven by cutting S with a suitable linear subspace and applying a Bertini argument, and the general case follows via Chow's lemma ([95], Chapter VI, §2.1). We conclude that the maps $\bar{s}^*\mathcal{E} \to \bar{s}^*\mathcal{F}$ have the same rank for all geometric points of S, and therefore by the criterion mentioned above both the kernel and the image are subbundles. Therefore the kernel and the cokernel of ϕ are locally free sheaves; they are semi-stable of degree 0 because the $\phi_{\widetilde{C}}$ have this property. Finally, (2) follows from Proposition 6.7.8 because the pullback of a finite locally free sheaf to every curve $\widetilde{C} \to X$ as above is again a finite locally free sheaf: polynomial equalities are preserved by base change. □

Now define a locally free sheaf $\mathcal{E}$ on S to be *essentially finite* if it is semi-stable of slope 0, and moreover there is a finite locally free sheaf $\mathcal{F}$ on S and subbundles $\mathcal{F}' \subset \mathcal{F}'' \subset \mathcal{F}$ with $\mathcal{E} \cong \mathcal{F}''/\mathcal{F}'$. To put briefly, an essentially finite sheaf is a subquotient of a finite locally free sheaf.

Proposition 6.7.11 *Assume moreover that the scheme S has a k-rational point $s : \mathrm{Spec}(k) \to S$. Then the full subcategory* EF_S *of the category of locally free sheaves on S spanned by essentially finite sheaves, together with the usual tensor product of sheaves and the functor $\mathcal{E} \mapsto s^*\mathcal{E}$, is a neutral Tannakian category over k.*

Proof By definition, the category EF_S is abelian. To see that it is stable by tensor product, take two essentially finite sheaves $\mathcal{E}_1$ and $\mathcal{E}_2$ which are subquotients of the finite locally free sheaves $\mathcal{F}_1$ and $\mathcal{F}_2$, respectively. Writing $\mathcal{E}_i \cong \mathcal{F}_i''/\mathcal{F}_i'$ with subbundles $\mathcal{F}_i' \subset \mathcal{F}_i'' \subset \mathcal{F}_i$, we see that $\mathcal{E}_1 \otimes_{\mathcal{O}_S} \mathcal{E}_2$ is isomorphic to a quotient of $\mathcal{F}_1'' \otimes_{\mathcal{O}_S} \mathcal{F}_2''$. It is thus a subquotient of $\mathcal{F}_1 \otimes_{\mathcal{O}_S} \mathcal{F}_2$ which is finite by Corollary 6.7.6. Therefore it remains to see that it is semi-stable of slope 0. It will suffice to show that its pullbacks to each proper normal curve $\pi : \widetilde{C} \to X$ as in the definition of semi-stability have slope 0, for then we may use the semi-stability of the pullback of $\mathcal{F}_1'' \otimes_{\mathcal{O}_S} \mathcal{F}_2''$ and apply the definition of semi-stability over curves via quotient bundles. But the $\pi^*\mathcal{E}_i$ are semi-stable of slope 0 as they are quotients of the $\pi^*\mathcal{F}_i''$ by the $\pi^*\mathcal{F}_i'$ which are all semi-stable of slope 0 by assumption. Therefore $\pi^*(\mathcal{E}_1 \otimes_{\mathcal{O}_S} \mathcal{E}_2) \cong \pi^*\mathcal{E}_1 \otimes_{\widetilde{C}} \pi^*\mathcal{E}_2$ has slope 0 by formula (6.19).

To show that the tensor structure on EF_S is rigid, observe that the dual of an essentially finite locally free sheaf is again essentally finite by a similar argument as above. Finally, consider the functor given by $\mathcal{E} \mapsto s^*\mathcal{E}$. It is a faithful exact tensor functor with values in the category of finite dimensional k-vector spaces, i.e. a fibre functor on EF_S. □

Remark 6.7.12 The point where the use of semi-stability was crucial in the construction was in the innocent-looking last part of the above proof. Indeed, for the purpose of embedding the category of finite locally free sheaves in an abelian category, the category of coherent sheaves on S is just as handy; it moreover carries a commutative tensor structure with unit. However, for duals to exist we need to know that the objects are locally free sheaves, and this is also needed for the construction of the fibre functor.

Definition 6.7.13 Let S be a proper integral scheme over a field k that has a k-rational point $s : \mathrm{Spec}(k) \to S$. The *fundamental group scheme* of S with base point s is the affine k-group scheme corresponding via Theorem 6.5.3 to the neutral Tannakian category EF_S and the fibre functor s^*. We denote it by $\pi_1^N(S, s)$ (the superscript N stands for 'Nori').

Proposition 6.7.14 *The group scheme $\pi_1^N(S, s)$ is an inverse limit of finite k-group schemes.*

Proof Let $\mathcal{F}$ be a finite locally free sheaf on S. Consider the direct sum $\widetilde{\mathcal{F}}$ of a system of representatives of $S(\mathcal{F} \oplus \mathcal{F}^\vee)$; by Proposition 6.7.4 this is a finite direct sum. As before, denote by $\langle\widetilde{\mathcal{F}}\rangle$ the full abelian subcategory of EF_S generated by the object $\widetilde{\mathcal{F}}$. It is stable by tensor product and duals because by definition $S(\mathcal{F} \oplus \mathcal{F}^\vee)$ contains all indecomposable direct summands of tensor products of the form $\mathcal{F}^{\otimes i} \otimes (\mathcal{F}^\vee)^{\otimes j}$. The category $\langle\widetilde{\mathcal{F}}\rangle$ thus equals the full Tannakian subcategory $\langle\mathcal{F}\rangle_\otimes$ of EF_S, and as such is equivalent to the category Rep_G for an affine group scheme G. The group scheme G is finite, because we know from Step 1 of the proof of Theorem 6.5.4 that $\langle\widetilde{\mathcal{F}}\rangle$ is equivalent to the category of finite dimensional comodules over a finite dimensional k-coalgebra $R_\mathcal{F}$ which must be the underlying coalgebra of the Hopf algebra of G. It remains to note that EF_S is the direct limit of the full Tannakian subcategories $\langle\mathcal{F}\rangle_\otimes$, as each object $\mathcal{E}$ is a subquotient of a finite locally free sheaf $\mathcal{F}$, and is therefore contained in $\langle\mathcal{F}\rangle_\otimes$. □

Remark 6.7.15 The arguments in the above proof also imply that *when k is of characteristic 0, every essentially finite locally free sheaf over S is finite.* Indeed, it follows from a theorem of Cartier (see [111], Theorem 11.4) that in characteristic 0 every finite group scheme is étale. On the other hand, for a finite étale group scheme G the regular representation $k[G]$ is manifestly semisimple (i.e. a direct sum of irreducible representations), hence each object in the category Rep_G is semisimple. Applying this to the category $\langle\widetilde{\mathcal{F}}\rangle$ of the above proof which is equivalent to Rep_G for some finite G, we see that each object is a direct sum of objects in $S(\mathcal{F} \oplus \mathcal{F}^\vee)$, and therefore a finite locally free sheaf by Corollary 6.7.6.

The next theorem shows why $\pi_1^N(S, s)$ is called a fundamental group scheme. To state it, introduce the category $\mathrm{Ftors}_{S,s}$ of triples (G, X, x), where G is a finite group scheme over k, X is a left G-torsor over S and x is a k-rational point in the fibre of X above s. A morphism $(G, X, x) \to (G', X', x')$ in this category is given by a pair of morphisms $\phi : G \to G'$, $\psi : X \to X'$ such that the G-action on X is compatible with the G'-action on X' via (ϕ, ψ), and moreover $\psi(x) = x'$.

Theorem 6.7.16 *There is an equivalence of categories between* $\mathrm{Ftors}_{S,s}$ *and the category of finite group schemes G over k equipped with a k-group scheme homomorphism $\pi_1^N(S, s) \to G$.*

Note that not all G-torsors over S have a k-point in the fibre above s. The proof will show that every G-torsor X gives rise to a homomorphism $\phi_X : \pi_1^N(S, s) \to G$ but one may recover X from ϕ_X only after securing a k-point in the fibre. We need the following general lemma.

Lemma 6.7.17 *Let G be a finite k-group scheme. Consider the neutral Tannakian category* Rep_G *of its finite dimensional representations, together with a fibre functor* $\omega : \mathrm{Rep}_G \to \mathrm{Vecf}_k$.

Given a non-neutral fibre functor η with values in the category LF_S *of locally free sheaves on S, the functor* $\mathbf{Hom}^{\otimes}(\eta, \omega_S)$ *given by sending T to* $\mathrm{Hom}^{\otimes}(\eta \otimes_{\mathcal{O}_S} \mathcal{O}_T, \omega \otimes_k \mathcal{O}_T)$ *on the category of schemes over S is representable by a left G-torsor over S.*

Proof By Lemma 6.3.3 the subfunctor $\mathbf{Isom}^{\otimes}(\eta, \omega_S) \subset \mathbf{Hom}^{\otimes}(\eta, \omega_S)$ given by tensor isomorphisms equals $\mathbf{Hom}^{\otimes}(\eta, \omega_S)$. Therefore for each $T \to S$ the natural map

$$\mathbf{Aut}^{\otimes}(\omega_S) \times \mathbf{Hom}^{\otimes}(\eta, \omega_S) \to \mathbf{Hom}^{\otimes}(\eta, \omega_S) \times \mathbf{Hom}^{\otimes}(\eta, \omega_S),$$
$$(g, \phi) \mapsto (\phi, g \circ \phi)$$

is bijective. As $\mathbf{Aut}^{\otimes}(\omega_S) \cong G_S$ by Proposition 6.3.5, if $\mathbf{Hom}^{\otimes}(\eta, \omega_S)$ is representable by a scheme X over S, we conclude that X carries the structure of a left G-torsor over S.

To show the representability of $\mathbf{Hom}^{\otimes}(\eta, \omega_S)$, we could invoke a general representability result for torsors under affine group schemes (e.g. [59], Proposition III.4.3. a)). But a direct reasoning is also available that generalizes our previous constructions. We only sketch it, leaving details to the readers. As G is finite, the regular representation $P = k[G]$ of G is a projective generator in Rep_G. In Steps 1–3 of the proof of Theorem 6.5.4 and the first step of the proof of Proposition 6.3.5 we have seen that $\omega(P)^* \otimes_{\mathrm{End}(P)} \omega(P)$ represents the functor $\mathbf{End}(\omega)$ on the category of k-algebras. A direct generalization of the argument shows that the coherent sheaf $\mathcal{A} := \eta(P)^* \otimes_{\mathrm{End}(P)} \omega_S(P)$ of $\mathcal{O}_S$-modules represents the functor $\mathbf{Hom}(\eta, \omega_S)$ on the category of quasi-coherent $\mathcal{O}_S$-algebras. (This is done by reduction to the affine case: for $S = \mathrm{Spec}\,(R)$ we have R-module structures on $(\omega \otimes R)(P)$ and $\eta(P)$; as before, the tensor product $\eta(P)^* \otimes_{\mathrm{End}(P)} (\omega \otimes R)(P)$ is defined and has an R-module structure, and we can glue these in the non-affine case.) Then as in the second step of the proof of Proposition 6.3.5 one shows that for each $T \to S$ endomorphisms in $\mathbf{Hom}^{\otimes}(\eta, \omega_S)(T)$ correspond to maps $\mathcal{A} \to \mathcal{O}_T$ that are $\mathcal{O}_S$-algebra homomorphisms for a canonical multiplicative structure on $\mathcal{A}$. With this structure $\mathcal{A}$ gives rise to an affine scheme over S by Remark 5.1.24. □

Remark 6.7.18 In fact, the construction of the lemma generalizes to an arbitrary affine group scheme G, and moreover yields a bijective correspondence between fibre functors with values in LF_S and left G-torsors over S; see [81], corollaire 3.2.3.3. The functor from torsors to fibre functors is given by 'twisting the fibre functor ω_S by the torsor X'.

Proof of Theorem 6.7.16 A G-torsor X gives rise to a finite locally free sheaf $\mathcal{E}_X$ on S by Lemma 6.7.1. As we have seen in the proof of Proposition 6.7.14, the full Tannakian subcategory $\langle \mathcal{E}_X \rangle_\otimes$ is equivalent to the category Rep_G for some finite group scheme G. Consider the fibre functor on Rep_G given by composing the equivalence $\mathrm{Rep}_G \xrightarrow{\sim} \langle \mathcal{E}_X \rangle_\otimes$ with the restriction of the functor s^* to $\langle \mathcal{E}_X \rangle_\otimes$. The inclusion functor $\langle \mathcal{E}_X \rangle_\otimes \to \mathrm{EF}_S$ then corresponds to a group scheme homomorphism $\phi_X : \pi_1^N(S, s) \to G$ by Corollary 6.3.6 and Proposition 6.7.11.

Conversely, a homomorphism $\phi : \pi_1^N(S, s) \to G$ induces a tensor functor $\phi^* : \mathrm{Rep}_G \to \mathrm{EF}_S$, whence a non-neutral fibre functor η with values in LF_S by composition with the inclusion of categories $\mathrm{EF}_S \to \mathrm{LF}_S$, and a neutral fibre functor ω by composing with s^*. Applying the lemma we obtain a G-torsor X_ϕ over S. However, for $\phi = \phi_X$ the G-torsor X_ϕ is not necessarily isomorphic to X. Indeed, the construction shows that we would get X back if instead of the above ω we applied the lemma to the forgetful fibre functor F sending an object of Rep_G to its underlying k-vector space. By fixing an isomorphism $\omega \cong F$ of fibre functors we can thus define a quasi-inverse to the functor of the previous paragraph. But such an isomorphism exists if and only if there is a k-rational point in the fibre of X above s. Indeed, the affine k-scheme $\mathbf{Isom}^\otimes(\eta, \omega_S)$ has a canonical k-rational point corresponding to s^*, so an isomorphism $\omega \cong F$ of functors yields a k-rational point of $\mathbf{Isom}^\otimes(\eta, F_S) \cong X$ above s. Conversely, such a point defines a trivialization of the G-torsor X_s over k, whence an isomorphism of functors $\omega \cong F$. □

Over an algebraically closed base field we may compare Nori's fundamental group scheme to Grothendieck's algebraic fundamental group.

Proposition 6.7.19 *Assume k is algebraically closed, and fix a k-valued geometric point $s = \bar{s}$ of S. The algebraic fundamental group $\pi_1(S, \bar{s})$ is canonically isomorphic to the group of k-points of the maximal pro-étale quotient of $\pi_1^N(S, s)$.*

Here 'maximal pro-étale quotient' means the inverse limit of those quotients of $\pi_1^N(S, s)$ that are finite étale group schemes over k.

Proof From Proposition 5.3.16 (1) and the fact that k is algebraically closed we conclude that a torsor under a finite étale k-group scheme G is one and the same thing as a finite étale Galois cover with group $G(k)$. In the proof of Proposition 5.4.6 we turned the set of finite étale Galois covers of S into an inverse system $(P_\alpha, \phi_{\alpha\beta})$ by specifying a point p_α in the geometric fibre $\mathrm{Fib}_{\bar{s}}(P_\alpha)$ and defining $\phi_{\alpha\beta}$ to be the unique morphism $P_\beta \to P_\alpha$ sending p_β to p_α. The group $\pi_1(S, \bar{s})$ was the automorphism group of this inverse system.

On the other hand, we have seen that each G_α-torsor P_α defines an object of EF_S. The full Tannakian subcategory $\langle P_\alpha\rangle_\otimes \subset \mathrm{EF}_S$ is equivalent to the category Rep_{G_α}. Let F be the neutral fibre functor on Rep_{G_α} given by the forgetful functor, and let η be the non-neutral fibre functor given by the composite $\mathrm{Rep}_{G_\alpha} \xrightarrow{\sim} \langle P_\alpha\rangle_\otimes \to \mathrm{EF}_S \to \mathrm{LF}_S$. By Lemma 6.7.17 the functor $\mathbf{Hom}^\otimes(\eta, F_S)$ on Rep_{G_α} is representable by a G_α-torsor over S and, as we have noted in the previous proof, fixing P_α yields an isomorphism between this G_α-torsor and P_α. Thus fixing points p_α as above turns the functors $\mathbf{Hom}^\otimes(\eta, F_S)$ on the various Rep_{G_α} into an inverse system as well. Passing to automorphisms defines an inverse system for the G_α whose inverse limit $\varprojlim G_\alpha$ is an affine group scheme over k whose group of k-points is $\pi_1(S, \bar{s})$ by construction. But $\varprojlim G_\alpha$ is also the maximal pro-étale quotient of $\pi_1^N(S, s)$ by Theorem 6.7.16. □

Corollary 6.7.20 *When k is algebraically closed of characteristic 0, there is a canonical isomorphism $\pi_1^N(S, s)(k) \xrightarrow{\sim} \pi_1(S, \bar{s})$ for each k-valued geometric point $s = \bar{s}$ of S.*

Proof This follows from the proposition above, Proposition 6.7.14 and the fact, already recalled in Remark 6.7.15, that in characteristic 0 all finite group schemes are étale. □

Remarks 6.7.21

1. Nori established in [73] a number of properties of $\pi_1^N(S, s)$. For instance, given two k-rational points s and t, one has $\pi_1(S, s) \times_k \bar{k} \cong \pi_1(S, t) \times_k \bar{k}$ (we have omitted the 'Spec' to ease notation). Also, one has an isomorphism of group schemes $\pi_1(S \times_k L) \cong \pi_1(S, s) \times_k L$ for a separable algebraic extension $L|k$. However, as Mehta and Subramanian have shown in [58], the corresponding property does not hold for extensions of algebraically closed fields, at least if one allows S to be singular. In the same paper they checked that $\pi_1^N(S, s)$ is compatible with direct products. This can be used to adapt the Lang–Serre arguments on the étale fundamental group of an abelian variety A, and obtain that $\pi_1^N(A, 0)$ is the inverse limit of the kernels of the multiplication-by-m maps $m : A \to A$ considered as finite group schemes. Nori proved this earlier in [74] by a direct method.
2. In their recent paper [21], Esnault and Hai develop a generalization of Nori's theory for smooth geometrically connected schemes S over a field k of characteristic 0, without assuming the existence of a k-point. They consider the category FConn_S of locally free sheaves on S equipped with the structure of a flat connection, and then define a connection $(\mathcal{E}, \nabla)$ to be finite if $f((\mathcal{E}, \nabla)) = g((\mathcal{E}, \nabla))$ for some polynomials $f \neq g$. They

show that in the proper case each object of Nori's category EF_S carries the structure of a flat connection lying in FConn_S. But in the absence of a k-rational point the category FConn_S has a priori only a non-neutral Tannakian structure. If ω is a non-neutral fibre functor coming from a $\bar{k}$-valued geometric point of S, the general Tannakian theory (Remark 6.5.13) shows that $\mathbf{Aut}_k^{\otimes}(\omega)$ is representable by a groupoid scheme Π over $\bar{k}$. Esnault and Hai identify the sections of the groupoid map $s : \Pi \to \mathrm{Spec}\,(\bar{k})$ with elements of the étale fundamental group $\pi_1(S \otimes_k \bar{k})$ and show that neutral fibre functors on FConn_S correspond to conjugacy classes of sections of the projection $\pi_1(S) \to \mathrm{Gal}\,(\bar{k}|k)$. This yields a Tannakian reinterpretation of Grothendieck's Section Conjecture: given a proper normal curve C of genus at least 2 over a finitely generated extension k of $\mathbf{Q}$, neutral fibre functors on FConn_C should come from k-rational points.

Exercises

1. Let V and W be infinite dimensional vector spaces. Show that the natural embedding $V^* \otimes_k W^* \to (V \otimes_k W)^*$ sending a tensor product $\phi \otimes \psi$ of functions to the function given by $v \otimes w \mapsto \phi(v)\psi(w)$ is not surjective.
   [*Hint:* Choose infinite sequences of linearly independent vectors $v_1, v_2, \dots$ and $w_1, w_2, \dots$ in V and W, respectively. Define a k-linear function on $V \otimes_k W$ by sending $v_i \otimes w_j$ to δ_{ij} (Kronecker delta), extending linearly to the span S of the $v_i \otimes w_j$, and extending by 0 outside S. Show that this function is not in the image of $V^* \otimes_k W^*$.]
2. Let B be a finite dimensional k-algebra, and let ω be the forgetful functor from the category of finitely generated left B-modules to the category Vecf_k of finite dimensional k-vector spaces.
   (a) Show that the set $\mathrm{End}(\omega)$ of functor morphisms $\omega \to \omega$ carries a natural k-algebra structure.
   (b) Verify that the map $B \mapsto \mathrm{End}(\omega)$ induced by mapping $b \in B$ to the functorial collection Φ^b of multiplication-by-b maps $V \to V$ on the underlying space of each finitely generated B-module V is an isomorphism of k-algebras.
   (c) State and prove a dual statement for a finite dimensional k-coalgebra A and the category of finite dimensional k-vector spaces equipped with a right A-comodule structure.
3. Denote by Comodf_A the category of finite dimensional right comodules over a k-coalgebra A. Show that the map $A' \mapsto \mathrm{Comodf}_{A'}$ yields a bijection between the sub-coalgebras $A' \subset A$ and the subcategories of Comodf_A closed under taking subobjects, quotients and finite direct sums.
4. Let A be a commutative ring, and $\mathcal{C}$ a full subcategory of the category of finitely generated A-modules containing A and stable by tensor product. Show that $(\mathcal{C}, \otimes)$ is a rigid tensor category if and only if its objects are projective A-modules.

5. Let k be a field. In the category GrVecf_k of finite dimensional *graded vector spaces* over k objects are families $(V_i)_{i\in\mathbf{Z}}$ of k-vector spaces indexed by $\mathbf{Z}$ such that the direct sum $\bigoplus V_i$ is finite dimensional, and morphisms are families $(\phi_i : V_i \to V_i')_{i\in\mathbf{Z}}$. Define a tensor product on GrVecf_k by $(V_i) \otimes (V_i') = (\bigoplus_{p+q=i} V_p \otimes_k V_q')$ and a functor $\mathrm{GrVecf}_k \to \mathrm{Vecf}_k$ by $(V_i) \mapsto \bigoplus V_i$. Check that GrVecf_k with the above additional structure is a neutral Tannakian category, and determine the Tannakian fundamental group.
6. Let (K, ∂) be a differential field.
   (a) Given a differential module (V, ∇) over K, show that a system of elements of V^∇ is k-linearly independent if and only if it is K-linearly independent.
   (b) Let $f_1, \dots, f_n$ be elements of K, and consider the *Wronskian matrix* $W(f_1, \dots, f_n) = [f_j^{(i-1)}]_{1\leq i,j\leq n}$, where $f_j^{(i-1)}$ denotes the $(i-1)$-st derivative of f_j and $f_j^{(0)} = f_j$ by convention. Show that $W(f_1, \dots, f_n)$ is invertible if and only if the f_j are k-linearly independent. [*Hint:* Define a differential module structure on K^n for which the vectors $(f_j, f_j', \dots, f_j^{n-1})$ are horizontal.]
7. Let (K, ∂) be a differential field with algebraically closed constant field k, and let $L|K$ be a finite Galois extension with group G (in the usual sense).
   (a) Show that the derivation $\partial : f \mapsto f'$ extends uniquely to L.
   (b) Let $f \in K[x]$ be a polynomial such that L is the splitting field of f over K. Index the roots α_i of f so that $\alpha_1, \dots, \alpha_n$ are k-linearly independent and the other α_i are in their k-span. Denote by W the Wronskian matrix $W(\alpha_1, \dots, \alpha_n)$; it is invertible by the previous exercise. Show that (L, ∂) is a Picard–Vessiot extension for the differential module on K^n with connection matrix $A = W'W^{-1}$, and the differential Galois group is G.
8. Let k be an algebraically closed field of characteristic 0, and $G \subset \mathrm{GL}_n(k)$ a closed subgroup. Show that there exists a Picard–Vessiot extension $(L, \partial) \supset (K, \partial)$ over k with differential Galois group isomorphic to G. [*Hint:* Treat first the case $G = \mathrm{GL}_n(k)$.]
   [*Remark.* C. and M. Tretkoff have shown that for $k = \mathbf{C}$ one may actually take $K = \mathbf{C}(t)$ with the usual derivation, thereby obtaining an analogue of Corolary 3.4.4. For a proof as well as a discussion of related work, see [107], Section 5.2.]
9. (suggested by Totaro) Let $\mathcal{E}$ and $\mathcal{F}$ be locally free sheaves on a scheme S, of respective ranks e and f.
   (a) Establish a canonical isomorphism of invertible sheaves
   $$\det(\mathcal{E} \otimes_{\mathcal{O}_S} \mathcal{F}) = \det(\mathcal{E})^{\otimes f} \otimes_{\mathcal{O}_S} \det(\mathcal{F})^{\otimes e}.$$
   (b) When $S = C$ is a normal curve, derive a proof of formula (6.19).
10. Let S be a proper integral scheme over a field k, and let $s : \mathrm{Spec}\,(k) \to S$ be a k-rational point. Consider the category of pairs (G, X) with G a finite group scheme over k and $X \to S$ a left G_S-torsor such that the fibre X_s over s has a k-rational point. Let F_s be the functor on this category sending a pair (G, X) to the set of k-rational points of the fibre X_s. Show that F_s is a pro-representable functor.

Bibliography

[1] Shreeram S. Abhyankar, Coverings of algebraic curves, *Amer. J. Math.* **79** (1957), 825–856.

[2] Michael F. Atiyah, Ian G. MacDonald, *Introduction to Commutative Algebra*, Addison-Wesley, Reading, 1969.

[3] Emil Artin, Otto Schreier, Eine Kennzeichnung der reell abgeschlossenen Körper, *Abh. Math. Sem. Hamburg* **5** (1927), 225–231.

[4] Arnaud Beauville, Monodromie des systèmes différentielles linéaires à pôles simples sur la sphère de Riemann [d'après A. Bolibruch], Séminaire Bourbaki, exposé 765, *Astérisque* **216** (1993), 103–119.

[5] Frits Beukers, Gauss' hypergeometric function, in R-P. Holzapfel *et al.* (eds.) *Arithmetic and Geometry Around Hypergeometric Functions*, Progress in Mathematics, vol. 260, Birkhäuser, Basel, 2007, 23–42.

[6] Enrico Bombieri, Walter Gubler, *Heights in Diophantine Geometry*, New Mathematical Monographs, vol. 4, Cambridge University Press, Cambridge, 2006.

[7] Niels Borne, Michel Emsalem, Note sur la détermination algébrique du groupe fondamental pro-résoluble d'une courbe affine, *J. Algebra* **320** (2008), 2615–2623.

[8] Jean-Benoît Bost, François Loeser, Michel Raynaud (eds.), *Courbes semi-stables et groupe fondamental en géométrie algébrique*, Progress in Mathematics, vol. 187, Birkhäuser, Basel, 2000.

[9] Vyjayanthi Chari, Andrew Pressley, *A Guide to Quantum Groups*, Cambridge University Press, Cambridge, 1994.

[10] Richard M. Crew, Etale p-covers in characteristic p, *Compositio Math.* **52** (1984), 31–45.

[11] Olivier Debarre, Variétés rationnellement connexes (d'après T. Graber, J. Harris, J. Starr et A. J. de Jong), Séminaire Bourbaki, exposé 905, *Astérisque* **290** (2003), 243–266.

[12] Pierre Deligne, *Equations différentielles à points singuliers réguliers*, Lecture Notes in Mathematics, vol. 163, Springer-Verlag, Berlin, 1970.

[13] Pierre Deligne, Le groupe fondamental de la droite projective moins trois points, in Y. Ihara *et al.* (eds.) *Galois Groups over* $\mathbf{Q}$, Math. Sci. Res. Inst. Publ., vol. 16, Springer-Verlag, New York, 1989, 79–297.

[14] Pierre Deligne, Catégories tannakiennes, in P. Cartier *et al.* (eds.) *The Grothendieck Festschrift II*, Progress in Mathematics, vol. 87, Birkhäuser, Boston, 1990, 111–195.

[15] Pierre Deligne, David Mumford, The irreducibility of the space of curves of given genus, *Publ. Math. IHES* **36** (1969), 75–109.

[16] Adrien Douady, Détermination d'un groupe de Galois. *C. R. Acad. Sci. Paris* **258** (1964), 5305–5308.

[17] Adrien and Régine Douady, *Algèbre. Théories galoisiennes*, Dunod, Paris, 1979.

[18] David Eisenbud, *Commutative Algebra with a View toward Algebraic Geometry*, Graduate Texts in Mathematics, vol. 150, Springer-Verlag, New York, 1995.

[19] Noam D. Elkies, ABC implies Mordell, *Internat. Math. Res. Notices* (1991), 99–109.

[20] Hélène Esnault, Phùng Hô Hai, Packets in Grothendieck's section conjecture, *Adv. Math.* **218** (2008), 395–416.

[21] Hélène Esnault, Phùng Hô Hai, The fundamental groupoid scheme and applications, preprint `arXiv:math/0611115`.

[22] Gerd Faltings, Curves and their fundamental groups (following Grothendieck, Tamagawa and Mochizuki), Séminaire Bourbaki, exposé 840, *Astérisque* **252** (1998), 131–150.

[23] Otto Forster, *Lectures on Riemann Surfaces*, Springer-Verlag, Berlin, 1981.

[24] Peter Freyd, *Abelian Categories*, Harper & Row, New York, 1964.

[25] Albrecht Fröhlich, Martin Taylor, *Algebraic Number Theory*, Cambridge Studies in Advanced Mathematics, vol. 27, Cambridge University Press, Cambridge, 1991.

[26] William Fulton, *Algebraic Topology: A First Course*, Graduate Texts in Mathematics, vol. 153, Springer-Verlag, New York, 1995.

[27] William Fulton, *Intersection Theory*, second edition, Springer-Verlag, Berlin, 1998.

[28] Phillip Griffiths, Joseph Harris, *Principles of Algebraic Geometry*, Wiley-Interscience, New York, 1978.

[29] Alexander Grothendieck, *Revêtements étales et groupe fondamental* (SGA 1), Lecture Notes in Mathematics, vol. 224, Springer-Verlag, Berlin and New York, 1971. New annotated edition: Société Mathématique de France, Paris, 2003.

[30] Alexander Grothendieck, *Cohomologie locale des faisceaux cohérents et théorèmes de Lefschetz locaux et globaux* (SGA 2), North-Holland, Amsterdam, 1968. New annotated edition: Société Mathématique de France, Paris, 2005.

[31] David Harbater, Galois coverings of the arithmetic line, in *Number Theory (New York, 1984–1985)*, Lecture Notes in Math., vol. 1240, Springer-Verlag, Berlin, 1987, 165–195.

[32] David Harbater, Abhyankar's conjecture on Galois groups over curves. *Invent. Math.* **117** (1994), 1–25.

[33] Robin Hartshorne, *Ample Subvarieties of Algebraic Varieties*, Lecture Notes in Mathematics, vol. 156, Springer-Verlag, Berlin and New York, 1970.

[34] Robin Hartshorne, *Algebraic Geometry*, Graduate Texts in Mathematics, vol. 52, Springer-Verlag, New York and Heidelberg, 1977.

[35] Edwin Hewitt, Kenneth A. Ross, *Abstract Harmonic Analysis II*, Grundlehren der Mathematischen Wissenschaften, vol. 152, Springer-Verlag, Berlin and Heidelberg, 1970.

[36] Luc Illusie, Grothendieck's existence theorem in formal geometry, in B. Fantechi *et al.*, *Fundamental Algebraic Geometry*, Mathematical Surveys and Monographs, vol. 123, American Mathematical Society, Providence, 2005, 179–233.

[37] Yulij Ilyashenko, Sergei Yakovenko, *Lectures on Analytic Differential Equations*, Graduate Studies in Mathematics, vol. 86, American Mathematical Society, Providence, 2008.

[38] André Joyal, Ross Street, An introduction to Tannaka duality and quantum groups, in *Category theory (Como, 1990)*, Lecture Notes in Math., vol. 1488, Springer-Verlag, Berlin, 1991, 413–492.

[39] Irving Kaplansky, *An Introduction to Differential Algebra*, Hermann, Paris, 1957.

[40] Nicholas M. Katz, An overview of Deligne's work on Hilbert's twenty-first problem, in *Mathematical Developments Arising from Hilbert Problems*, Proceedings of Symposia in Pure Mathematics, vol. 28/2, American Mathematical Society, Providence, 1976, 537–557.

[41] Nicholas M. Katz, Serge Lang, Finiteness theorems in geometric class field theory, *L'Enseign. Math.* **27** (1981), 285–314.

[42] Masaki Kashiwara, Pierre Schapira, *Categories and Sheaves*, Grundlehren der Mathematischen Wissenschaften, vol. 332, Springer-Verlag, Berlin, 2006.

[43] Moritz Kerz, Alexander Schmidt, On different notions of tameness in aritmetic geometry, preprint `arxiv:0807.0979`, 2008.

[44] Steven Kleiman, The Picard scheme, in B. Fantechi *et al.*, *Fundamental Algebraic Geometry*, Mathematical Surveys and Monographs, vol. 123, American Mathematical Society, Providence, 2005, 237–329.

[45] Max-Albert Knus, Alexander S. Merkurjev, Markus Rost, Jean-Pierre Tignol, *The Book of Involutions*, American Mathematical Society

Colloquium Publications, vol. 44, American Mathematical Society, Providence, 1998.

[46] Jochen Koenigsmann, On the 'section conjecture' in anabelian geometry, *J. reine angew. Math.* **588** (2005), 221–235.

[47] Wolfgang Krull, Galoissche Theorie der unendlichen algebraischen Erweiterungen, *Math. Ann.* **100** (1928), 687–698.

[48] Serge Lang, *Algebra*, third edition, Addison-Wesley, Reading, 1993.

[49] Serge Lang, *Number Theory III*, Encyclopaedia of Mathematical Sciences, vol. 60, Springer-Verlag, Berlin, 1991.

[50] Serge Lang, *Algebraic Number Theory*, second edition, Springer-Verlag, New York, 1994.

[51] Serge Lang, Jean-Pierre Serre, Sur les revêtements non ramifiés des variétés algébriques, *Amer. J. Math.* **79** (1957), 319–330.

[52] Hendrik W. Lenstra, Jr, *Galois Theory of Schemes*, course notes available from the server of the Universiteit Leiden Mathematics Department.

[53] Joseph Le Potier, *Lectures on Vector Bundles*, Cambridge Studies in Advanced Mathematics, vol. 54, Cambridge University Press, Cambridge, 1997.

[54] Max Lieblich, Martin Olsson, Generators and relations for the étale fundamental group, to appear in *Quart. J. of Pure and Appl. Math.*

[55] Shahn Majid, *Foundations of Quantum Group Theory*, Cambridge University Press, Cambridge, 1995.

[56] Gunter Malle, B. Heinrich Matzat, *Inverse Galois Theory*, Springer-Verlag, Berlin, 1998.

[57] Hideyuki Matsumura, *Commutative Ring Theory*, Cambridge Studies in Advanced Mathematics, vol. 8, Cambridge University Press, Cambridge, 1989.

[58] Vikhram Mehta, Swaminathan Subramanian, On the fundamental group scheme, *Invent. Math.* **148** (2002), 143–150.

[59] James S. Milne, *Étale Cohomology*, Princeton University Press, 1980.

[60] John Milnor, *Morse Theory*, Annals of Mathematics Studies, vol. 51, Princeton University Press, 1963.

[61] Shinichi Mochizuki, The profinite Grothendieck conjecture for closed hyperbolic curves over number fields, *J. Math. Sci. Univ. Tokyo* **3** (1996), 571–627.

[62] Shinichi Mochizuki, The local pro-p anabelian geometry of curves. *Invent. Math.* **138** (1999), 319–423.

[63] Shinichi Mochizuki, Absolute anabelian cuspidalizations of proper hyperbolic curves, *J. Math. Kyoto Univ.* **47** (2007), 451–539.

[64] David Mumford. *The Red Book of Varieties and Schemes*, Lecture Notes in Mathematics 1358, Springer-Verlag, Berlin, 1988.

[65] David Mumford, *Abelian Varieties*, Oxford University Press, Oxford, 1970.

[66] James Munkres, *Topology*, second edition, Prentice Hall, Englewood Cliffs, 2000.

[67] Masayoshi Nagata, *Local Rings*, Interscience, New York and London, 1962.

[68] Jürgen Neukirch, Kennzeichnung der p-adischen und der endlichen algebraischen Zahlkörper, *Invent. Math.* **6** (1969), 296–314.

[69] Jürgen Neukirch, *Algebraic Number Theory*, Grundlehren der Mathematischen Wissenschaften, vol. 322, Springer-Verlag, Berlin, 1999.

[70] Jürgen Neukirch, Alexander Schmidt, Kay Wingberg, *Cohomology of Number Fields,* Grundlehren der Mathematischen Wissenschaften, vol. 323, Springer-Verlag, Berlin, 2000.

[71] Nikolay Nikolov, Dan Segal, On finitely generated profinite groups I. Strong completeness and uniform bounds, *Ann. of Math.* **165** (2007), 171–238.

[72] Madhav Nori, On the representations of the fundamental group, *Compositio Math.* **33** (1976), 29–41.

[73] Madhav Nori, The fundamental group-scheme, *Proc. Indian Acad. Sci. Math. Sci.* **91** (1982), 73–122.

[74] Madhav Nori, The fundamental group-scheme of an abelian variety, *Math. Ann.* **263** (1983), 263–266.

[75] Florian Pop, 1/2 Riemann existence theorem with Galois action, in G. Frey and J. Ritter (eds.) *Algebra and Number Theory*, de Gruyter, Berlin, 1994.

[76] Florian Pop, Étale Galois covers of affine smooth curves. The geometric case of a conjecture of Shafarevich. On Abhyankar's conjecture, *Invent. Math.* **120** (1995), 555–578.

[77] Florian Pop, Alterations and birational anabelian geometry, in H. Hauser *et al.* (eds.) *Resolution of Singularities (Obergurgl, 1997)*, Progress in Mathematics, vol. 181, Birkhäuser, Basel, 2000, 519–532.

[78] Wayne Raskind, Abelian class field theory of arithmetic schemes, in B. Jacob, A. Rosenberg (eds.) *K-Theory and Algebraic Geometry: Connections with Quadratic Forms and Division Algebras*, Proc. Symp. Pure Math., vol. 58/1, American Mathematical Society, Providence, 1995, 85–187.

[79] Michel Raynaud, Revêtements de la droite affine en caractéristique $p > 0$ et conjecture d'Abhyankar, *Invent. Math.* **116** (1994), 425–462.

[80] Walter Rudin, *Real and Complex Analysis*, third edition, McGraw-Hill, New York, 1987.

[81] Neantro Saavedra Rivano, *Catégories tannakiennes*, Lecture Notes in Mathematics 265, Springer-Verlag, Berlin, 1972.

[82] Peter Schauenburg, Tannaka duality for arbitrary Hopf algebras, *Algebra Berichte*, vol. 66, Verlag Reinhard Fischer, Munich, 1992.

[83] Alexander Schmidt, Tame class field theory for arithmetic schemes. *Invent. Math.* **160** (2005), 527–565.

[84] Alexander Schmidt, Michael Spieß, Singular homology and class field theory for varieties over finite fields, *J. reine angew. Math.* **527** (2000), 13–36.

[85] Leila Schneps, Pierre Lochak (eds.), *Geometric Galois Actions 1*, London Mathematical Society Lecture Note Series, vol. 242. Cambridge University Press, Cambridge, 1997.

[86] Jean-Pierre Serre, Morphismes universels et variété d'Albanese, Séminaire Chevalley, année 1958/58, exposé 10, reprinted in *Exposés de séminaires 1950-1999*, Société Mathématique de France, Paris, 2001, 141–160.

[87] Jean-Pierre Serre, Morphismes universels et différentielles de troisième espèce, Séminaire Chevalley, année 1958/59, exposé 11, reprinted in *Exposés de séminaires 1950-1999*, Société Mathématique de France, Paris, 2001, 161–168.

[88] Jean-Pierre Serre, On the fundamental group of a unirational variety, *J. London Math. Soc.* **34** (1959), 481–484.

[89] Jean-Pierre Serre, Zeta and L functions, in O. Schilling (ed.) *Arithmetical Algebraic Geometry*, Harper and Row, New York, 1965, 82–92.

[90] Jean-Pierre Serre, *Cours d'arithmétique,* Presses Universitaires de France, Paris, 1970. English edition: *A course in arithmetic*, Graduate Texts in Mathematics, vol. 7, Springer-Verlag, New York, 1973.

[91] Jean-Pierre Serre, *Lectures on the Mordell–Weil Theorem*, Vieweg, Braunschweig, 1989.

[92] Jean-Pierre Serre, Construction de revêtements étales de la droite affine en caractéristique p, *C. R. Acad. Sci. Paris Sér. I Math.* **311** (1990), 341–346.

[93] Jean-Pierre Serre, *Topics in Galois Theory*, Jones and Bartlett, Boston, 1992.

[94] Igor R. Shafarevich, On p-extensions (in Russian), *Mat. Sbornik* 20(62) (1947), 351–363. English translation: *Amer. Math. Soc. Translation Series* 4 (1956), 59–72.

[95] Igor R. Shafarevich, *Basic Algebraic Geometry*, second edition, Springer-Verlag, Berlin, 1994.

[96] Stephen S. Shatz, *Profinite Groups, Arithmetic and Geometry*, Annals of Mathematics Studies, No. 67, Princeton University Press, 1972.

[97] Michael Spieß, Tamás Szamuely, On the Albanese map for smooth quasi-projective varieties, *Math. Ann.* **325** (2003), 1–17.

[98] Jakob Stix, Projective anabelian curves in positive characteristic and descent theory for log étale covers, *Bonner Mathematische Schriften*, vol. 354, Bonn, 2002.

[99] Jakob Stix, On cuspidal sections of algebraic fundamental groups, preprint `arXiv:0809.0017`.

[100] Joseph H. Silverman, *The Arithmetic of Elliptic Curves*, Graduate Texts in Mathematics, vol. 106, Springer-Verlag, New York, 1986.

[101] Tamás Szamuely, Groupes de Galois de corps de type fini [d'après Pop], Séminaire Bourbaki, exposé 923, *Astérisque* **294** (2004), 403–431.

[102] Akio Tamagawa, The Grothendieck conjecture for affine curves, *Compositio Math.* **109** (1997), 135–194.

[103] Akio Tamagawa, Finiteness of isomorphism classes of curves in positive characteristic with prescribed fundamental groups, *J. Algebraic Geom.* **13** (2004), 675–724.

[104] Domingo Toledo, Projective varieties with non-residually finite fundamental group, *Publ. Math. IHES* **77** (1993), 103–119.

[105] Karl-Heinz Ulbrich, On Hopf algebras and rigid monoidal categories, *Israel J. Math.* **72** (1990), 252–256.

[106] Bartel Leendert van der Waerden, *Algebra* I (7te Auflage), II (5te Auflage), Springer-Verlag, Berlin, 1966–67. English translation: Springer-Verlag, New York, 1991.

[107] Marius van der Put, Michael F. Singer, *Galois Theory for Linear Differential Equations*, Grundlehren der mathematischen Wissenschaften, vol. 328, Springer-Verlag, Berlin, 2003.

[108] Veeravalli S. Varadarajan, Meromorphic differential equations, *Exposition. Math.* **9** (1991), 97–188.

[109] Angelo Vistoli, Grothendieck topologies, fibered categories and descent theory, in B. Fantechi *et al.*, *Fundamental Algebraic Geometry*, Mathematical Surveys and Monographs, vol. 123, American Mathematical Society, Providence, 2005, 1–104.

[110] Claire Voisin, *Hodge Theory and Complex Algebraic Geometry I*, Cambridge Studies in Advanced Mathematics, vol. 76, Cambridge University Press, Cambridge, 2002.

[111] William C. Waterhouse, *Introduction to Affine Group Schemes*, Graduate Texts in Mathematics, vol. 66, Springer-Verlag, New York and Berlin, 1979.

[112] André Weil, Généralisation des fonctions abeliennes, *J. Math. Pures Appl.* **17** (1938), 47–87.

Index